THOMAS, JOHN JAMES
AN INTRODUCTION TO STATISTICAL
000223879

QA276.T45

KU-205-911

179

WITHDRAWN
FROM STOCK

# an introduction to statistical analysis for economists

London School of Economics
*Handbooks in Economic Analysis*

EDITORS:    S.G.B.  Henry    *Lecturer in Economics, University College, London*

A.A.  Walters    *Cassel Professor of Money and Banking at the London School of Economics and Political Science*

# an introduction to statistical analysis for economists

## J.J. Thomas

*Lecturer in Economics, London School of Economics and Political Science*

**Weidenfeld and Nicolson**
**London**

© 1973 J.J. Thomas

First published in Great Britain in 1973 by
Weidenfeld & Nicolson Ltd, 11 St John's Hill,
London SW11.

All rights reserved. No part of this publication
may be reproduced, stored in a retrieval system,
or transmitted, in any form or by any means,
electronic, mechanical, photocopying, recording
or otherwise, without the prior permission of the
Copyright owner.

ISBN 0 297 76620 1 cased

ISBN 0 297 76621 x paperback

*Printed in Great Britain by*
REDWOOD BURN LIMITED
*Trowbridge & Esher*

# contents

202299

# preface

This book is intended for students of economics and provides an introduction to statistical theory which develops basic ideas from first principles and, given its audience, provides a much fuller coverage of regression analysis than is usual in an elementary statistical text. It has evolved out of a lecture course that I have taught to M.Sc. students at the London School of Economics over the last six years as a prerequisite to an elementary course in econometric theory. The students in the course have been fairly typical economics students in that they do not have an extensive knowledge of mathematics and statistics. As a result, the mathematical requirements for this book are not high: some algebra is used throughout and some elementary differential calculus is used in chapters 8 and 9. Matrix algebra is not required, as the use of this technique is confined to the appendices to chapters 9 and 11.

Given the intended audience, the mathematical constraints and the objective of brevity, some topics have had to be omitted from the text. In particular there is no discussion of Bayesian inference and decision theory or of maximum likelihood methods of estimation. The former topic is excluded because the value of this approach in empirical work in economics has not yet been fully established, while the latter is excluded mainly on mathematical grounds.

In writing a textbook in statistics the choice of notation is always a problem, since it is extremely difficult to choose a convention which is completely consistent. I have chosen as follows: (i) Parameters and variables to be estimated are denoted by Greek letters while the corresponding estimators of the parameters and variables are denoted by the corresponding roman letters (so that $\varepsilon_i$ is the error in the regression model while $e_i$ is the corresponding residual in the model). (ii) For the numbering within chapters of sections, equations, tables and figures I have used a simple decimal system in which, for example, the second equation in the seventh chapter is numbered 7.2. (iii) Some equations appear several times in different chapters and in this case when they reappear they continue to be labelled with their original numbers. The footnotes are grouped together at the end of the book.

This text has evolved out of teaching a course over a number of years and so it is difficult to make full acknowledgements to individuals. I have benefited from the reactions of many students, whose questions and comments sharpened up the course. Also during this period I have benefited from discussions with my colleagues Meghnad Desai, David Hendry and Kenneth F. Wallis, some of whose ideas and comments have affected the evolution of the course and are reflected in this book. I am very grateful to Susan Taylor, who read an earlier draft of the first five chapters and whose detailed comments were extremely useful in the revision of this part of the book. Finally, I should like to express my gratitude to the secretaries whose good humour and accurate typing helped enormously in the production of the book. Susan Swan typed most of the manuscript, Luba Mumford typed chapter 6 and Ronnie Morse typed most of the final revisions. Needless to say I am responsible for any errors that remain.

J. J. THOMAS

*London*
*October 1972*

# part one

# single variable statistics

The equanimity of your average tosser of coins depends on a law, or rather a tendency, or let us say a probability, or at any rate a mathematically calculable chance, which ensures that he will not upset himself by losing too much nor upset his opponent by winning too often. This made for a kind of harmony and a kind of confidence. It related the fortuitous and the ordained into a reassuring union which we recognize as nature. The sun came up about as often as it went down, in the long run, and a coin showed heads about as often as it showed tails.

TOM STOPPARD                              *Rosencrantz and Guildenstern are Dead**

* Published by Faber and Faber, London 1967.

# chapter 1
# introduction

## 1.1 introduction

The aim of this book is to provide an introduction to statistical theory for economists. Interest in and the use of statistical methods by economists has increased enormously in recent years and it is convenient here to discuss some of the uses of statistical analysis and to outline the structure of the rest of the book.

At the outset we shall distinguish between descriptive statistics and statistical inference:

*Descriptive statistics.* In many situations which are of interest to economists, large quantities of numerical data are generated. For example, the annual filing of income-tax returns and the decennial population census provide data at regular intervals over time, and there are many special surveys carried out from time to time to collect information on the behaviour of consumers and producers. Such data are potentially of great use, but often the sheer quantity may make interpretation difficult. Comprehension can be aided by means of tabulation, grouping and graphical presentation of the data and also by developing summary statistics, which can be used to highlight particular features of the data which are of interest. These topics will be discussed in this chapter.

*Statistical inference.* In some situations the summary of large quantities of data may represent the object of the exercise, but in others it may be no more than the first step in the analysis. For example, the data may represent a sample which has been drawn from some population: a selection of firms chosen from all the firms in a particular industry, perhaps. In such a situation it is of interest to see what can be said about the characteristics of the population on the basis of the data in the sample. This question will be explored in the next five chapters, concentrating on problems involving a single variable. For example, given a sample of family incomes in London, can we say anything about London family incomes in general?

While this may be an interesting question, there is one which is usually of more interest to economists. Economic theories are concerned with relationships

3

between variables, and one important use of statistical methods is the investigation of relationships between variables. For example, if we have data on expenditure as well as incomes for our sample of London families, we might well ask whether there is evidence of a relationship between expenditure and income. If we have an economic theory which suggests the form of a relationship between economic variables, it is of interest to obtain some idea of the order of magnitude involved. This question is very important in policy discussions, where the choice between alternatives may hinge on the relative speeds and magnitudes of reaction. Given the importance of this subject, the discussion of relationships between variables occupies chapters 7 to 11.

The main characteristic of problems of statistical inference is that we have to say something about a population on the basis of incomplete data, in the sense that we only examine a sample and not every member of the population. The result is that we cannot be absolutely certain that our inference is correct. For example, suppose we select 100 schoolchildren in a particular city and discover that 40 of them are suffering from dental decay. Can we conclude that exactly 40 per cent of all the schoolchildren in the city are suffering from dental decay? The answer here is no. However, on the basis of the statistical theory which will be developed in chapters 2 to 6, we shall see how we may choose a range of values for the proportion of schoolchildren in the city suffering from dental decay with a specified probability that our chosen range will contain the true proportion.

The underlying theory involves us first in a formal discussion of probability (chapter 2). Then we shall develop a number of mathematical models (probability distributions) which may serve to describe the main characteristics of a range of populations, both human and otherwise (chapter 3). Methods of sampling and sample characteristics will be treated in chapter 4, leading on to the discussion of statistical inference in chapters 5 and 6.

Numerical data form the grist for the statistical mill, and the data available to economists may be classified into two main types. Firstly, and most commonly, we have *time series* data: the information collected at regular intervals of time and published by governments and international bodies and usually relating to the main aggregates of macro-economics; consumption, investment, government expenditure, etc. Secondly, we may have data obtained by taking a sample of individual consumers or families, firms or industries, and collecting information over some short interval of time. This is known as *cross-section* data and micro-economic information is often collected in this form.

The principles involved in constructing a logical, consistent set of national income accounts and the difficulties and compromises which result in practice will not be discussed in this text. Nor shall we discuss the important question of errors of measurement and the accuracy of economic data. On economic statistics

and the national income accounts the reader should consult Blyth (1960)[1] or
Nicholson (1969), while on the accuracy of economic statistics he may find
Morgenstern (1950) illuminating.

## 1.2 frequency distributions

We shall illustrate the statistical techniques for summarizing numerical data by
means of the following example. Suppose that income data have been collected
from two groups of men, each group consisting of 100 individuals. The groups
are drawn from two different cities in England with the object of examining the
distribution of income in the two cities and making some kind of comparison
between them. The data for the two groups are presented in Table 1.1, incomes
being rounded to the nearest £10.

*Table 1.1.* Hypothetical income data

City A

| | | | | | | | | | |
|---|---|---|---|---|---|---|---|---|---|
| 1820 | 350 | 6830 | 1490 | 2470 | 3900 | 2560 | 8440 | 6240 | 7580 |
| 2660 | 9780 | 7460 | 6800 | 6230 | 4060 | 3790 | 5210 | 6660 | 7480 |
| 2340 | 1680 | 2740 | 2050 | 8600 | 2820 | 6830 | 7870 | 2700 | 9550 |
| 5240 | 1260 | 2990 | 6420 | 5640 | 9440 | 9890 | 5900 | 6440 | 1280 |
| 3780 | 5560 | 1690 | 530 | 7790 | 1710 | 8270 | 9370 | 3950 | 1340 |
| 7030 | 1620 | 1130 | 800 | 1750 | 1840 | 9480 | 1090 | 3160 | 3600 |
| 5660 | 8440 | 3520 | 6870 | 6380 | 5720 | 1020 | 8680 | 5920 | 7790 |
| 9950 | 6300 | 3820 | 2700 | 1230 | 1500 | 1690 | 6160 | 2980 | 7230 |
| 1610 | 3320 | 3200 | 1460 | 8630 | 9330 | 2590 | 5270 | 4410 | 4350 |
| 3120 | 5760 | 6670 | 4370 | 790 | 3010 | 4070 | 8640 | 4640 | 7060 |

City B

| | | | | | | | | | |
|---|---|---|---|---|---|---|---|---|---|
| 4700 | 7980 | 3490 | 4310 | 2590 | 6510 | 1560 | 4910 | 6880 | 5280 |
| 4250 | 750 | 5230 | 9320 | 3620 | 4110 | 7890 | 4950 | 5560 | 7650 |
| 3290 | 4230 | 4230 | 5070 | 3470 | 3330 | 6720 | 2600 | 5510 | 1480 |
| 1510 | 1660 | 6170 | 740 | 3700 | 1080 | 3820 | 5410 | 2620 | 2200 |
| 3160 | 4560 | 7330 | 4610 | 7880 | 5160 | 6290 | 1100 | 5540 | 4210 |
| 7700 | 2020 | 4210 | 8620 | 3870 | 6300 | 6210 | 2910 | 1210 | 3320 |
| 9810 | 4420 | 2680 | 4630 | 5700 | 5070 | 2420 | 3030 | 3520 | 4320 |
| 5240 | 7780 | 3240 | 3210 | 3730 | 6310 | 3880 | 7460 | 7630 | 7210 |
| 4920 | 5380 | 7980 | 7630 | 4500 | 3230 | 8850 | 4140 | 8670 | 3350 |
| 8310 | 6720 | 7300 | 5100 | 6680 | 5310 | 4770 | 9100 | 5100 | 5820 |

When presented in this form, comparison between the two sets of data is
extremely difficult. An inspection of the data reveals a great deal of overlapping
of incomes between the two cities, so that we do not have a clear-cut and obvious
difference in incomes between the two cities. However, the data in Table 1.1 have
not been arranged in any special order and the detail can be made clearer if the
incomes are rearranged and tabulated by increasing size for the two groups. This
is shown in Table 1.2.

In this table the incomes have been tabulated by putting all those less than
£1000 in the first column, all those between £1000 and £1999 in the second, and

*Table 1.2.* Income data retabulated by increasing size

City A

| | | | | | | | | | |
|---|---|---|---|---|---|---|---|---|---|
| 350 | 1020 | 2050 | 3010 | 4060 | 5120 | 6160 | 7030 | 8270 | 9330 |
| 530 | 1090 | 2340 | 3120 | 4070 | 5240 | 6230 | 7060 | 8440 | 9370 |
| 790 | 1130 | 2470 | 3160 | 4350 | 5270 | 6240 | 7230 | 8440 | 9440 |
| 800 | 1230 | 2560 | 3200 | 4370 | 5560 | 6300 | 7460 | 8600 | 9480 |
| | 1260 | 2590 | 3320 | 4410 | 5640 | 6380 | 7480 | 8630 | 9550 |
| | 1280 | 2660 | 3520 | 4640 | 5660 | 6420 | 7580 | 8640 | 9780 |
| | 1340 | 2700 | 3600 | | 5720 | 6440 | 7790 | 8680 | 9890 |
| | 1460 | 2700 | 3780 | | 5760 | 6660 | 7790 | | 9950 |
| | 1490 | 2740 | 3790 | | 5900 | 6670 | 7870 | | |
| | 1500 | 2820 | 3820 | | 5920 | 6800 | | | |
| | 1610 | 2980 | 3900 | | | 6830 | | | |
| | 1620 | 2990 | 3950 | | | 6830 | | | |
| | 1680 | | | | | 6870 | | | |
| | 1690 | | | | | | | | |
| | 1690 | | | | | | | | |
| | 1710 | | | | | | | | |
| | 1750 | | | | | | | | |
| | 1820 | | | | | | | | |
| | 1840 | | | | | | | | |

City B

| | | | | | | | | | |
|---|---|---|---|---|---|---|---|---|---|
| 740 | 1080 | 2020 | 3030 | 4110 | 5070 | 6170 | 7210 | 8310 | 9100 |
| 750 | 1100 | 2200 | 3160 | 4140 | 5070 | 6210 | 7300 | 8620 | 9320 |
| | 1210 | 2420 | 3210 | 4210 | 5100 | 6290 | 7330 | 8670 | 9810 |
| | 1480 | 2590 | 3230 | 4210 | 5100 | 6300 | 7460 | 8850 | |
| | 1510 | 2600 | 3240 | 4230 | 5160 | 6310 | 7630 | | |
| | 1560 | 2620 | 3290 | 4230 | 5230 | 6510 | 7630 | | |
| | 1660 | 2680 | 3320 | 4250 | 5240 | 6680 | 7650 | | |
| | | 2910 | 3330 | 4310 | 5280 | 6720 | 7700 | | |
| | | | 3350 | 4320 | 5310 | 6720 | 7780 | | |
| | | | 3470 | 4420 | 5380 | 6880 | 7880 | | |
| | | | 3490 | 4500 | 5410 | | 7890 | | |
| | | | 3520 | 4560 | 5510 | | 7980 | | |
| | | | 3620 | 4610 | 5540 | | 7980 | | |
| | | | 3700 | 4630 | 5560 | | | | |
| | | | 3730 | 4700 | 5700 | | | | |
| | | | 3820 | 4770 | 5820 | | | | |
| | | | 3870 | 4910 | | | | | |
| | | | 3880 | 4920 | | | | | |
| | | | | 4950 | | | | | |

so on. Arranged in this way, the visual impression of the two groups is that, despite the overlap, they do differ. The first group has a relatively large number of low incomes while a relatively large number of incomes in the second group lie between £3000 and £8000.

So far we have merely rearranged the data and have preserved all the detail that was contained in Table 1.1. However, in attempting to compare the two sets of income data this degree of detail may be unnecessary. For example, it is more interesting to note that 35 of the incomes in the first group are less than £3000 as compared with 17 in the second group than it is to look at the individual incomes.

By suppressing some of the detail, the data may be presented as a *frequency distribution* in which we record the number of incomes falling in certain ranges for the two groups. The frequency distributions are presented in Table 1.3. This summary contains less detailed information than did Table 1.1, but the difference between the two groups is easier to see. It is clear that the rearrangement of the income data in the form presented in Table 1.2 is not necessary if our object is to construct the frequency distribution. This can be done directly from Table 1.1 by recording the number of incomes in each of the income intervals.

*Table 1.3*. Frequency distributions of incomes

| Income interval | City A Frequency (cum.) | | City B Frequency (cum.) | |
|---|---|---|---|---|
| 0 to 999 | 4 | (4) | 2 | (2) |
| 1000 to 1999 | 19 | (23) | 7 | (9) |
| 2000 to 2999 | 12 | (35) | 8 | (17) |
| 3000 to 3999 | 12 | (47) | 18 | (35) |
| 4000 to 4999 | 6 | (53) | 19 | (54) |
| 5000 to 5999 | 10 | (63) | 16 | (70) |
| 6000 to 6999 | 13 | (76) | 10 | (80) |
| 7000 to 7999 | 9 | (85) | 13 | (93) |
| 8000 to 8999 | 7 | (92) | 4 | (97) |
| 9000 to 9999 | 8 | (100) | 3 | (100) |
| Total | 100 | | 100 | |

Table 1.2 did provide a visual impression of the two sets of data, but this can be achieved more conveniently by plotting the frequency distribution in a diagram. For example, Figure 1.1 illustrates one way in which this may be done by plotting the frequencies against the corresponding income intervals. Such a diagram is called a *histogram*, and in this example the two frequency distributions have been drawn in the same diagram to facilitate the comparison of incomes. An alternative way to compare the two sets of income data is to plot the *cumulative frequency distribution*, which shows the proportion (percentage) of people in each sample who have incomes less than the various levels of income. This is illustrated in Figure 1.2, which brings out very clearly the contrast between the two groups over the lower income ranges.

While these visual methods of presentation bring out many of the distinctive features of the data, they may still carry more information than is useful for some purposes and not enough for others. For example, the reader could write a verbal summary of the pictures presented in Figures 1.1 and 1.2, but the aspects which were stressed would probably vary from reader to reader. It would be useful, therefore, to develop some numerical measures which would capture the main features of a set of data and might also make comparisons between sets of data

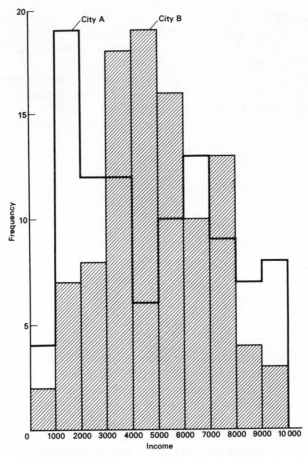

Figure 1.1

easier to perform. We shall consider two types of measures here, one being a 'representative' or average value for the set of data, the other being a measure of how the data are dispersed around the representative value.

### 1.3 the representative value
There is no single, obvious interpretation of 'representative' and we shall consider a number of alternative measures. In general it will not be the case that one of these is right and the others wrong; rather the choice will depend on the nature of the data and which features are being highlighted.

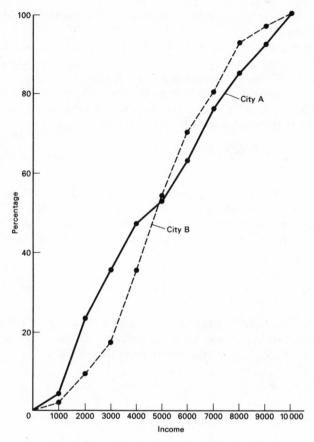

Figure 1.2 Percentage of individuals with incomes below given level

*The arithmetic mean*

This measure is most likely to be familiar to the reader when he thinks of an average value. It is defined as the sum of the values of the observations divided by the number of observations and is the way in which batting averages are calculated in many sports. Thus, from the income data for City A presented in Table 1.1, we would calculate the value of the arithmetic mean as

$$\frac{1820 + 2660 + 2340 + \ldots + 7230 + 4350 + 7060}{100},$$

which equals £4748·4. From the data for City B we calculate that the arithmetic mean is £4907·4. A comparison of these figures confirms the visual impression gained from Figures 1.1 and 1.2.

The arithmetic mean will figure frequently in what follows and we shall find it convenient to develop a general notation for dealing with summations. The first step is to let $X$ denote income, using a subscript to denote which income observation we are considering. For example, if we sum down adjacent columns, then $X_1 = 1820$, $X_2 = 2660$, and so on until $X_{99} = 4350$ and $X_{100} = 7060$. We shall denote the arithmetic mean of the set of $X$s by $\bar{X}$, so that

$$\bar{X} = \frac{X_1 + X_2 + \dots + X_{99} + X_{100}}{100}.$$

The operation we are performing is to add together the values of variable $X$ from $X_1$ to $X_{100}$, and we may indicate this operation by writing $\sum_{i=1}^{100} X_i$, where $\sum$ is to be interpreted as an instruction to add together what follows. What follows is $X_i$, where $i$ goes in integer steps from $i = 1$ to $i = 100$. Using this notation (often called 'sigma' notation), we may write

$$\bar{X} = \frac{1}{100} \sum_{i=1}^{100} X_i$$

to represent the arithmetic mean of our income data. The definition of the arithmetic mean can be written quite generally in this notation, so that if we have $n$ observations $X_1, \dots, X_n$, then their arithmetic mean is

$$\bar{X} = \frac{1}{n} \sum_{i=1}^{n} X_i. \tag{1.1}$$

As further examples of this notation, the reader may consider

(a) $\sum_{i=1}^{4} i = 1 + 2 + 3 + 4 = 10,$

(b) $\sum_{i=1}^{3} (X_i - Y_i) = (X_1 - Y_1) + (X_2 - Y_2) + (X_3 - Y_3)$

$$= (X_1 + X_2 + X_3) - (Y_1 + Y_2 + Y_3) = \sum_{i=1}^{3} X_i - \sum_{i=1}^{3} Y_i.$$

We shall use this notation extensively below, so that the reader who has not met this convention before is advised to attempt the summation problems at the end of this chapter.

To return to our discussion of the arithmetic mean, it is easy to compute and easy to comprehend, but it may not always serve as a 'representative' figure. For example, suppose our data consisted of the observations 1, 4, 5, 6, 499. Here the arithmetic mean is 103, which is not representative of the data since it is dominated by one large value. We shall now discuss two 'representative' figures which are not affected by extreme observations.

*The median*
The median is that value of the variable which divides the data so that half of the values are less than and half greater than the median. Suppose we have $n$ observations on some variable $X$ and we arrange them in order of increasing magnitude, so that $X_{(1)} \leq X_{(2)} \leq \ldots \leq X_{(n)}$, where the parentheses around the subscript will be used to indicate that the values have been ranked in this fashion. First, suppose that $n$ is an odd number; for example, let $n = 2m + 1$. Then the median value is $X_{(m+1)}$. The data given in the previous paragraph, 1, 4, 5, 6, 499, are already ordered. Here $n = 5$ and $m = 2$, so that $X_{(3)} = 5$ is the median value. When $n$ is an even number ($n = 2m$), then the median falls between $X_{(m)}$ and $X_{(m+1)}$ and the convention usually followed is to take the average of these two values, so that

$$\text{Median } (M) = \frac{X_{(m)} + X_{(m+1)}}{2}.$$

For our income data $n = 100$ and for City A we have $X_{(1)} = 350$, $X_{(2)} = 530, \ldots, X_{(50)} = 4350$, $X_{(51)} = 4370, \ldots, X_{(100)} = 9950$. Here

$$M = \tfrac{1}{2}(4350 + 4370) = £4360.$$

For City B the median income is £4735 and again a comparison of the medians confirms the visual impression obtained from Figures 1.1 and 1.2. It is clear that, since (at most) only $X_{(m)}$ and $X_{(m+1)}$ enter into the calculation of the median, this statistic will not be affected by extreme values.

*The mode*
The mode is that value of the variable which appears most frequently among the data. Thus if the observations are 1, 2, 3, 3, 4, the mode is 3. However, there may be a problem in defining the mode for small quantities of data. Thus in our earlier example, the observations were 1, 4, 5, 6, 499 and there is no mode, since all the values appear once and once only. Another problem is that there may be multiple modes, as is the case with the income data for City A, where all values appear only once except for £1690, £2700, £6830, £7790 and £8440. These five values each appear twice and are all modes by the definition given above. Hence the mode must be used with caution as a representative value and is most useful

with very large quantities of data, especially when they have been grouped into a frequency distribution.

So far we have noted that for the income data a comparison of either the two arithmetic means or the two medians confirms the general impression obtained from the visual representation of the data. It will not always be convenient to present the data in graphical form, in which case we may wish to summarize the salient features of the data in terms of statistics such as the arithmetic mean or the median. If this is the objective, we shall need more than one kind of statistic to achieve a good summary, since the dispersion of the distribution is also of interest. For example, both the following sets of data have the same arithmetic mean and median, but differ very much in other respects: (a) 1, 33, 50, 66, 100; (b) 48, 49, 50, 51, 52. It would not be very informative if we merely reported that both sets of data had the same mean. We shall therefore consider some possible measures of dispersion.

## 1.4  measures of dispersion
### The range
The simplest measure of dispersion is the range, which is defined as the difference between the largest and smallest observations. This measure would bring out the difference between the two sets of data given above very sharply, since for the first set the range is $100 - 1 = 99$, while for the second set the range is $52 - 48 = 4$. In the case of our income data we find that the range for City A is $9950 - 350 = £9600$, while for City B the range is $9810 - 740 = £9070$.

For small quantities of data the range may be a useful summary statistic, but it can be criticized on the grounds that it only makes use of two observations, regardless of the amount of data, and that these are the two most extreme observations. Given we have the other observations, it would seem reasonable to make use of this information. Two measures of dispersion which make use of this information will now be considered.

### The quartile deviation
This measure of dispersion can be thought of as the most natural addition to the information presented by the median. The median divides the data so that half the values lie above and half below the median value. We now define $Q_1$ as the lower quartile, so that 25 per cent of the observations lie below this value, and define $Q_3$ as the upper quartile, so that 25 per cent of the observations lie above this value. Then the *interquartile range*, $Q_3 - Q_1$, includes 50 per cent of the observations and gives an indication of the dispersion of the middle section of the data. By concentrating on the central 50 per cent, the interquartile range is not affected by extreme observations. The quartile deviation is defined as

Q.D. $= \frac{1}{2}(Q_3 - Q_1)$

and it provides a measure of the average deviation of the quartiles from the median, since

$$\text{Q.D.} = \frac{1}{2}[(Q_3 - M) + (M - Q_1)]. \tag{1.2}$$

To illustrate the calculation of this statistic we shall consider the income data again. For City A we have $Q_3 = £6830$ and $Q_1 = £2340$, giving an interquartile range of £4490 and a quartile deviation of £2245. For City B, $Q_3 = £6310$ and $Q_1 = £3330$, so that the interquartile range is £2980 and the quartile deviation is £1490. A comparison of the quartile deviations for the two sets of data suggests that incomes are somewhat more closely concentrated around the median in City B than in City A, which agrees with the visual impression given by Figures 1.1 and 1.2.

*The root mean square deviation*
The expression for the quartile deviation given in equation 1.2 relates this measure of dispersion to the median as the representative value. This linkage is reasonable since both measures are concerned with percentages of the data. The arithmetic mean has also been discussed as a representative value and we shall now discuss measures of dispersion which may be associated with the arithmetic mean. Since we are interested in measuring the dispersion of the data around the arithmetic mean, a reasonable measure would be to average the dispersion from the mean. Suppose we have a set of observations $X_1, \ldots, X_n$ from which we calculate the arithmetic mean $\bar{X} = \sum X_i / n$. Then we might define our measure of dispersion as

$$\frac{1}{n} \sum_{i=1}^{n} (X_i - \bar{X}). \tag{1.3}$$

Unfortunately this measure will not do, since if we expand 1.3 we see that

$$\frac{1}{n} \sum (X_i - \bar{X}) = \frac{1}{n} \left( \sum X_i - n\bar{X} \right) = \sum X_i / n - \bar{X} = \bar{X} - \bar{X} = 0.$$

The fact that the positive and negative deviations from the arithmetic mean exactly cancel out presents a minor problem which may be overcome in a number of ways. One way is to ignore the signs of the deviations and treat them all as being positive. We may then define the *mean absolute deviation* as

$$\frac{1}{n} \sum_{i=1}^{n} |X_i - \bar{X}|,$$

where the vertical lines are used to indicate that we are considering the 'modulus' (that is, the positive value) of the expression $X_i - \overline{X}$. An alternative way of dealing with this problem is to square the deviations before averaging and then take the square root at the end of the calculation. We define the *root mean square deviation* (RMSD) as[2]

$$s' = \sqrt{\frac{1}{n} \sum_{i=1}^{n} (X_i - \overline{X})^2}.$$  (1.4)

(As we shall meet the concept below, the reader may note that the square of the root mean square deviation is called the *mean square deviation* (MSD).)

Of the two measures of dispersion around the arithmetic mean that we have introduced, the mean absolute deviation may appear the easier to understand and interpret, but in what follows we shall concentrate on the root mean square deviation. One reason for this is computational: in any situation involving more than a few observations it is easier to calculate the root mean square deviation than the mean absolute deviation. The root mean square deviation also has some theoretical advantages over the mean absolute deviation, but we must defer a discussion of this point until chapter 4.

As it stands, equation 1.4 involves calculating the arithmetic mean, calculating the deviations from $\overline{X}$, squaring these deviations and averaging the squared deviations. However, the equation can be expressed in a form which is considerably more efficient for computational purposes. Consider the expansion[3]

$$\sum_{i=1}^{n} (X_i - \overline{X})^2 = \sum (X_i^2 - 2\overline{X}X_i + \overline{X}^2) = \sum X_i^2 - 2\overline{X}\sum X_i + n\overline{X}^2$$

$$= \sum X_i^2 - \left(\sum X_i\right)^2 / n = \sum X_i^2 - n\overline{X}^2,$$

since $\sum X_i = n\overline{X}$.

Then we can rewrite the root mean square deviation in either of the forms

$$s' = \sqrt{\frac{1}{n}\left(\sum X_i^2 - \left(\sum X_i\right)^2 / n\right)} = \sqrt{\frac{1}{n}\left(\sum X_i^2 - n\overline{X}^2\right)}.$$  (1.5)

Clearly, summing the original data or the squares of the observations and performing one subtraction is computationally easier than squaring and summing the results of $n$ subtractions.

For our income data we calculate that for City A the root mean square deviation is £2745, while for City B the RMSD is £2137. A comparison of these values confirms that incomes in City A are more dispersed than is the case in City B.

This is true not only in absolute terms, but even more so in relation to the means of the two groups. Incomes in City A show a greater dispersion about a lower mean than is the case for City B. One way in which we may combine this relative information is to express the root mean square deviation as a percentage of the arithmetic mean. We define the *coefficient of variation* as

$$V = 100 \left( \frac{\text{root mean square deviation}}{\text{arithmetic mean}} \right)$$

and for our income data we calculate $V_A = 57 \cdot 8$ per cent and $V_B = 43 \cdot 5$ per cent.

## 1.5 conclusions

We have now shown that it is possible to compress a considerable amount of numerical information into a small number of summary statistics, such as the arithmetic mean and the root mean square deviation. These statistics were illustrated by means of the hypothetical income data for two cities, whose means and root mean square deviations were compared above. The reader will have noticed that in making these comparisons we were careful not to say anything about incomes in general in the two cities. Our aim will be to draw inferences about the population from which the sample is drawn, but before we can do so we must develop some statistical theory starting with the basic ideas of probability. This is the subject we shall turn to in the next chapter.

# problems

**1.1** Evaluate the following summations:

(a) $\sum_{i=0}^{4} 2^i$, (b) $\sum_{i=0}^{4} 2i$, (c) $\sum_{i=1}^{3} aX_i$, (d) $\sum_{i=1}^{4} (a + bX_i)$,

(e) $\sum_{i=1}^{3} (2X_i + 3)^2$, (f) $\sum_{i=1}^{4} (a + bX_i)(c + dY_i)$.

**1.2** Table 1.4 contains data on 100 incomes collected in City C. Given these data:

(a) Form a frequency distribution for the data.

(b) Plot a histogram for the data and the cumulative frequency distribution.

(c) Calculate the mode, the median and the arithmetic mean for the data.

(d) Calculate the range, the quartile deviation and the root mean square deviation.

(e) Compare the results obtained for City C with those reported earlier for City A and City B.

*Table 1.4.* Income data for City C

| | | | | | | | | | |
|---|---|---|---|---|---|---|---|---|---|
| 5700 | 1790 | 3710 | 9730 | 4810 | 7100 | 8090 | 3790 | 5460 | 5450 |
| 8950 | 3690 | 1800 | 4290 | 2750 | 7940 | 1650 | 670 | 5330 | 6030 |
| 940 | 1560 | 880 | 9290 | 2990 | 7680 | 3270 | 8500 | 6390 | 4020 |
| 1630 | 4550 | 2640 | 8860 | 3940 | 2220 | 9450 | 6040 | 9280 | 2950 |
| 8050 | 7270 | 6640 | 7340 | 1860 | 8700 | 846 | 4530 | 6190 | 5870 |
| 4180 | 670 | 4900 | 5060 | 2790 | 9870 | 6720 | 990 | 3320 | 8670 |
| 6620 | 9570 | 3140 | 950 | 8520 | 1970 | 1880 | 7090 | 5410 | 8710 |
| 4230 | 9620 | 590 | 9840 | 8170 | 2740 | 7460 | 8500 | 1950 | 5100 |
| 9080 | 3500 | 1580 | 4140 | 7720 | 8180 | 9070 | 5820 | 7520 | 6250 |
| 3340 | 720 | 3360 | 1950 | 4920 | 5250 | 1930 | 9860 | 2570 | 7320 |

# chapter 2
# probability theory

## 2.1 introduction

We shall develop the basic ideas of probability theory in the context of simple experiments which involve tossing coins or throwing dice. The particular characteristics of an experiment to note are that (i) a number of different outcomes to the experiment exist and (ii) before the experiment is performed we are uncertain which outcome will occur. For example, if our experiment involves throwing a die and observing the score obtained, there are six outcomes – the numbers 1, 2, 3, 4, 5, 6. Before the die is cast we do not know what the outcome will be, but we shall assume that the coins and dice are physically constructed in such a way that there is no particular bias towards some outcomes rather than others. For example, we rule out two-headed pennies! We shall also assume that the tossing or throwing process is fair and does not bias the experiment towards any particular outcome. These assumptions enable us to regard the set of outcomes as being symmetrical, with none of the outcomes being more likely to occur than any of the others.

The outcomes of an experiment may be thought of as the basic building blocks, and we may now define an *event* to consist of a group of outcomes having a given characteristic. For example, in throwing a die, if the event is that we obtain an even score, then there are three outcomes, 2, 4, 6, which correspond to this event. In some cases an event may consist of a single outcome; for example, in tossing a coin only one outcome corresponds to the event 'getting a head'.

Let us now introduce some notation. Suppose an experiment has $n$ outcomes, which we denote as $e_1, \ldots, e_n$. Events will be denoted by $E_1, E_2$, say $E_i, i = 1, \ldots, k$. We now define two important properties of events. Suppose we have a set of events $E_1, \ldots, E_k$ defined for a given experiment and we allocate the outcomes of the experiment to the events. Then if the events account for all the outcomes they are said to be *exhaustive*. For example, in throwing a die, if $E_1$ is the event 'obtaining an even score' and $E_2$ 'obtain an odd score', then the events are exhaustive. However, if $E_1$ is the event 'get an even score' and $E_3$ is 'get a score $\leq 4$' the events are not exhaustive, since the outcome 'score 5' is not included in either event. The second case to consider is that in which we define $E_i, i = 1, \ldots, k$, in such a way that each outcome corresponds to one and only one event. In such a case the events are said to be *mutually exclusive*. For example, in throwing the die $E_1$ and $E_2$ are mutually exclusive, but $E_1$ and $E_3$ are not, since the outcomes 'score 2' and 'score 4' correspond to both events.

We may illustrate these concepts diagrammatically. For this example we have six outcomes which are represented by points inside the rectangle. Events are shown by drawing a line to include the corresponding outcomes. Figure 2.1(a) makes clear that $E_1$ and $E_2$ are exhaustive and mutually exclusive, while (b) shows that $E_1$ and $E_3$ are neither exhaustive nor mutually exclusive. These diagrams, which are called Venn diagrams after their inventor, will often be used in the discussion that follows.

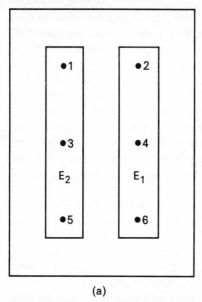
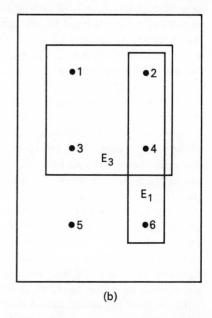

(a)                                              (b)

Figure 2.1

## 2.2 the definition of probability

Given our framework of experiments, outcomes and events, we are now ready to define the probability of an event occurring. If, out of the total of $n$ outcomes to an experiment, $n_i$ corresponds to the event $E_i$, then the probability that $E_i$ occurs, $p(E_i)$, is defined as

$$p(E_i) = \frac{n_i}{n}. \qquad\qquad 2.1$$

For example, if $E_1$ is 'getting an even score', then three outcomes correspond to this event and $p(E_1) = \frac{3}{6} = \frac{1}{2}$. Similarly, if we toss a coin and define $E_1$ as the event 'get a head', then $p(E_1) = \frac{1}{2}$.

Given this definition of a probability, we may obtain the following results. Firstly, the limits on the outcomes corresponding to $E_i$ are that there may be no outcomes that correspond to the event or that all the outcomes of the experiment correspond to $E_i$. In the first case $p(E_i) = 0$ and in the second $p(E_i) = 1$. Thus we conclude that

$$0 \leq p(E_i) \leq 1. \tag{2.2}$$

For example, if we are throwing a die and $E_1$ is the event 'score 7' and $E_2$ is the event 'score a number between 1 and 6 inclusive', then $p(E_1) = 0$, while $p(E_2) = 1$.

Secondly, consider the case in which an experiment is divided into $k$ events which are exhaustive and mutually exclusive, and let the number of outcomes corresponding to $E_i$ be $n_i (i = 1, \ldots, k)$. Since the events are exhaustive and mutually exclusive,

$$n_1 + n_2 + \ldots + n_k = n,$$

or, dividing through by $n$,

$$\frac{n_1}{n} + \frac{n_2}{n} + \ldots + \frac{n_k}{n} = 1,$$

that is,

$$p(E_1) + p(E_2) + \ldots + p(E_k) = 1. \tag{2.3}$$

Thirdly, consider any one of these events, say $E_1$. Denoting the event 'not $E_1$' by $\bar{E}_1$, we have $n_2 + n_3 + \ldots + n_k$ outcomes corresponding to $\bar{E}_1$. Since

$$n_1 = n - (n_2 + n_3 + \ldots + n_k)$$

$$\frac{n_1}{n} = 1 - \frac{n_2 + \ldots + n_k}{n}$$

$$p(E_1) = 1 - p(\bar{E}_1). \tag{2.4}$$

These three results define the most important characteristics of probabilities, and any set of numbers which satisfy 2.2, 2.3 and 2.4 can represent the probabilities of a set of events corresponding to some experiment. For example, $\frac{1}{8}, \frac{3}{8}, \frac{3}{8}, \frac{1}{8}$ satisfy 2.2 and 2.3 and hence could be a set of probabilities. We shall show in the next section that these are in fact the probabilities of obtaining 0, 1, 2 or 3 heads in tossing three unbiased coins.

### 2.3 some rules of probability

Working with the definition of a probability given in 2.1, it is possible to answer many questions easily from first principles. We shall now extend the discussion of probability to deal with more complex situations.

*The addition rule*

Let $E_i$ and $E_j$ be two events corresponding to a given experiment and consider the probability of observing either event $E_i$ or event $E_j$, that is, $p(E_i \text{ or } E_j)$. Here we need to consider two cases, which are illustrated in Figure 2.2.

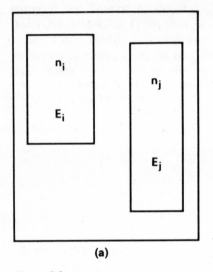
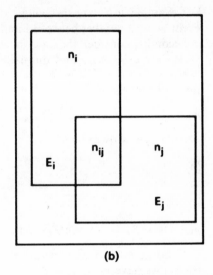

(a)                                    (b)

Figure 2.2

In the example shown in part (a), the events $E_i$ and $E_j$ are mutually exclusive. Out of the total of $n$ outcomes we assume $n_i$ correspond to $E_i$ and $n_j$ correspond to $E_j$, so that $(n_i + n_j)$ correspond to either $E_i$ or $E_j$. Hence, using 2.1, we may define the required probability as

$$p(E_i \text{ or } E_j) = \frac{n_i + n_j}{n} = \frac{n_i}{n} + \frac{n_j}{n},$$

that is,

$$p(E_i \text{ or } E_j) = p(E_i) + p(E_j). \tag{2.5}$$

For example, if the experiment involves throwing a fair die and $E_1$ is the event 'score six' and $E_2$ is the event 'score 4 or less', then $n = 6$, $n_1 = 1$ and $n_2 = 4$.

$P(E_1 \text{ or } E_2) = \frac{1}{6} + \frac{4}{6} = \frac{5}{6}$.

Suppose, however, that in the same experiment $E_3$ is 'score an even number' and $E_2$ is 'score 4 or less'. Then

$p(E_3 \text{ or } E_2) = \frac{3}{6} + \frac{4}{6} = \frac{7}{6}$,

giving a probability which is greater than unity! The reason for this result may be seen if we consider Figure 2.1(b) and note that the outcomes 'score 2' and 'score 4' have been credited both to $E_i$ and $E_j$, so that we have double-counted these two outcomes. Clearly, only five of the six outcomes correspond to $E_i$ or $E_j$, so that the correct result is

$p(E_i \text{ or } E_j) = \frac{5}{6}$.

The double-counting, which arises because the events $E_i$ and $E_j$ are not mutually exclusive, may be avoided if we modify the formula given in 2.5 by subtracting those outcomes which correspond to *both* $E_i$ and $E_j$. Thus

$$p(E_i \text{ or } E_j) = p(E_i) + p(E_j) - p(E_i \text{ and } E_j), \tag{2.6}$$

and in the example given above we have

$p(E_i \text{ or } E_j) = \frac{3}{6} + \frac{4}{6} - \frac{2}{6} = \frac{5}{6}$.

The result expressed in 2.5 and 2.6 is often called the *addition* rule of probability. It may easily be extended for events that are mutually exclusive. If events $E_i$, $E_j$, $E_k$ with outcomes $n_i$, $n_j$ and $n_k$ are mutually exclusive, then

$$p(E_i \text{ or } E_j \text{ or } E_k) = \frac{n_i + n_j + n_k}{n}$$
$$= p(E_i) + p(E_j) + p(E_k). \tag{2.7}$$

If the events are not mutually exclusive, the extension becomes more complicated as there are now more opportunities for double- or triple-counting. Figure 2.3(a) shows one possibility for which

$$p(E_i \text{ or } E_j \text{ or } E_k) = p(E_i) + p(E_j) + p(E_k) - p(E_i \text{ and } E_j)$$
$$- p(E_i \text{ and } E_k) - p(E_j \text{ and } E_k) + p(E_i \text{ and } E_j \text{ and } E_k). \tag{2.8}$$

For example, in throwing a fair die, let $E_1$ be the event 'score 2', $E_2$ be the event 'score 4 or less' and $E_3$ be 'score an even number'. Then $p(E_1 \text{ or } E_2 \text{ or } E_3) = \frac{1}{6} + \frac{4}{6} + \frac{3}{6} - \frac{1}{6} - \frac{1}{6} - \frac{2}{6} + \frac{1}{6} = \frac{5}{6}$. This is illustrated in Figure 2.3(b).

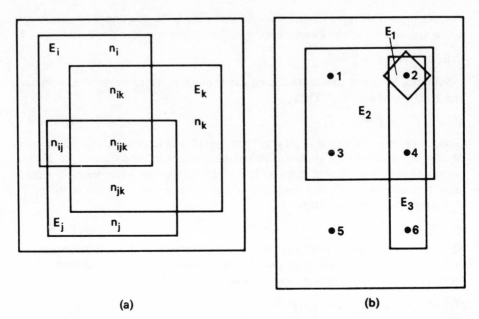

(a)                                                              (b)

Figure 2.3

*The multiplication rule*
The other case of interest is the probability that two events may both occur, that is, $p(E_i$ and $E_j)$. We shall investigate this probability now, but shall find it convenient to consider a further concept first, that of a *conditional* probability. By a conditional probability we mean, for example, the probability that event $E_i$ will occur given event $E_j$ has occurred. This probability we shall denote as $p(E_i|E_j)$ and it may be explored by considering Figure 2.2(b).

There are $n$ outcomes to the experiment, but since we know that $E_j$ has occurred, the set of outcomes which are relevant is reduced from $n$ to $n_j$. Of this set of outcomes, $n_{ij}$ satisfy the requirement that they are in both $E_i$ and $E_j$. Hence, from 2.1,

$$p(E_i|E_j) = \frac{n_{ij}}{n_j}.$$

Multiplying this ratio by $n/n$ and rearranging:

$$p(E_i|E_j) = \frac{n_{ij}}{n} \cdot \frac{n}{n_j} = \frac{n_{ij}}{n} \div \frac{n_j}{n}$$

or

$$p(E_i|E_j) = p(E_i \text{ and } E_j)/p(E_j). \tag{2.9}$$

From 2.9 we obtain

$$p(E_i \text{ and } E_j) = p(E_i|E_j)\,p(E_j). \tag{2.10}$$

We could also consider $p(E_j|E_i)$, which from Figure 2.2(b) we see is

$$p(E_j|E_i) = \frac{n_{ij}}{n_i} = \frac{p(E_i \text{ and } E_j)}{p(E_i)},$$

so that

$$p(E_i \text{ and } E_j) = p(E_j|E_i)\,p(E_i).$$

For example, suppose we have a bag containing nine discs which are numbered from 1 to 9, but which are physically identical, and draw two out without replacement, shaking the bag well before each draw. What is the probability that both will be even numbers? If $E_1$ is even number on the first draw and $E_2$ is even number on the second draw, then we want $p(E_1 \text{ and } E_2)$. Since the bag is well shaken before draws and the discs and physically identical, we may assume that all outcomes are equally likely. From 2.10 we have

$$p(E_1 \text{ and } E_2) = p(E_2|E_1)\,p(E_1).$$

On the first draw there are nine equally likely outcomes, of which four correspond to $E_1$. Hence $p(E_1) = \frac{4}{9}$. Given $E_1$ has occurred, one even-numbered disc has been removed from the bag, so that there are now eight outcomes, three of which correspond to $E_2$. Hence $p(E_2|E_1) = \frac{3}{8}$. Combining these results yields $p(E_1 \text{ and } E_2) = \frac{3}{8} \cdot \frac{4}{9} = \frac{1}{6}$.

Suppose now, however, that the first disc drawn is replaced before the second draw is made (and that the process of making the draw does not affect the physical properties of the disc that was drawn first). Does this affect $p(E_1 \text{ and } E_2)$? Clearly, $p(E_1)$ is not affected, but what of $p(E_2|E_1)$? If the first disc is replaced and the bag shaken, there are now nine equally likely outcomes of which four correspond to $E_2$. Hence $p(E_2|E_1) = \frac{4}{9}$ and $p(E_1 \text{ and } E_2) = \frac{4}{9} \cdot \frac{4}{9} = \frac{16}{81}$. The point to note here is that if the first disc is replaced, $p(E_2|E_1) = p(E_2)$, that is, it is the probability of getting an even number *regardless of what has happened on the first draw*. This result is important, and if $p(E_2|E_1) = p(E_2)$ the events $E_1$ and $E_2$ are said to be *statistically independent*. For such events we may formalize the result demonstrated in the last example and rewrite 2.10 as

$$p(E_i \text{ and } E_j) = p(E_i)\,p(E_j), \tag{2.11}$$

when $E_i$ and $E_j$ are statistically independent.[1] For statistically independent events this result may be extended to

$$p(E_i \text{ and } E_j \text{ and } E_k) = p(E_i)\,p(E_j)\,p(E_k). \tag{2.12}$$

## 2.4 counting outcomes. permutations and combinations

Given the addition and multiplication rules, we are now in a position to solve a wide range of probability problems. As an example, consider the following problem. Suppose that a fair coin is tossed three times and the outcomes (heads or tails) are recorded. What is the probability of getting either three heads or three tails? We first note that since we either get heads or tails, but not both, the events are mutually exclusive. Secondly, we may assume that what happens on one toss of the coin will not affect what happens on later tosses, that is, the events are independent. Let $E_1$ be the event 'three heads' and $E_2$ be 'three tails'; then we have

$$p(E_1 \text{ or } E_2) = p(E_1) + p(E_2).$$

$E_1$ in turn requires (head on first toss) and (head on second toss) and (head on third toss), and since the events are independent we have $p(E_1) = \frac{1}{2} \cdot \frac{1}{2} \cdot \frac{1}{2} = \frac{1}{8}$. Similarly, $E_2$ requires (tail on first toss) and (tail on second toss) and (tail on third toss), and $p(E_2) = \frac{1}{2} \cdot \frac{1}{2} \cdot \frac{1}{2} = \frac{1}{8}$. Hence

$$p(E_1 \text{ and } E_2) = \frac{1}{8} + \frac{1}{8} = \frac{1}{4}.$$

Keeping to the same experiment, let us now consider the probability of getting either one head and two tails $(E_3)$ or two heads and one tail $(E_4)$. Following the argument presented above, we have

$$p(E_3 \text{ or } E_4) = p(E_3) + p(E_4)$$

and we need to evaluate $p(E_3)$ and $p(E_4)$. This is more involved than finding $p(E_1)$ or $p(E_2)$ as there are more configurations of outcomes corresponding to $E_3$ and $E_4$. For example, corresponding to $E_3$ we have [(head on first) and (tail on second) and (tail on third)] or [(tail on first) and (head on second) and (tail on third)] or [(tail on first) and (tail on second) and (head on third)]. Hence

$$p(E_3) = \frac{1}{8} + \frac{1}{8} + \frac{1}{8} = \frac{3}{8}.$$

From the symmetry of the experiment $p(E_4) = \frac{3}{8}$ and so $p(E_3 + E_4) = \frac{3}{8} + \frac{3}{8} = \frac{3}{4}$. (The reader may confirm these results by noting that $E_1$, $E_2$, $E_3$ and $E_4$ are exhaustive and mutually exclusive, so that

$$p(E_3) + p(E_4) = 1 - p(E_1) - p(E_2) = 1 - \frac{1}{4} = \frac{3}{4}.)$$

This example raises a new problem, namely the need to count the number of outcomes corresponding to a given event. One way in which this can be done is to write out all the outcomes to an experiment and then group up the outcomes corresponding to the events which are being considered. For example, in our problem of tossing three coins there are eight different outcomes:

| 1st toss | 2nd toss | 3rd toss |
|----------|----------|----------|
| T | T | T |
| T | T | H |
| T | H | T |
| H | T | T |
| T | H | H |
| H | T | H |
| H | H | T |
| H | H | H |

and it is easy to see that three outcomes correspond to one head and two tails. In this example it is fairly simple to list all the possible outcomes, but if we extend this experiment to consider tossing more coins, the number of outcomes rises rapidly. For example, if we toss six coins, each coin can land in one of two ways so that there are $2^6 = 64$ different outcomes to the experiment, and listing them becomes a tedious business. For example, as we shall show, there are twenty different outcomes corresponding to three heads and three tails. We shall now develop some rules which will enable us to count the number of outcomes without having to write them out completely.

First, suppose we have $n$ different objects; in how many different ways can we arrange them in a line? For example, if we have four books, in how many ways can they be arranged upright, the right way up with their titles facing out, on a shelf? The first object may be chosen in $n$ ways, leaving $(n-1)$ objects to be arranged. The second object can be chosen in $(n-1)$ ways, so that there are $n(n-1)$ ways of choosing the first two objects. The third object can be chosen in $(n-2)$ ways, so that the first three objects can be chosen in $n(n-1)(n-2)$ different ways. If we continue this process it is clear that the total number of arrangements is $n(n-1)(n-2)\dots(3)(2)(1)$. We shall denote this expression as

$$n! = n(n-1)(n-2)\dots(2)(1), \tag{2.13}$$

where $n!$ is called '$n$ factorial'. This is defined for positive integer values, but in addition we shall find it useful to define

$$0! \equiv 1. \tag{2.14}$$

As an example we have $4! = (4)(3)(2)(1) = 24$. Thus we could arrange our four books in twenty-four different ways.

If, instead of arranging all $n$ objects, we choose a subset of $r$ objects, how many different sets of $r$ can we choose? From the argument developed above, the number of arrangements is $n(n - 1)(n - 2) \ldots (n - r + 1)$. If we notice that

$$n! = n(n - 1)(n - 2) \ldots (n - r + 1) \times (n - r)(n - r - 1) \ldots (2)(1)$$

$$= n(n - 1)(n - 2) \ldots (n - r + 1) \times (n - r)!$$

we may write

$$n(n - 1) \ldots (n - r + 1) = \frac{n!}{(n - r)!}.$$

The left-hand side of this expression, the number of permutations of $r$ objects chosen from $n$, is denoted by

$$P_r^n = \frac{n!}{(n - r)!}. \tag{2.15}$$

In deriving the result presented in 2.15 it has been assumed that the order of the objects within the subset matters, but this is not always the case. For example, if we want to know how many subcommittees of three can be chosen from a committee of six, then $P_3^6 = 6!/3! = 120$ overstates the number of different groups of three. It overstates by the factor of how many ways each group of three may be arranged among themselves, which is $3!$ Hence there are in fact $120/3! = 120/6 = 20$ different subcommittees which may be chosen. The result given in 2.15 may be modified if the permutations of the $r$ objects do not matter to give

$$C_r^n = \frac{n!}{(n - r)! \, r!}, \tag{2.16}$$

where $C_r^n$ denotes the number of combinations of $r$ objects chosen from $n$ objects.

So far we have discussed the number of arrangements of $n$ objects when all of them are different, but our results are modified in an important way if some of the objects are the same. We may see this in the following example, in which we consider first the $3! = 6$ permutations of the letters A, B and C. These are listed in the first column, while the second column shows what happens if two

| | |
|---|---|
| ABC | AAC |
| ACB | ACA |
| BAC | AAC |
| BCA | ACA |
| CAB | CAA |
| CBA | CAA |

of the three letters are the same. There are only three different arrangements and 3! overstates the number of arrangements by the factor 2! In general terms the number of arrangements of $n$ objects, $r$ of which are the same, is

$$\frac{n!}{r!}.$$

This argument can be extended if there is more than one group of similar objects. In particular, if $r$ objects are of one kind and the remaining $(n - r)$ are all of another kind, the number of different arrangements of $n$ objects is

$$C_r^n = \frac{n!}{(n - r)!\, r!},$$

the same formula as that obtained in 2.16 and an important result for what follows. The reader should note that, since $0! \equiv 1$,

$$C_0^n = C_n^n = \frac{n!}{0!\, n!} = 1.$$

As an example, suppose a coin is tossed six times. In how many ways can we get three heads and three tails? The answer is

$$C_3^6 = \frac{6!}{3!\, 3!} = 20.$$

One further important mathematical result the reader should note is the Binomial Theorem, which will be given here without proof.[2] The theorem is concerned with the expansion of an expression such as $(a + b)^n$ and states that

$$(a + b)^n = a^n + C_1^n a^{n-1} b + C_2^n a^{n-2} b^2 + \ldots + C_r^n a^{n-r} b^r + \ldots$$
$$+ C_{n-1}^n ab^{n-1} + b^n. \tag{2.17}$$

The right-hand side of 2.17 may be written in an alternative form:

$$(a + b)^n = a^n + na^{n-1} b + \frac{n(n - 1)}{2!} a^{n-2} b^2 + \ldots$$

$$+ \frac{n(n - 1) \ldots (n - r + 1)}{r!} a^{n-r} b^r + \ldots + nab^{n-1} + b^n.$$

One special case of the Binomial Theorem is of particular interest to us. Suppose we put $b = p$, where $0 \leq p \leq 1$, and $a = (1 - p)$. Then

$$(a + b)^n = [(1 - p) + p]^n = 1.$$

The expansion corresponding to 2.17 is

$$[(1 - p) + p]^n = (1 - p)^n + C_1^n (1 - p)^{n-1} p + C_2^n (1 - p)^{n-2} p^2 + \ldots$$
$$+ C_r^n (1 - p)^{n-r} p^r + \ldots + C_{n-1}^n (1 - p) p^{n-1} + p^n. \qquad (2.18)$$

Given its definition, $p$ may be interpreted as a probability, and we shall use this form of the Binomial Theorem to derive some important results in the next chapter.

# problems

**2.1** If a card is drawn from a standard pack of 52 playing cards which has been well shuffled, what is the probability that the card is
    (a) either a Jack or a Queen,
    (b) either a black King or a red Queen,
    (c) either a club or less than ten,
    (d) either a red card or an ace?

**2.2** If two cards are drawn from a well-shuffled pack, find the probability that both cards are black
    (a) when the first card is replaced between draws, and
    (b) when the first card is not replaced.

**2.3** If we throw two unbiased dice and record the numbers shown, what is the probability that
    (a) the numbers shown on the two faces sum to 5,
    (b) the number shown on one face equals twice the number shown on the other,
    (c) one face shows 5 and the other face shows a number less than five,
    (d) the two faces show different scores?

**2.4** An urn contains four physically identical discs which are numbered 1, 2, 3, 4. Two discs are chosen at random. Find the probability that the numbers on the discs are consecutive integers if
    (a) the first disc is replaced before the second is drawn, and
    (b) the first is not replaced.

**2.5** The probabilities that three drivers will be able to drive home safely after drinking are 1/3, 1/4 and 1/5 respectively. If they set out to drive home after a party, what is the probability that all three drivers will have accidents? What is the probability that at least one driver will drive home safely?

**2.6** A section of a computer consists of 100 identical components and the section will fail if any of the components fails. If the probability that a component will fail is 0.01, how reliable is the computer section?

**2.7** A bag contains five physically identical discs, numbered 1, ..., 5. If the bag is well shaken and the discs are drawn out one by one to form a five-digit number,

what is the probability that
    (a) the number obtained is 12345, and
    (b) the number ends in 5?

**2.8** A bag contains four physically identical discs, two of which are numbered 1 and two of which are numbered 2. If the four discs are drawn one by one to form a four-digit number and the last digit is 1, what is the probability that the two 2's occur in consecutive positions?

**2.9** An experiment involves three identical black bags, the first of which contains two identical red balls, the second contains one red ball and one white ball which are physically identical, while the third bag contains two identical white balls. The experiment involves choosing one of the bags at random and then drawing one of the balls from the selected bag without looking at the other. If the ball drawn is white, what is the probability that the other ball in this bag is white?

# chapter 3

# random variables and frequency distributions

## 3.1 random variables

In the previous chapter we characterized an experiment as consisting of a number of outcomes, each of which has some probability of occurring. For example, if we are tossing a number of coins, say $n$, we may be interested in the number of heads we obtain. Here we may denote the number of heads observed in the experiment as $X$, where $X$ may take any one of the integer values $0, \ldots, n$. Before the experiment is carried out we do not know which value of $X$ will occur, but for each value of $X$ we may specify the probability it will take a particular value, say $p(X = X_0)$. We shall define $X$ as a *random variable* and the set of probabilities which correspond to the values $X$ may take as a theoretical *probability distribution, $p(X)$*.

In the experiments we have considered involving coin tossing or dice throwing, the random variables have only been defined to take on integer values: for example, the *number* of heads, the *sum* of the scores observed in throwing three dice. Here $X$ is a *discrete* random variable and is the outcome of a counting exercise. However, when we consider experiments involving measurement, $X$ is not confined to take only integer values. For example, if $X$ denotes the height of a man aged between 25 and 30 and living in the United Kingdom, our measurement need not be an integer value. In practice there will be constraints on how small a difference in height our instruments can detect, but in principle we could go on subdividing any fraction of an inch into ever finer measurements and we may think of $X$ as a *continuous* random variable. We shall first consider discrete random variables.

Given we have defined a discrete random variable, $X$, we are interested in specifying the corresponding probabilities, $p(X)$. We could do this by calculating the probabilities using the results developed in the previous chapter. For example, if $X$ denotes the number of heads obtained in tossing three unbiased coins, where $X = 0, \ldots, 3$, we have calculated that $p(0) = \frac{1}{8}$, $p(1) = \frac{3}{8}$, $p(2) = \frac{3}{8}$ and $p(3) = \frac{1}{8}$. This information could be presented in the form

| X | p(X) |
|---|---|
| 0 | $\frac{1}{8}$ |
| 1 | $\frac{3}{8}$ |
| 2 | $\frac{3}{8}$ |
| 3 | $\frac{1}{8}$ |

As an alternative to listing the probabilities in full, we shall explore the possibility of defining a theoretical probability distribution as a general mathematical formula, a function of X from which probabilities may be derived by substituting particular numerical values of X into the formulae. These mathematical formulae may be thought of as a set of models which have different characteristics and which may describe interesting experimental situations. Since the probability distribution gives the set of probabilities corresponding to the range of values of X; the models are subject to certain logical constraints. In particular, if a discrete random variable X has a range $0, \ldots, n$, then the probability distribution should have the properties that

$$0 \leq p(X) \leq 1 \tag{3.1}$$

and

$$\sum_{X=0}^{n} p(X) = 1. \tag{3.2}$$

We shall examine a number of theoretical distributions in the sections that follow, beginning with two discrete distributions.

### 3.2 the rectangular distribution
If X is a discrete random variable defined over the range $0, \ldots, n$, then we may define the rectangular distribution as[1]

$$p(X) = \frac{1}{n+1}. \tag{3.3}$$

It is easy to see that this distribution satisfies the constraints imposed by 3.1 and 3.2, since

$$0 < \frac{1}{n+1} \leq 1, \quad \text{if } n \text{ is non-negative, and}$$

$$\sum_{X=0}^{n} p(X) = \sum_{X=0}^{n} \frac{1}{n+1} = (n+1)\left(\frac{1}{n+1}\right) = 1.$$

The rectangular distribution may be represented graphically by the histogram in Figure 3.1. Here a rectangle with base of unit length and height $1/(n + 1)$ corresponds to the probability of each value of $X$. The graphical representation of the probability distribution obeys the constraints, since each probability may be represented by an area under the probability distribution which is less than one and the total area under the probability distribution is equal to one. We shall represent other theoretical distributions graphically below and shall interpret the areas under the curve corresponding to the distribution as probabilities and require the total area under the curve to equal unity.

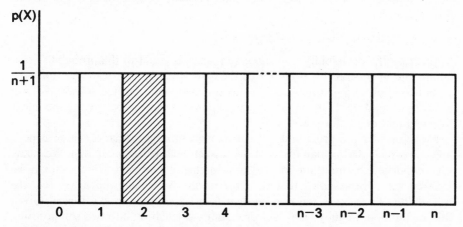

Figure 3.1 Shaded area corresponds to $p(2) = 1/(n + 1)$

The rectangular distribution would provide a theoretical model for the following experiment. Suppose ten identical discs which are numbered $0, ..., 9$ are put into a bag which is shaken well before we draw out a disc without looking and record the number on the disc. Here the discrete random variable $X = 0, ..., 9$ and

$$p(X) = \tfrac{1}{10}.$$

We shall return to the discussion of this distribution in the next chapter when we discuss random sampling and random numbers.

### 3.3 the binomial distribution
Earlier in the chapter we discussed the probabilities of obtaining various numbers of heads when tossing three unbiased coins. This is one example of a type of ex-

periment which has many applications. To obtain a general formulation of this experiment, we note the following properties of our coin-tossing experiment. Firstly, we are only interested in one result of our experiment, the number of heads. Thus we may generalize here by calling those results we are interested in *successes*, the residual set of results *failures*, and letting the discrete random variable $X = 0, 1, \ldots, n$ represent the number of successes in $n$ performances of the experiment being considered. Secondly, we may recall that the evaluation of the probability of two heads in our coin-tossing experiment resolved itself into two components: the probability of getting two heads and one tail multiplied by the number of ways in which we could permute two heads and one tail. In this case we had

$$p(\text{two heads and one tail}) = C_2^3(\tfrac{1}{2})^2\,(\tfrac{1}{2}) = \tfrac{3}{8}.$$

In this case the probability of a success (a head) is equal to the probability of a failure, but in general this need not be the case. For example, if we are throwing an unbiased die and getting a six is a success, then $p(\text{success})$ is $1/6$ and $p(\text{failure})$ is $5/6$. In general we shall denote the probability of a success in a given experiment by $p$ and that of a failure by $(1 - p)$.

Now consider the probability of $m$ successes in $n$ performances of an experiment in which the probability of a success in each experiment is $p$. We made one important assumption in the coin-tossing experiment which should be stressed here: we assumed that the experiments were independent, *so that the probability of a success remains constant from experiment to experiment.* Given this assumption, we may break our probability calculation into two components: the probability of $m$ successes and $(n - m)$ failures ignoring the order in which they occur is, if the events are independent,

$$p^m(1 - p)^n,$$

while the number of different arrangements of $n$ outcomes where $m$ are of one kind and $(n - m)$ are of another is given by

$$C_m^n = \frac{n!}{m!(n - m)!}.$$

Combining these results, we obtain

$$p(X = m) = C_m^n p^m(1 - p)^{n-m},$$

that is,

$$p(X) = C_X^n p^X(1 - p)^{n-X}, \qquad X = 0, \ldots, n. \tag{3.4}$$

This formula defines the *binomial distribution*.[2]

As an example of the use of equation 3.4 we may consider the following example. Suppose we throw three unbiased dice and call getting a six a success. If we let $X$ be a discrete random variable representing the number of successes, we may calculate $p(X)$ for $X = 0, ..., 3$. In this case $n = 3$, $p = 1/6$ and $(1 - p) = 5/6$. Substituting these values into 3.4, we obtain

$p(0) = C_0^3 (\frac{1}{6})^0 (\frac{5}{6})^3 = \frac{125}{216}$

$p(1) = C_1^3 (\frac{1}{6}) (\frac{5}{6})^2 = \frac{75}{216}$

$p(2) = C_2^3 (\frac{1}{6})^2 (\frac{5}{6}) = \frac{15}{216}$

$p(3) = C_3^3 (\frac{1}{6})^3 (\frac{5}{6})^0 = \frac{1}{216}$.

Figure 3.2 provides a graphical illustration of some binomial distributions. In 3.2(a) we have three examples where $p = \frac{1}{2}$, while $n = 1$, 2 and 3, while in 3.2(b) we have three cases where $p = 1/6$, with $n = 1$, 2, 3. We observe that the histograms vary considerably and that there is no single basic shape for the binomial distribution. While equation 3.4 is a general definition of the binomial distribution, what distinguishes the form of one binomial distribution from another is the pair of values we give to $n$ and $p$. Thus if we know we are dealing with a binomial distribution, we can calculate the values of $p(X)$ once $n$ and $p$ are specified. Once our experiment is defined, $n$ and $p$ are defined and remain constants for that experiment. We shall call these constants which define the characteristics of a particular distribution the *parameters* of the distribution. Parameters are a feature of all theoretical distributions, although the number of parameters necessary to define a distribution varies.

We note from Figure 3.2 that when $p = \frac{1}{2}$ the distribution is *symmetrical*, while for $p = \frac{1}{6}$ the binomial distribution is *skewed*, with the bulk of the area under the curve being close to zero. Had we taken a value of $p$ which was close to unity we would have found the binomial distribution skewed towards unity. Beyond these general results, however, forming an impression of the distribution from $n$ and $p$ may not be easy without calculating probabilities and perhaps plotting the histogram, a process which may be very labour-intensive. Suppose we consider the binomial distribution with $n = 100$ and $p = 0.36$. We know the distribution will be skew, but unless we calculate some values for $p(X)$ we may have very little feeling for the properties of this particular binomial distribution. We shall now therefore explore ways of presenting this information in a form which summarizes the main characteristics of the distribution.

## 3.4 expectations
In tackling this problem the reader may recall the discussion of statistics which summarize the main characteristics of a body of empirical data. There it was

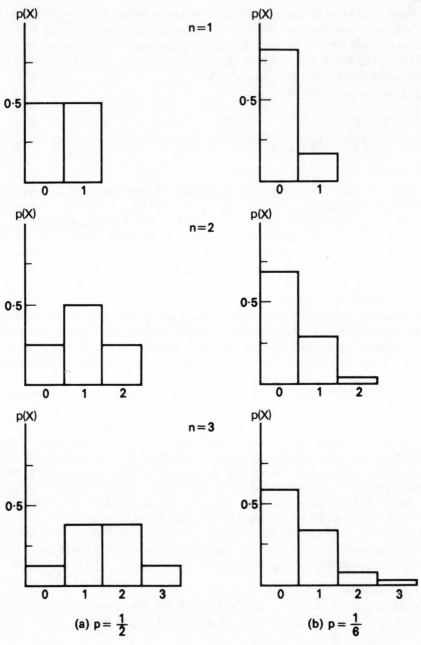

(a) $p = \frac{1}{2}$    (b) $p = \frac{1}{6}$

Figure 3.2

shown that a representative value and a measure of the dispersion around the average could provide a useful summary of a body of data. We shall consider the problem of developing similar measures in the case of frequency distributions, starting with the population analogy to the arithmetic mean, the population mean.[3] The reader will recall that the arithmetic mean is defined as $\overline{X} = 1/n \sum_{i=1}^{n} X_i$, that is, we calculate a weighted sum of the $n$ values $X_1, \ldots, X_n$, the weights being the relative frequencies with which each value appears in the set of data, that is, $1/n$. We may define the average value for the probability distribution in the same way: by weighting each of the values of $X$ in the range $0, \ldots, n$ by the relative frequency with which it occurs, which of course is $p(X)$. We therefore defined the *expected value* as[4]

$$E(X) = \sum_{X=0}^{n} p(X)\, X. \tag{3.5}$$

As an example of the calculation of an expected value, consider the binomial distribution with $n = 3$ and $p = \frac{1}{2}$. In this case we have

$$E(X) = \tfrac{1}{8}(0) + \tfrac{3}{8}(1) + \tfrac{3}{8}(2) + \tfrac{1}{8}(3) = 1\cdot 5.$$

This result is what we might expect, since we have already noticed that, when $p = \frac{1}{2}$, the binomial distribution is symmetrical. In Figure 3.3(a) we assume that each of the columns in the histogram has a base of length one unit, so that 1·5 corresponds to the centre of the distribution and the area to the left of the expected value of 1·5 equals the area to the right.

We may interpret this result as follows. Suppose we could perform the hypothetical experiment of tossing three unbiased coins and record the number of heads obtained, $X = 0, \ldots, 3$. The probabilities $p(X)$ are those presented in Figure 3.3(a), and while we would not know what value of $X$ to expect in any given per-

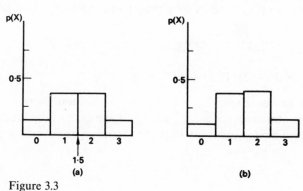

Figure 3.3

formance of the experiment, we would expect that in a very long run of replications of the experiment (tending in the limit to infinity) the relative frequencies with which the values of $X$ occur would be the probabilities, $p(X)$. Hence, if $X = 0$ occurs one-eighth of the time, etc., we would expect the average number of successes to be 1·5. Figure 3.3(b) shows part of an empirical experiment in which three coins were tossed 500 times and the number of heads recorded. Even with only 500 replications the relative frequencies obtained are close to the theoretical values of $p(X)$, and the average value of $X$ calculated from these data is 1·54, which is close to the expected value of 1·5.

We shall be using expected values in a number of places below, so that it is important to develop some general properties of expected values. Suppose we consider some discrete random variable, $X = 0, ..., n$ with probability distribution $p(X)$; what is the expected value of $a + bX$?

$$E(a + bX) = \sum_{X=0}^{n} p(X)(a + bX) = \sum ap(X) + p(X)(bX).$$

Expanding the two summations, we have

$$\sum ap(X) = ap(0) + ap(1) + ... + qp(n) = a[p(0) + ... + p(n)] = a \sum p(X) = a$$

and

$$\sum p(X)(bX) = bp(0)(0) + bp(1)(1) + ... + bp(n)(n) = b \sum p(X) X = bE(X),$$

so that

$$E(a + bX) = a + bE(X). \tag{3.6}$$

We also note that

$$E[(a + bX)^2] = E(a^2 + 2abX + b^2X^2) = \sum p(X)(a^2 + 2abX + b^2X^2)$$

$$= a^2 \sum p(X) + 2ab \sum p(X) X + b^2 \sum p(X) X^2$$

$$= a^2 + 2abE(X) + b^2E(X^2). \tag{3.7}$$

As the measure of dispersion around the population mean we shall define the population standard deviation (and its square, the population variance).[5] We recall that for a set of empirical data the mean square deviation was defined as

$$\frac{1}{n} \sum_{i=1}^{n} (X_i - \bar{X})^2,$$

that is, we calculate the squared deviation of each value of $X$ from the mean, $\overline{X}$, and form a weighted sum of the squared deviations, where the weights are the relative frequencies with which the $X$s appear, namely $1/n$. For the probability distribution, the relative frequencies are given by $p(X)$ and the population mean, $E(X)$, corresponds to $\overline{X}$. We defined the population variance as

$$E[X - E(X)]^2 = \sum_{X=0}^{n} p(X)[X - E(X)]^2. \tag{3.8}$$

Expanding this expression, we obtain

$$E[X - E(X)]^2 = \sum p(X)[X^2 - 2E(X).X + (E(X))^2]$$

$$= \sum p(X)X^2 - 2E(X)\sum p(X)X + [E(X)]^2,$$

since $\sum p(X)[E(X)]^2 = [E(X)]^2 \sum p(X)$

$$= E(X^2) - 2[E(X)]^2 + [E(X)]^2$$

$$= E(X^2) - [E(X)]^2. \tag{3.9}$$

If we denote the population mean by $\mu$ and the population variance by $\sigma^2$, then 3.9 may be rewritten more compactly as

$$\sigma^2 = E[(X - \mu)^2] = E(X^2) - \mu^2. \tag{3.10}$$

To illustrate the calculation of the population variance, we return to the example of the binomial distribution with $n = 3$ and $p = \frac{1}{2}$, for which we have calculated $\mu = 1.5$. Here we have

$$E(X^2) = \sum p(X)X^2 = \tfrac{1}{8}(0) + \tfrac{3}{8}(1^2) + \tfrac{3}{8}(2^2) + \tfrac{1}{8}(3^2) = 3$$

so that

$$\sigma^2 = E(X^2) - \mu^2 = 3 - (1.5)^2 = \tfrac{3}{4}.$$

The standard deviation $\sigma$ is therefore $\sqrt{3}/2$, that is, 0·8660. It will be left as an exercise for the reader to confirm that for the binomial distribution with $n = 3$ and $p = \frac{1}{6}$ we have $\mu = \frac{1}{2}$ and $\sigma^2 = \frac{5}{12}$ (so that $\sigma = 0.6455$).

We calculate the expected value and the variance from first principles by making use of equations 3.5 and 3.9, but in the case of the binomial distribution we may calculate general formulae for $\mu$ and $\sigma^2$ in terms of $n$ and $p$. For the binomial distribution $p(X) = C_X^n p^X (1 - p)^{n-X}$, $X = 0, ..., n$, so that if we substitute for $p(X)$ in equation 3.5,

$$\mu = \sum_{X=0}^{n} p(X) . X = \sum_{X=0}^{n} C_X^n p^X (1 - p)^{n-X} (X)$$

$$= (1 - p)^n (0) + n(1 - p)^{n-1} p(1) + \frac{n(n-1)}{2!} (1 - p)^{n-2} p^2 (2)$$

$$+ \frac{n(n-1)(n-2)}{3!} (1 - p)^{n-3} p^3 (3) + ... + n(1 - p) p^{n-1} + p^n (n)$$

$$= np \left[ (1 - p)^{n-1} + (n - 1)(1 - p)^{n-2} p + \frac{(n-1)(n-2)}{2!} (1 - p)^{n-3} p^2 \right.$$

$$\left. + ... + (n - 1)(1 - p) p^{n-2} + p^{n-1} \right]$$

$$= np \sum_{X=0}^{n-1} C_X^{n-1} p^X (1 - p)^{n-1-X} = np,$$

since $\sum C_X^{n-1} p^X (1 - p)^{n-1-X} = 1$. Thus, for the binomial distribution,

$$\mu = np. \tag{3.11}$$

It can also be shown that for the binomial distribution the variance is

$$\sigma^2 = np(1 - p), \tag{3.12}$$

although this result will not be proved here.[6]

To summarize the calculation for our binomial distributions,

|  | $\mu$ | $\sigma$ |
|---|---|---|
| $n = 3, p = \frac{1}{2}$ | 1·5 | 0·866 |
| $n = 3, p = \frac{1}{6}$ | 0·5 | 0·646 |

we see that while both distributions have the same range, the first is somewhat more widely dispersed around its expected value than is the second. We shall return to the standard deviation in the context of continuous random variables and their theoretical frequency distributions. However, one general result may be noted here concerning the standard deviation and the dispersion of any distribution. This is *Tchebycheff's inequality*,[7] and it states that if we consider a range of $\pm k$ standard deviations around the mean of the distribution, then the proportion of the distribution lying *outside* this range is less than or equal to $1/k^2$. Thus at least three-quarters of any distribution lies inside a range $\mu \pm 2\sigma$. As we shall see when we investigate particular distributions, this inequality often considerably understates the proportion of the distribution falling within a given range, but it does provide a lower bound.

### 3.5 continuous random variables
We have already noted that continuous random variables can arise when we are concerned with taking measurements rather than counting outcomes. We may also obtain a continuous random variable as the limit of a discrete random variable by letting the number of outcomes, $n$, tend to infinity. For example, consider a rectangular distribution defined over the range 0–1 with $X = 0, 0\cdot1, \ldots,$ $0\cdot9$ and $p(X) = \frac{1}{10}$. If we now increase $n$ so that $X = 00, 0\cdot01, \ldots, 0\cdot99$, then $p(X) = \frac{1}{100}$; if $X = 000, 0\cdot001, \ldots, 0\cdot999$, then $p(X) = \frac{1}{1000}$ and so on. As we continue to subdivide the range 0–1, $p(X)$ gets smaller and smaller and, in the limit, as $n$ tends to infinity $p(X)$ tends to zero. Thus one property of a continuous random variable is that if $X$ can take any one of an infinite number of values, then the probability that it will take any particular value must tend to zero.

Geometrically we may think of this result as follows: we have shown that for a discrete random variable we may represent a probability as an area under the histogram – a rectangle whose base is centred on the value of $X$ being considered. As we let $n$ increase, the length of the base of the rectangle decreases relative to the range of $X$ and, in the limit when $n$ tends to infinity and $X$ becomes continuous, the base of the rectangle tends to a point and hence the area of the rectangle tends to zero.

As a consequence of this result, when we deal with continuous random variables we shall not attempt to evaluate the probability that $X$ takes some particular value $X_0$, since $p(X_0) = 0$. Rather we shall think in terms of the probability that $X$ falls within some given range; e.g., if $X$ has the range $-\infty \leq X \leq \infty$, we shall consider probabilities of the form $p(X_1 \leq X \leq X_2)$, $p(X \leq X_1)$ and $p(X \leq X_2)$. In analysing the probabilities corresponding to a continuous random variable, we shall find that the mathematical formula which represents the theoretical frequency distribution is usually difficult to work with directly and it will be much easier to evaluate probabilities as areas under the curve of the frequency distribution. In our notation we shall distinguish between the probability function $p(X)$ for a discrete random variable and the frequency function $f(X)$ for a continuous random variable.

In discussing continuous random variables we shall concentrate on one particular distribution, the normal distribution. In practice this distribution has been found to provide a good model for many measurements, such as human heights, weights, girths, etc., while it provides an important theoretical basis for the development of statistical inference.

### 3.6 the normal distribution

This frequency distribution is defined for a continuous random variable $X$ in the range $-\infty \leqq X \leqq \infty$ and is given by the formula

$$f(X) = \frac{1}{\sqrt{2\pi}\,\sigma} \exp\left( -\frac{1}{2}\left( \frac{X-\mu}{\sigma} \right)^2 \right) dX. \tag{3.13}$$

It is shown geometrically in Figure 3.4.

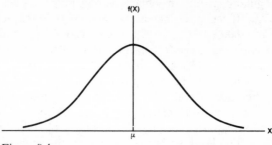

Figure 3.4

Here we note the following:

(i) The term $dX$ in equation 3.13 may be interpreted as indicating that we are considering the value of $f(X)$ not at one point, but in some range $dX$ around the point. (The analogy to summation for a continuous variable is integration.)

(ii) The symbols $\pi$ and $e$ denote mathematical constants, where $\pi$ is the ratio of the circumference of a circle to its diameter and equals 3·1412, while $e$ represents the sum of the convergent infinite series $(1 + 1/1! + 1/2! + ...)$ and equals 2·17828.

(iii) The constant term $1/\sqrt{2\pi}\,\sigma$ has to be included to ensure the area under the frequency distribution (that is, the sum of the probabilities) equals unity.

(iv) Since $X$ only enters equation 3.13 in the squared term in the exponent of $e$, the value of $f(X)$ is the same whether we substitute $X$ or $-X$. This means that the normal distribution is symmetrical for corresponding positive and negative values of $X$.

(v) If we evaluate the expected values for the normal distribution we find that the mean $E(X) = \mu$ and the variance $E[(X-\mu)^2] = \sigma^2$. It is clear that equation 3.13 represents a family of distributions, the values of which will vary only with $\mu$ and $\sigma^2$. For this reason $\mu$ and $\sigma^2$ are referred to as the parameters of the normal distributions, and since the normal distribution depends only on $\mu$ and $\sigma^2$ we shall convey the relevant information by means of the notation $N(\mu, \sigma^2)$. Thus $N(25, 100)$ will denote a normal distribution with a mean of 25 and a variance of 100.

We could evaluate the probability that $X$ will fall inside some range, say $X_1 \leqq X \leqq X_2$, by integrating equation 3.13 over the relevant range, but this is not an easy integral to work with and the possibility of tabulating the relevant areas under the curve is made difficult by the fact that different combinations of both $\mu$ and $\sigma^2$ would need to be tabulated, a formidable printing task. However, this problem may be avoided by noting one important property of the normal distribution. Suppose we define a new continuous random variable, $Z$, in terms of $X$ as

$$Z = \frac{X - \mu}{\sigma}, \tag{3.14}$$

so that $-\infty \leqq Z \leqq \infty$, and note the result that $dZ = dX/\sigma$. Then substituting 3.14 in equation 3.13 gives us

$$f(Z) = \frac{1}{\sqrt{2\pi}} \exp(-\tfrac{1}{2}Z^2)\,dZ, \tag{3.15}$$

as the transformed normal distribution. If we evaluate the mean and the variance for this normal distribution we find that $\mu = 0$ and $\sigma^2 = 1$, so that this distribution may be denoted as $N(0, 1)$. The mathematical form of the normal distribution is such that if we are concerned with $p(X_1 \leqq X \leqq X_2)$ and define for the variable $Z$

$$Z_1 = \frac{X_1 - \mu}{\sigma} \quad \text{and} \quad Z_2 = \frac{X_2 - \mu}{\sigma},$$

then we have

$$p(X_1 \leqq X \leqq X_2) = p(Z_1 \leqq Z \leqq Z_2).$$

This transformation is illustrated in Figure 3.5. The distribution $N(0, 1)$ is known as the *standard* normal distribution.

This presents an escape from the tabulation problem, since by means of the formula defined in equation 3.14 we may transform any normal random variable into a particular normal distribution. Thus if we evaluate and tabulate the probabilities corresponding to $f(Z)$, they can be made relevant to any normal distribution by means of the transformation given in equation 3.14.

(i) The effect of subtracting μ is to shift the centre of the distribution from μ to 0.

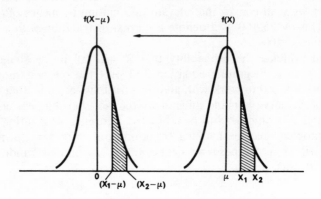

(ii) Dividing by σ produces the standard normal distribution, N (0, 1).

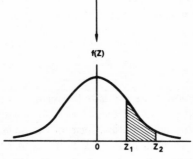

Figure 3.5 (a) The effect of subtracting $\mu$ is to shift the centre of the distribution from $\mu$ to 0

(b) Dividing by $\sigma$ produces the standard normal distribution, $N(0, 1)$

This solution depends on the evaluation and tabulation of 3.15, a task which has been performed and whose results are presented in the Appendix, Table A1. The simplest way of presenting the data is in the form of the areas under the normal distribution within given regions, and Table A1 presents the data in the form $p(Z \leq Z_1)$, as shown in Figure 3.6(a).

Use is made of the symmetry of the normal distribution, so that since $p(Z \leq -Z_1) = p(Z \geq Z_1) = 1 - p(Z \leq Z_1)$, only areas corresponding to positive values of $Z$ are tabulated. This is illustrated in Figure 3.6(b).

As some examples of the areas under the standard normal distribution we first note that $p(Z \leq 0) = 0.5000$, a result which corresponds to the symmetry

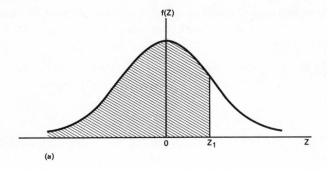

(a)

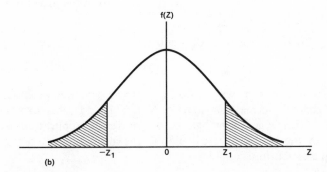

(b)

Figure 3.6

of the normal distribution. Secondly, we find from Table A1 that $p(Z < 1) = 0.8413$ and $p(0 \leq Z \leq 1) = p(Z \leq 1) - p(Z \leq 0) = 0.8413 - 0.5000 = 0.3413$. Thus $p(-1 \leq Z \leq 1) = 2p(0 \leq Z \leq 1) = 2(0.3413) = 0.6826$. Since the standard deviation is unity, the result we have just obtained is that the area under the standard normal distribution within the range $\pm \sigma$ around the mean, $\mu = 0$, is $0.6826$, that is, just over two-thirds of the total area under the curve. The reader may confirm that $p(Z < 2) = 0.9772$, so that $p(Z > 2) = 1 - 0.9773 = 0.0228$ and hence $p(-2 \leq Z \leq 2) = 1 - 2(0.0228) = 1 - 0.0456 = 0.9544$. That is, about 95 per cent of the area under the curve lies in a range of two standard deviations around the mean. The reader may note that these proportions fall well inside the limits set by Tchebycheff's inequality.

We have now given some examples of how to evaluate probabilities in connection with the standard normal distribution $N(0, 1)$, but given the transformation defined in equation 3.14, the areas contained in Table A1 may be used in connection with any normal distribution. For example, suppose we are concerned with a continuous random variable, $X$, which follows a particular normal distribution $N(158, 25)$, and we want to evaluate $p(153 < X < 168)$. By means of 3.14 we may calculate

$$Z_1 = \frac{153 - 158}{5} = -1,$$

and

$$Z_2 = \frac{168 - 158}{5} = 2.$$

Hence $p(153 \leq X \leq 168) = p(-1 \leq Z \leq 2)$, and from the probabilities we have calculated above,

$$p(-1 \leq Z \leq 2) = p(-1 \leq Z \leq 0) + p(0 \leq Z \leq 2)$$

$$= 0.3413 + 0.4772 = 0.8185.$$

This example illustrates the way in which the transformation provides probabilities for the general normal distribution $N(\mu, \sigma^2)$, and other examples will be given below. At this stage we shall go on to consider the application of the normal distribution to the problem of calculating the probabilities corresponding to the binomial distribution.

### 3.7  the normal approximation to the binomial distribution

In an earlier section of this chapter we discussed how the rectangular distribution could be made to tend to a continuous distribution by further subdividing the range 0–1. We may now consider what happens to the shape of the binomial distribution if we let $n$ become large. Figure 3.7 shows the histograms for binomial distributions for which $p = \frac{1}{2}$ and $n$ equals (a) 2, (b) 10 and (c) 20. From the definition of $X$ as the number of successes in the binomial distribution, it is a discrete variable and hence cannot really tend to a continuous variable as $n$ tends to infinity, but as $n$ increases the unit strip corresponding to any value of $X$ is tending to zero relative to the range of $X$, and hence the properties of the binomial distribution tend to those of a continuous distribution as $n$ tends to infinity. The interesting mathematical result here is that if $p = \frac{1}{2}$, the binomial distribution tends to the normal distribution as $n$ tends to infinity. This result might seem to be of academic interest, but in practice the normal distribution provides a very good approximation to the binomial distribution long before $n$ tends to infinity.

In fitting a normal distribution to a binomial distribution $n$ and $p$ are known, and hence we know $\mu = np$ and $\sigma^2 = np(1 - p)$ for the binomial distribution,

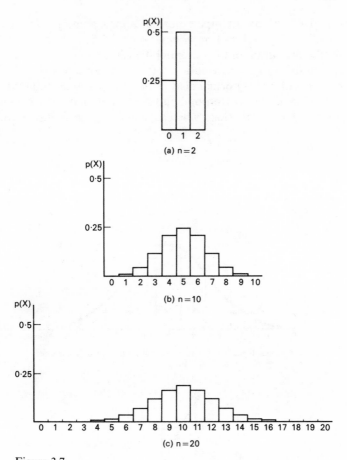

Figure 3.7

but we need a rule for choosing the values of $\mu$ and $\sigma^2$ for the normal distribution. The obvious one is to choose the same mean and variance for the normal distribution. This will ensure that the two distributions are centred on the same point and have the same dispersion. As an example we take a binomial distribution with $n = 10$ and $p = \frac{1}{2}$ and consider the probability $p(X = 6 \text{ or } 7)$ which we may calculate directly as

$$p(6 \text{ or } 7) = p(6) + p(7) = C_6^{10} \left(\tfrac{1}{2}\right)^6 \left(\tfrac{1}{2}\right)^4 + C_7^{10} \left(\tfrac{1}{2}\right)^7 \left(\tfrac{1}{2}\right)^3$$

$$= 0.2051 + 0.1172 = 0.3223.$$

As an alternative we may attempt a normal approximation. Here with $n = 10$ and $p = \frac{1}{2}$ and $\mu = np = 5$ and $\sigma^2 = np(1 - p) = 2.5$, so that $\sigma = 1\cdot58$. Figure 3.8(a) shows the histogram of the binomial distribution corresponding to $n = 10$ and $p = \frac{1}{2}$, with the normal distribution $N(5, 2.5)$ superimposed.

Matching the continuous normal distribution to the discrete binomial distribution poses a problem, since corresponding to $X = 6$ we have a strip of unit length under the binomial distribution, whereas for the normal distribution $X = 6$

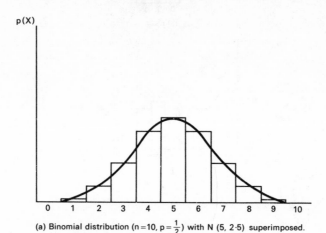

(a) Binomial distribution ($n = 10$, $p = \frac{1}{2}$) with N (5, 2·5) superimposed.

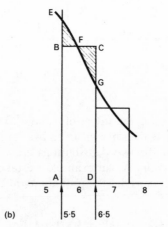

(b)

Figure 3.8  (a) Binomial distribution ($n = 10$, $p = \frac{1}{2}$) with $N(5, 2.5)$ superimposed
(b) Continuity connection

is a point. The solution to this problem is to make a 'continuity connection', which is illustrated in Figure 3.8(b). The 'true' area under the binomial distribution corresponding to $p(6)$ is the rectangle $ABCD$. Visually, the area under the normal distribution which best approximates this rectangle is $AEGD$, since we may hope that areas $BEF$ and $CFG$ will be approximately equal. We may find the co-ordinates of points $A$ and $D$ by assuming that the point corresponding to $X = 6$ under the normal distribution lies at the centre of the unit strip corresponding to $X = 6$ under the binomial distribution. This gives $A = 5.5$ and $D = 6.5$ as the corresponding points under the normal distribution. Our normal approximation then becomes

$$\underset{\text{binomial}}{p(X = 6 \text{ or } 7)} \approx \underset{\text{normal}}{p(5{\cdot}5 \leq X \leq 7{\cdot}5)}.$$

To evaluate the approximation, we make use of the transformation to the standard normal distribution given in 3.14. First we calculate

$$Z_1 = \frac{5{\cdot}5 - 5}{1{\cdot}58} = 0{\cdot}316 \quad \text{and} \quad Z_2 = \frac{7{\cdot}5 - 5}{1{\cdot}58} = 1{\cdot}581.$$

Then

$$p(5{\cdot}5 \leq X \leq 7{\cdot}5) = p(0.316 \leq Z \leq 1{\cdot}581)$$

$$= p(Z \leq 1{\cdot}581) - p(Z \leq 0{\cdot}136)$$

$$= 0{\cdot}9430 - 0{\cdot}6240 = 0{\cdot}3190.$$

A comparison of this value with $0{\cdot}3223$, the value obtained by evaluating the binomial probabilities, shows that the approximation is good; the two values differ by $0{\cdot}0033$, which is about 1 per cent of the correct value.

The normal approximation is particularly useful if we wish to calculate the probability of '$m$ or more' successes, which might involve calculating a large number of binomial probabilities, but only involves one calculation using the normal approximation. For example, in our present example

$$p(6 \text{ or more}) = p(6) + p(7) + p(8) + p(9) + p(10)$$

$$= 0{\cdot}2051 + 0{\cdot}1172 + 0{\cdot}0439 + 0{\cdot}0098 + 0{\cdot}0010$$

$$= 0{\cdot}3770.$$

Using the normal approximation,

$$\underset{\text{binomial}}{p(6 \text{ or more})} \approx \underset{\text{normal}}{p(X \geq 5{\cdot}5)} = p(Z \geq 0{\cdot}316)$$

$$= 1 - 0{\cdot}6240 = 0{\cdot}3760.$$

The reader may observe the importance of the continuity correction in providing a good approximation if we calculate

$$p(\text{6 or more}) \underset{\text{binomial}}{\approx} p(Z \geq 6),$$
$$\phantom{p(\text{6 or more})}\underset{\text{normal}}{\phantom{\approx p(Z \geq 6)}}$$

where now

$$Z = \frac{6 - 5}{1 \cdot 58} = 0 \cdot 633$$

and

$$p(Z \geq 0 \cdot 633) = 0 \cdot 2633.$$

It was mentioned that for $p = \frac{1}{2}$ the binomial distribution tends to the normal distribution as $n$ tends to infinity, and we have shown that the normal distribution provides a good approximation for the binomial distribution for quite small values of $n$. It also holds when $p \neq \frac{1}{2}$, provided $p$ is not close either to zero or to unity. For example, if we have a binomial distribution with $n = 12$ and $p = \frac{1}{3}$, then using the binomial distribution we may calculate

$$p(6) = C_6^{12} (\tfrac{1}{3})^6 (\tfrac{2}{3})^6 = 0 \cdot 108$$

Here $\mu = np = 4$ and $\sigma = \sqrt{np(1 - p)} = 1 \cdot 63$ and

$$p(X = 6) \underset{\text{binomial}}{\approx} p(5 \cdot 5 \leq X \leq 6 \cdot 5).$$
$$\phantom{p(X = 6)}\underset{\text{normal}}{\phantom{\approx p(5 \cdot 5 \leq X \leq 6 \cdot 5)}}$$

Here

$$Z_1 = \frac{5 \cdot 5 - 4}{1 \cdot 63} = 0 \cdot 92 \quad \text{and} \quad Z_2 = \frac{6 \cdot 5 - 4}{1 \cdot 63} = 1 \cdot 53$$

and

$$p(0 \cdot 92 \leq Z \leq 1 \cdot 53) = 0 \cdot 1158,$$

which compares reasonably well with the correct value of $0 \cdot 108$, since the error of $0 \cdot 0078$ is about 7 per cent of the correct value.

# problems

**3.1** A fair coin is tossed five times. Use the binomial distribution to calculate the values of $p(X)$, where $X$ equals the number of heads obtained. Plot the corresponding histogram.

**3.2** A fair coin is tossed five times and three heads are obtained. What is the probability that a head was obtained on each of the first two tosses?

**3.3** If you had to choose between attempting to throw one six in two throws of a die or two sixes in four throws of the die, which would you choose?

**3.4** A firm plans to bid £30 per ton for a contract to supply 1000 tons of a metal. It has two competitors A and B and it assumes that the probability that A will bid less than £30 per ton is 0·3 and that B will bid less than £30 per ton is 0·7. If the lowest bidder gets all the business and the firms bid independently, what is the expected value of the contract to the first firm?

**3.5** A man carrying a purse containing 3 $2\frac{1}{2}$p pieces, 3 5p pieces and 3 10p pieces, goes to a shop to buy typing paper. The paper comes in packets of three thicknesses, namely 10 sheets for $2\frac{1}{2}$p, 25 sheets for 5p, and 60 sheets for 10p. The man shakes two coins out of his purse without looking and buys the corresponding quantity of paper. If $X$ denotes the number of sheets of paper which can be bought with two coins, evaluate $p(X)$, the probability distribution of $X$. What is the expected value of $X$?

**3.6** For the rectangular distribution $p(X) = 1/n, X = 1, \ldots, n$, show that $E(X) = (n + 1)/2$ and $\sigma^2 = (n^2 - 1)/12$. (*Hint*. Note that the sum of the first $n$ natural numbers $1, \ldots, n$, is $n(n + 1)/2$ and the sum of their squares is $\dfrac{n(n + 1)(2n + 1)}{6}$.)

**3.7** Given the areas under the standard normal distribution $N(0, 1)$ presented in Table A1, find the following probabilities:

    (a) $p(Z > 1\cdot5)$,
    (b) $p(Z < -1\cdot25)$,
    (c) $p(0 \leq Z \leq 0\cdot94)$,
    (d) $p(-0\cdot42 \leq Z \leq 0)$,

(e) $p(0.31 \leq Z \leq 0.32)$,

(f) $p(0.16 \leq Z \leq 1.96)$.

**3.8**  By using a suitable transformation, find the probabilities of the values of $X$ given below for the following normal distributions $N(\mu, \sigma^2)$:

(a) $N(100, 100)$: $p(90 \leq X \leq 110)$; $p(X \geq 130)$; $p(105 \leq X \leq 110)$,

(b) $N(-20, 16)$: $p(X \leq -30)$; $p(-24 \leq X \leq 20)$; $p(X \geq 0)$,

(c) $N(0, 4)$: $p(0 \leq X \leq 3)$; $p(X \leq -6)$; $p(-3 \leq X \leq 3)$.

(d) $N(300, 25)$: $p(X \geq 292)$; $p(289 \leq X \leq 311)$; $p(X \leq 304)$.

**3.9**  A set of examination marks in economics is approximately normally distributed with a mean of 75 and standard deviation of 5. If the top 20 per cent of students get A's and the bottom 20 per cent get F's, what mark is the lowest A and what mark is the highest F?

**3.10**  A firm making metal rods aims at a length of 4 cm. The lengths of the rods it produces follow a normal distribution with mean 4·01 cm and standard deviation 0·03 cm. Each rod costs £6 to produce and may be used immediately if its length lies between 3·98 and 4·02 cm. If its length is less than 3·98 cm, the rod is useless, but has a scrap value of £1. If its length exceeds 4·02 cm it may be shortened and used, but at a cost of £2. Find the average cost per usable rod and the average profit per rod if the firm sells the rods at a price of

(a) £6, and

(b) £8.

**3.11**  If the probability of an individual customer buying a commodity from a shop is 0·64 and 25 consumers enter the shop on a particular day, use the normal approximation to the binomial distribution to evaluate the probability that

(a) 19 or more customers buy the commodity,

(b) 17 customers buy this commodity.

**3.12**  If the probability of a student passing an examination in statistics is 0·8 and 25 students take the examination, use the normal approximation to the binomial distribution to evaluate the probability that

(a) at least 10 students pass,

(b) 20 or more students pass,

(c) all the students pass.

**3.13**  A company selects a random sample of 100 individuals from a telephone directory for a survey concerning one of its products. If 20 per cent of the people in the phone book use the product, what is the probability that 25 or more of the members of the sample will use the product? How large must the sample be for the probability to be 0·95 that less than 25 per cent of the sample members use the product?

**3.14**  In a particular maternity hospital, it is found that the probability that a baby is a girl is 0·55. Use the normal approximation to the binomial distribution

to calculate the probability that out of 100 births on a given day
    (a) between 50 and 60 will be boys,
    (b) over 60 will be girls, and
    (c) not more than 50 will be girls.
**3.15** Given the following probability distributions for the discrete random variables $X$ and $Y$, calculate $E(X)$, $E(Y)$ and $E(X)/E(Y)$. If $Z = X/Y$, calculate $p(Z)$ and $E(Z)$. Comment on your results.

(a)

| $X$ | $-1$ | $\frac{2}{2}$ | $2$ |
|---|---|---|---|
| $p(X)$ | $\frac{1}{2}$ | $\frac{3}{8}$ | $\frac{1}{8}$ |

,

| $.Y$ | $1$ | $3$ |
|---|---|---|
| $p(Y)$ | $\frac{1}{2}$ | $\frac{1}{2}$ |

(b)

| $X$ | $-1$ | $3$ |
|---|---|---|
| $p(X)$ | $\frac{1}{2}$ | $\frac{1}{2}$ |

,

| $Y$ | $-1$ | $\frac{1}{3}$ |
|---|---|---|
| $p(Y)$ | $\frac{1}{4}$ | $\frac{3}{4}$ |

(c)

| $X$ | $-1$ | $\frac{2}{3}$ | $2$ |
|---|---|---|---|
| $p(X)$ | $\frac{1}{2}$ | $\frac{3}{8}$ | $\frac{1}{8}$ |

,

| $Y$ | $-1$ | $\frac{1}{3}$ |
|---|---|---|
| $p(Y)$ | $\frac{1}{4}$ | $\frac{3}{4}$ |

# chapter 4

# sampling and sampling theory

## 4.1 introduction

In the last chapter we discussed theoretical frequency distributions and suggested that they might be regarded as models for experimental situations. For example, the binomial distribution can describe the outcome of tossing coins, or the number of female births in a given hospital, while the normal distribution could describe the heights or weights of a human population and is used by some educators as a model for the distribution of human intelligence. The next step in our development of the theory of statistical inference is to see what can be deduced about the properties of a sample which is drawn from some population. In practice we may be concerned with a sample drawn from an actual population, such as the expenditures of a sample of London families, but in order to develop sample properties we shall consider the conceptual experiment of drawing samples from populations which are theoretical and are represented by the frequency distributions which were discussed in the previous chapter.

Since our aim is to infer something about the population from the information contained in the sample, we would like the sample to be representative of the population. For example, if we are trying to advise legislators on public attitudes towards equal pay for women, a sample consisting only of men (or only of women) is unlikely to give a complete or balanced picture of the whole population.

There are a number of ways in which we might attempt to make a sample representative of the population. For example, if we know the relevant characteristics of the population, we could use this information to determine the characteristics of the sample. In the United Kingdom the decennial census, which has been a complete enumeration of the population, provides information on age, sex, income, education and many other variables. Since from these data we know the proportion of the population falling into various categories, we may aim to select a sample so that the proportion of sample members with given characteristics correspond to those in the population.

This idea is the basis of the sampling method known as *quota sampling*, a method which is often used in market research surveys. Given the information to be collected, knowledge of the population being sampled is used to determine the quotas of respondents in the various age/sex/income/religious/educational categories that the interviewers should contact. Selecting the members of the population with the required characteristics is then left to the interviewers. At first sight this method of sampling might seem to present an easy way of obtaining a sample which is a miniature replica of the population, but it can suffer from the disadvantage that filling the quotas is left to the discretion of the interviewers, thus introducing a subjective element into the process. Thus we may find that a quota sample contains a smaller proportion of bad-tempered, ugly or uncooperative people than we would find in the total population.

An alternative way of getting a sample which is representative of the population is to ensure that every member of the population has a known (and usually equal) probability of being included in the sample, but then making the method of selection non-subjective. This is the aim of the method which is known as *random sampling*, that is, the use of some random process to select the members of the sample. An example of a sample drawn by a random process from a population whose members have a known probability of being selected is the prize draw in a raffle, where all the numbers are put into a drum and well stirred before a number is drawn. If there are $n$ numbers, $X = 1, ..., n$, then $p(X) = 1/n$ for the first draw, and the rectangular distribution provides the theoretical model for this sampling situation.

In practice it is difficult to find a physical method of sampling which does ensure $p(X) = 1/n$ when a member of the sample is selected, since it is not always easy to shuffle a drum full of raffle tickets adequately. For this reason it is often convenient to use *random numbers* to draw a sample by a random method. The phrase 'random numbers' is somewhat misleading, since it is not the numbers which are random but the method of choosing them. Some device (nowadays often an electronic computer[1]) is used which simulates the physical process of pulling numbers out of a hat, but which is designed to ensure that $p(X) = 1/n$. Sets of numbers generated in this way have been tabulated and a small set may be found in Table A6 of the Appendix. They may be used as follows. If we want to select twelve people at random from an alphabetical list of 100 names, we begin by numbering the names consecutively 00, 01, ..., 98, 99. Then we use a table of random numbers to choose twelve names at random. For example, reading down the first two columns of Table A6 we obtain 12, 30, 59, 99, 27, 93, 84, 68, 60, 97, 58. Given these numbers, we select the corresponding names from our list. If any number had turned up again after being selected, we would have ignored it and continued selecting random numbers until we had obtained twelve different numbers.

We may hope that a sample selected by a random process will be representative of the population, but this does not mean that *every* random sample will be a miniature replica of the population being sampled. For example, if our list of 100 names consisted of equal numbers of men and women, a representative sample of twelve would consist of six men and six women, but there is no reason why our sample based on twelve random numbers should reproduce this breakdown exactly. It would be possible to choose a random sample of twelve from this list consisting only of men, although the probability of doing so $\left(\left(\frac{1}{2}\right)^{12} = 0\cdot000244\right)$ is quite low. One way of avoiding this kind of unrepresentativeness would be to draw a *stratified* random sample, obtained by listing the fifty men and fifty women separately and drawing a random sample of six from each group.[2] Stratification may help to reduce the probability that a random sample will be extremely unrepresentative, but cannot in general make this probability zero. However, if we were to carry out an experiment in which we draw a large number of random samples, then in the long run they may be expected to reproduce the characteristics of the population from which they are drawn. We shall explore this idea more fully below.

The random selection process can obviously avoid the subjective bias that may result if the interviewer selects the members of the sample. It is also important in ensuring that a sample of a given size contains the maximum amount of information. Suppose that a student group decides to poll a sample of 200 from its membership on a particular issue. Two alternative sampling methods have been suggested, both based on the student refectory, which contains 1000 tables, each holding four people. The first method involves numbering the 4000 seating places and selecting 200 of them by means of a table of random numbers. The students in these seats on a given day will then be polled. The second method is to work in terms of the tables: to number the 1000 tables, select 50 tables at random, and then poll all four students at each table. (We are assuming that all the tables are always full.) Both methods yield 200 replies and the second is easier to carry out, but will both samples yield the same amount of information? They may do, but the second method is likely to yield less information than the first, since there is a tendency for friends to eat together, and friends often have similar views, interests and opinions. Hence if all four students at the same table are in favour of some issue, we know less about the views of the student body than if we find that four randomly chosen students (almost certainly at different tables) are in favour of the issue.

The reason for this statement is that while we expect the four students chosen at random to provide *independent* pieces of information, this may not be the case for our students sitting at the same table. To maximize the information that a sample contains, we should aim at making the observations independent of one another, hence the importance of random sampling methods.

## 4.2 random variables and independence

We have already met the concept of independence in our discussion of independent events, and we shall now extend the idea to cover random variables. We shall do so for discrete random variables in the context of the simple dice-throwing experiments we discussed in chapter 2. Suppose we throw two unbiased dice and record the score shown on each. There are thirty-six equally likely outcomes to this experiment, each of which may be represented by the pair of numbers $(i, j)$, $i, j = 1, \ldots, 6$, where $i$ represents the score on the first die, and $j$ the score on the second. The full set of outcomes is

| | | | | | |
|---|---|---|---|---|---|
| (1, 1) | (2, 1) | (3, 1) | (4, 1) | (5, 1) | (6, 1) |
| (1, 2) | (2, 2) | (3, 2) | (4, 2) | (5, 2) | (6, 2) |
| (1, 3) | (2, 3) | (3, 3) | (4, 3) | (5| 3) | (6, 3) |
| (1, 4) | (2, 4) | (3, 4) | (4, 4) | (5, 4) | (6, 4) |
| (1, 5) | (2, 5) | (3, 5) | (4, 5) | (5, 5) | (6, 5) |
| (1, 6) | (2, 6) | (3, 6) | (4, 6) | (5, 6) | (6, 6) |

First we consider the case in which we define two discrete random variables; $X = 1, \ldots, 6$ is the score on the first die and $Y = 1, \ldots, 6$ is the score on the second. Having introduced two random variables, we shall now discuss probabilities concerning both of them. To do so we define the *joint probability distribution* as providing the probabilities that $X = i$ and $Y = j$, that is, $p(X = i$ and $Y = j)$, $i, j = 1, \ldots, 6$. Where the range of variation for $X$ and $Y$ is clear we shall abbreviate this expression to $p(X, Y)$. For this experiment, each $(X, Y)$ combination corresponds to one and only one of the thirty-six outcomes listed above. Hence $p(X, Y) = \frac{1}{36}$ and the joint probability distribution may be represented in tabular form as Table 4.1.

*Table 4.1.*

| X \ Y | 1 | 2 | 3 | 4 | 5 | 6 | $p(X)$ |
|---|---|---|---|---|---|---|---|
| 1 | $\frac{1}{36}$ | $\frac{1}{36}$ | $\frac{1}{36}$ | $\frac{1}{36}$ | $\frac{1}{36}$ | $\frac{1}{36}$ | $\frac{1}{6}$ |
| 2 | $\frac{1}{36}$ | $\frac{1}{36}$ | $\frac{1}{36}$ | $\frac{1}{36}$ | $\frac{1}{36}$ | $\frac{1}{36}$ | $\frac{1}{6}$ |
| 3 | $\frac{1}{36}$ | $\frac{1}{36}$ | $\frac{1}{36}$ | $\frac{1}{36}$ | $\frac{1}{36}$ | $\frac{1}{36}$ | $\frac{1}{6}$ |
| 4 | $\frac{1}{36}$ | $\frac{1}{36}$ | $\frac{1}{36}$ | $\frac{1}{36}$ | $\frac{1}{36}$ | $\frac{1}{36}$ | $\frac{1}{6}$ |
| 5 | $\frac{1}{36}$ | $\frac{1}{36}$ | $\frac{1}{36}$ | $\frac{1}{36}$ | $\frac{1}{36}$ | $\frac{1}{36}$ | $\frac{1}{6}$ |
| 6 | $\frac{1}{36}$ | $\frac{1}{36}$ | $\frac{1}{36}$ | $\frac{1}{36}$ | $\frac{1}{36}$ | $\frac{1}{36}$ | $\frac{1}{6}$ |
| $p(Y)$ | $\frac{1}{6}$ | $\frac{1}{6}$ | $\frac{1}{6}$ | $\frac{1}{6}$ | $\frac{1}{6}$ | $\frac{1}{6}$ | 1 |

We may first note that the joint probability distribution satisfies the two criteria for a probability distribution that we discussed in the previous chapter, namely

$0 \leq p(X, Y) \leq 1$   and   $\sum_{X} \sum_{Y} p(X, Y) = 1.$

The joint probability function provides the probability connected with any $(X, Y)$ combination, but we may also be interested in the probabilities associated with $X$ regardless of the value taken by $Y$. To correspond to this we may define the *marginal* probability distribution, $p(X)$, as

$$p(X) = \sum_{Y} p(X, Y). \tag{4.1}$$

For example, $p(X = 1) = p(1, 1) + p(1, 2) + \ldots + p(1, 6) = 6(\frac{1}{36}) = \frac{1}{6}$. By symmetry the marginal probability distribution for $Y, p(Y)$, may be defined as

$$p(Y) = \sum_{X} p(X, Y). \tag{4.2}$$

The two marginal distributions are presented as the final column and row in Table 4.1, and the reader may note that

$$\sum_{X} p(X) = \sum_{Y} p(Y) = \sum_{X} \sum_{Y} p(X, Y) = 1.$$

We may now provide a formal definition of independence for random variables. If $X$ and $Y$ are two discrete random variables, with joint probability distribution $p(X, Y)$ and marginal probability distributions $p(X)$ and $p(Y)$, then $X$ and $Y$ are statistically independent if

$$p(X, Y) = p(X) p(Y), \tag{4.3}$$

for all $X$ and $Y$. Conversely, if $p(X, Y) \neq p(X) p(Y)$, the random variables $X$ and $Y$ are statistically dependent. Equation 4.3 may be compared with our definition of independent events in equation 2.11.

Applying 4.3 to our dice-throwing experiment we have $p(X, Y) = p(X) p(Y)$ for all $X$ and $Y$, so that the two random variables are independent, a result we would expect from the nature of the experiment. However, in contrast to this result consider another experiment involving two unbiased dice, but now we define $X$ as the number of 1's obtained in the two throws and $Y$ as the number of odd scores obtained. Both $X$ and $Y$ may take the values 0, 1, 2, and the joint probability distribution and marginal probability distributions may be obtained most easily by an appropriate retabulation of the thirty-six equally likely outcomes to the experiment. This is presented in Table 4.2.

If we look at the values of $p(X, Y)$, $p(X)$ and $p(Y)$, it is clear that $p(X, Y) \neq p(X) p(Y)$ for any $X$ and $Y$ and we conclude that the two random variables are dependent. Again, this result may be expected from the nature of the experiment:

if there are zero odd scores in the two throws ($Y = 0$), then there cannot be either one or two 1's scored ($X = 1, 2$). Conversely, if the scores are both 1's ($X = 2$), then there cannot be only one odd score in the two throws ($Y = 1$).

*Table 4.2* (a) Outcomes corresponding to $(X, Y)$

| X \ Y | 0 | 1 | 2 |
|---|---|---|---|
| 0 | (2, 2) (2, 4) (2, 6) (4, 2) (4, 4) (4, 6) (6, 2) (6, 4) (6, 6) | (2, 3) (2, 5) (4, 3) (4, 5) (6, 3) (6, 5) (3, 2) (3\| 4) (3, 6) (5, 2) (5, 4) (5, 6) | (3, 3) (3, 5) (5, 3) (5, 5) |
| 1 | | (1, 2) (1, 4) (1, 6) (2, 1) (4, .1) (6, 1) | (1, 3) (1, 5) (3, 1) (5, 1) |
| 2 | | | (1, 1) |

(b) Relative frequency of outcomes, $p(X, Y)$

| X \ Y | 0 | 1 | 2 | $p(X)$ |
|---|---|---|---|---|
| 0 | $\frac{9}{36}$ | $\frac{12}{36}$ | $\frac{4}{36}$ | $\frac{25}{36}$ |
| 1 | 0 | $\frac{6}{36}$ | $\frac{4}{36}$ | $\frac{10}{36}$ |
| 2 | 0 | 0 | $\frac{1}{36}$ | $\frac{1}{36}$ |
| $p(Y)$ | $\frac{9}{36}$ | $\frac{18}{36}$ | $\frac{9}{36}$ | 1 |

In considering two random variables, there is a further probability distribution of interest, the *conditional* probability distribution, which provides the probability that one of the variables (say $X$) takes some value, given the other variable has some specified value (say $Y = Y_j$). We may denote the conditional distributions for $X$ given $Y$ as $p(X|Y)$ and, using the same argument that was used to derive the conditional probability in equation 2.9 of chapter 2, we may write

$$p(X|Y) = \frac{p(X, Y)}{p(Y)}.$$

(4.4)

We note that rearrangement of equation 4.3 gives

$$\frac{p(X, Y)}{p(Y)} = p(X),$$

so that when $X$ and $Y$ are *independent*,

$$p(X|Y) = p(X).$$

(4.5)

The conditional distribution may be obtained for $Y$ by substituting $Y$ for $X$ in equations 4.4 and 4.5.

As an example, consider our second experiment and the conditional probability distribution for $X$ when $Y = 1$, that is, $p(X|1)$, $X = 0, 1, 2$. We may evaluate these probabilities from first principles by examining part (a) of Table 4.1, where we note that there are 18 outcomes corresponding to $Y = 1$, of which 12 correspond to $X = 0$, 6 correspond to $X = 1$, and 0 to $X = 2$. Hence,

$$p(X = 0|1) = \tfrac{12}{18} = \tfrac{2}{3}; \qquad p(X = 1|1) = \tfrac{6}{18} = \tfrac{1}{3}; \qquad p(X = 2|1) = \tfrac{0}{18} = 0.$$

Alternatively, using equation 4.4,

$$p(0|1) = \tfrac{12}{36} \div \tfrac{18}{36} = \tfrac{2}{3}; \qquad p(1|1) = \tfrac{6}{36} \div \tfrac{18}{36} = \tfrac{1}{3}; \qquad p(2|1) = 0 \div \tfrac{18}{36} = 0.$$

In our earlier discussion of probability distributions we discussed the parameters of the distribution, the mean and variance. We may similarly define parameters for the joint probability distribution. For example, if we denote the expected value of the $X$s by $\mu_X$, then

$$\mu_X = \sum_X \sum_Y X p(X, Y) = \sum_X X p(X) \tag{4.6}$$

and, by symmetry, the expected value of the $Y$s is

$$\mu_Y = \sum_X \sum_Y Y p(X, Y) = \sum_Y Y p(Y).$$

Similarly, the variances of the two random variables may be defined as

$$\sigma_X^2 = \sum_X \sum_Y (X - \mu_X)^2 \, p(X, Y) = \sum_X (X - \mu_X)^2 \, p(X)$$

and

$$\sigma_Y^2 = \sum_X \sum_Y (Y - \mu_Y)^2 \, p(X, Y) = \sum_Y (Y - \mu_Y)^2 \, p(Y). \tag{4.7}$$

For the joint probability distribution, a new parameter may be defined which is of interest, that is, the *covariance*, which we define as

$$\sigma_{XY} = E\left[ (X - \mu_X)(Y - \mu_Y) \right]. \tag{4.8}$$

This expected value may be written as

$$\sigma_{XY} = \sum_X \sum_Y (X - \mu_X)(Y - \mu_Y) p(X, Y) \tag{4.9}$$

and if $X$ and $Y$ are independent random variables we may use 4.3 to write equation 4.9 as

$$\sigma_{XY} = \sum_X \sum_Y (X - \mu_X)(Y - \mu_Y) p(X) p(Y)$$

$$= \sum_X (X - \mu_X) p(X) \cdot \sum_X (Y - \mu_Y) p(Y)$$

$$= \left( \sum_X Xp(X) - \mu_X \right) \left( \sum_Y Yp(Y) - \mu_Y \right) = 0.$$

That is, if $X$ and $Y$ are independent variables, the covariance $\sigma_{XY}$ will be zero. In other words, the covariance reflects any tendency of the two variables to vary together. For example, to the extent that the quantity of a commodity demanded $(Y)$ is not independent of, but varies inversely with, the price of the commodity $(X)$, we would expect a negative covariance between $X$ and $Y$. We shall consider this result in the next section. When $X$ and $Y$ are not independent, 4.9 may be expanded to give

$$\sigma_{XY} = \sum_X \sum_Y (XY - X\mu_Y - \mu_X Y + \mu_X \mu_Y) p(X, Y)$$

$$= \sum_X \sum_Y XYp(X, Y) - \mu_Y \sum_X \sum_Y Xp(X, Y) - \mu_X \sum_X \sum_Y Yp(X, Y)$$

$$+ \mu_X \mu_Y \sum_X \sum_Y p(X, Y)$$

$$= \sum_X \sum_Y XYp(X, Y) - \mu_Y \mu_X - \mu_X \mu_Y + \mu_X \mu_Y$$

$$= E(XY) - \mu_X \mu_Y. \tag{4.10}$$

As an example, consider our second dice-throwing experiment as summarized in Table 4.2. Here we have

$$\mu_X = \sum_X Xp(X) = 0(\tfrac{25}{36}) + 1(\tfrac{10}{36}) + 2(\tfrac{1}{36}) = \tfrac{12}{36} = \tfrac{1}{3},$$

$$\mu_Y = \sum_Y Yp(Y) = 0(\tfrac{9}{36}) + 1(\tfrac{18}{36}) + 2(\tfrac{9}{36}) = 1,$$

$$\sum_X \sum_Y XYp(X, Y) = 0(\tfrac{9}{36}) + 0(\tfrac{12}{36}) + 0(\tfrac{12}{36}) + 0(0) + 1(\tfrac{6}{36}) + 2(\tfrac{4}{36})$$

$$+ 0(0) + 1(0) + 4(\tfrac{1}{36}) = \tfrac{18}{36} = \tfrac{1}{2}.$$

Hence $\sigma_{XY} = E(XY) - \mu_X \mu_Y = \tfrac{1}{2} - (1)(\tfrac{1}{3}) = \tfrac{1}{6}.$

It is left as an exercise for the reader to check that $\sigma_{XY} = 0$ for our first dice-throwing experiment.

### 4.3 sampling distributions

Having discussed randomness and independence, we shall now discuss some properties of random samples. Consider some discrete probability distribution $p(X)$, $X = 0, \ldots, N$, which has a mean $\mu$ and a variance of $\sigma^2$, from which we select a random sample (with replacement between draws) of $n$ observations, $X_i, i = 1, \ldots, n$. We form the sum $\sum_{i=1}^{n} X_i$ and consider its expected value

$$E\left( \sum_{i=1}^{n} X_i \right) = E(X_1 + \ldots + X_n)$$

$$= E(X_1) + \ldots + E(X_n) = n\mu,$$

since $E(X_i) = \mu$. Thus

$$E(X_1 + \ldots + X_n) = n\mu. \tag{4.11}$$

From this result it is easy to see that if we consider not $\sum_{i=1}^{n} X_i$, but the sample mean $\bar{X}$, then

$$E(\bar{X}) = E\left( \frac{1}{n} \sum X_i \right) = \frac{1}{n} E\left( \sum X_i \right) = \frac{1}{n}(n\mu)$$

$$= \mu. \tag{4.12}$$

That is, on average the mean of a random sample is equal to the mean of distribution from which the sample is drawn. This result is of great importance in what follows.

Now consider the variance of the sum $\sum_{i=1}^{n} X_i$,

$$E\left[ \sum_{i=1}^{n} X_i - E\left( \sum X_i \right) \right]^2 = E\left[ \sum_{i=1}^{n} X_i - n\mu \right]^2$$

$$= E\left[ \sum (X_i - \mu) \right]^2$$

$$= E[(X_1 - \mu) + \ldots + (X_n - \mu)]^2$$

$$= E\left[ \sum_{i=1}^{n} (X_i - \mu)^2 + \sum_{i \neq j} (X_i - \mu)(X_j - \mu) \right]$$

$$= \sum_{i=1}^{n} E(X_i - \mu)^2 + \sum_{i \neq j} E(X_i - \mu)(X_j - \mu).$$

Now $E(X_i - \mu)^2$ is the definition of the variance $\sigma^2$ and, *if the observations are independent* as we expect in a random sample, the covariances $E(X_i - \mu)(X_j - \mu)$

will be zero, so that

$$\text{variance} \left( \sum_{i=j}^{n} X_i \right) = \sum_{i=1}^{n} \sigma^2 = n\sigma^2. \tag{4.13}$$

If we consider the variance of the mean of the random sample, $\sigma_{\bar{X}}^2$,

$$E[\bar{X} - E(\bar{X})]^2 = E\left[ \frac{1}{n} \sum X_i - \mu \right]^2$$

$$= E\left[ \frac{1}{n} \sum (X_i - \mu) \right]^2$$

$$= \frac{1}{n^2} E\left[ \sum (X_i - \mu) \right]^2$$

$$= \frac{1}{n^2} (n\sigma^2) = \frac{\sigma^2}{n}. \tag{4.14}$$

To summarize these two important results, we have shown that if we draw a random sample of size $n$ from a discrete probability distribution with mean $\mu$ and variance $\sigma^2$, the expected value of the sample mean is $\mu$ and the variance of the sample mean is $\sigma^2/n$.

Suppose we are interested in the mean, $\mu$, of some distribution and we imagine a situation in which we can draw repeated random samples of size $n$ from the distribution. Then while we would not expect the arithmetic mean of any particular sample to be equal to $\mu$, we know now that on average $\bar{X}$ will equal $\mu$. Further, we also have some information concerning the dispersion of values of $\bar{X}$ around $\mu$ in repeated samples: given $\sigma_X = \sigma/\sqrt{n}$, we know from Tchebycheff's inequality that at least $(1 - k^2)$ of the values of $\bar{X}$ will lie in the range

$$\mu \pm k \frac{\sigma}{\sqrt{n}}.$$

These results are quite general, but to calculate probabilities we require a more detailed knowledge of the sampling distribution of $\bar{X}$. This distribution depends on the population being sampled, and we begin by considering sampling from a normal distribution.

## 4.4 sampling from a normal distribution

We stated earlier that the normal distribution was important for the theory of statistical inference. One reason for this is the following result. Suppose we have a normal distribution $N(\mu, \sigma^2)$ and repeated samples of size $n$ are drawn by a random process (so that the observations are independent). From the results obtained above we know the $E(\overline{X}) = \mu$ and $\sigma_{\overline{X}}^2 = \sigma^2/n$, *and when we sample from a normal population, the sampling distribution of $\overline{X}$ is a normal distribution.*[3] That is, the sampling distribution of $\overline{X}$ for random samples of size $n$ from $N(\mu, \sigma^2)$ is

$$N\left(\mu, \frac{\sigma^2}{n}\right).$$    (4.15)

This is an important result, since given values of $\mu$, $\sigma^2$ and $n$, we may now make probability statements about $\overline{X}$. If we apply the transformation given in 3.14 to 4.15, we have

$$Z = \frac{\overline{X} - \mu}{\sigma/\sqrt{n}},$$    (4.16)

where $Z$ follows the standard normal distribution, $N(0, 1)$. For example, suppose we draw a random sample of 25 observations from $N(25, 25)$ and we want to find the probability that the mean of the random sample, $\overline{X}$, will be greater than 26. From 4.15 we have the distribution of $\overline{X}$ as $N(25, 1)$, and using 4.16 we calculate

$$Z = \frac{26 - 25}{1} = 1\cdot0,$$

so that $p(\overline{X} > 26) = p(Z > 1\cdot0)$. From Table A1 we obtain $p(Z > 1\cdot0) = 0\cdot1587$, so that if we draw repeated random samples of 25 observations from $N(25, 25)$ and calculate $\overline{X}$, we would expect nearly 16 per cent of the calculated values of $\overline{X}$ to exceed 26.

Suppose that with the same normal population $N(25, 25)$ we now draw a random sample of size 100 and calculate $p(X > 26)$. In this case the relevant normal distribution for $\overline{X}$ is $N(25, 0\cdot25)$, and using 4.16 we have

$$Z = \frac{26 - 25}{0\cdot5} = 2\cdot0,$$

so that $p(\overline{X} > 26) = p(Z > 2\cdot0) = 0\cdot0228$. That is, with repeated random samples of size 100 we would now expect just over 2 per cent of values of $\overline{X}$ to exceed 26. This decrease in $p(\overline{X} > 26)$ follows from 4.5, since for a given $\sigma^2$ the dispersion of the distribution of $\overline{X}$ decreases in proportion to $(1/\sqrt{n})$.

The result that the mean of random samples from a normal population has a normal sampling distribution is obviously important if we are sampling from a normal distribution. What happens if the distribution we are sampling from is not normal? Suppose, for example, we sample from a rectangular distribution?

The answer is that the distribution of $\overline{X}$ is no longer given exactly by 4.15, but it may be shown that 4.15 holds as an approximate result. The proof of this result is known as the *Central Limit Theorem* (CLT), which states that if we draw random samples of size $n$ from a distribution with mean $\mu$ and variance $\sigma^2$ and calculate the sample mean $\overline{X}$, then the variable

$$\left( \frac{\overline{X} - \mu}{\sigma/\sqrt{n}} \right)$$

tends to follow the standard normal distribution as $n$ tends to infinity.

This result may only appear to be of interest for very large samples, but as we saw earlier in the case of the normal approximation to the binomial distribution which worked for small values of $n$, the result given by the Central Limit Theorem provides a good approximation for the distribution of $\overline{X}$ in small samples. As an example, we have used a table of random numbers to draw 100 samples of size 10 from the rectangular distribution $X = 0, ..., 9, p(X) = \frac{1}{10}$, and have calculated $\overline{X}$ for each sample. Table 4.3(a) presents the 100 values of $\overline{X}$, which are summarized in a frequency distribution in Table 4.3(b). The histogram corresponding to this frequency distribution is shown in Figure 4.1.

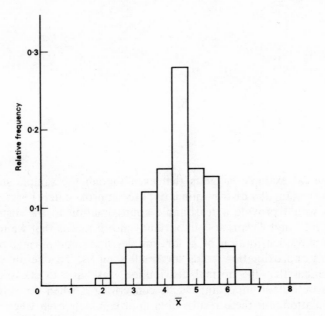

Figure 4.1

*Table 4.3*   (a) Sample mean, $\overline{X}$, for 100 samples of size 10 drawn from
a table of random numbers ($X = 0, \ldots, 9$)

| | | | | | | | | | |
|---|---|---|---|---|---|---|---|---|---|
| 4·2 | 5·0 | 4·6 | 5·3 | 4·2 | 4·4 | 5·6 | 5·2 | 4·0 | 4·3 |
| 3·3 | 4·6 | 4·9 | 5·8 | 5·2 | 4·5 | 5·5 | 4·9 | 3·5 | 3·9 |
| 2·6 | 4·1 | 4·3 | 5·7 | 5·4 | 3·6 | 3·0 | 3·3 | 3·3 | 5·3 |
| 4·7 | 5·3 | 4·5 | 5·4 | 4·5 | 3·7 | 5·5 | 4·3 | 3·2 | 4·4 |
| 4·8 | 2·3 | 2·7 | 4·6 | 4·7 | 5·0 | 3·9 | 2·1 | 5·1 | 3·7 |
| 3·0 | 3·7 | 4·3 | 5·8 | 6·6 | 5·2 | 5·2 | 6·1 | 4·9 | 5·5 |
| 5·1 | 3·4 | 4·0 | 4·0 | 4·4 | 4·6 | 4·6 | 5·3 | 4·7 | 4·3 |
| 5·9 | 4·4 | 4·0 | 4·6 | 3·3 | 4·0 | 4·4 | 4·3 | 4·4 | 4·2 |
| 5·3 | 5·3 | 4·2 | 2·8 | 5·9 | 6·7 | 4·3 | 4·5 | 3·8 | 3·0 |
| 3·7 | 5·1 | 3·7 | 5·7 | 4·0 | 4·7 | 5·0 | 4·1 | 5·2 | 4·6 |

(b) Frequency distribution of 100 sample means presented
in Table 4.3(a)

| Class interval | Frequency |
|---|---|
| 1·75–2·25 | 1 |
| 2·25–2·75 | 3 |
| 2·75–3·25 | 5 |
| 3·25–3·75 | 12 |
| 3·75–4·25 | 15 |
| 4·25–4·75 | 28 |
| 4·75–5·25 | 15 |
| 5·25–5·75 | 14 |
| 5·75–6·25 | 5 |
| 6·25–6·75 | 2 |
| | |
| Total | 100 |

This empirical example suggests that even though the sample size of 10 is small and the rectangular distribution is not even approximately normal, a normal distribution would provide a reasonable approximation to the distribution of $\overline{X}$. The Central Limit Theorem is important, since it means that we may use the normal sampling distribution of $\overline{X}$ for samples from non-normal populations and that this approximation holds for small samples. This result explains an earlier statement that the normal distribution was basic to statistical theory: not only do we obtain important results concerning $\overline{X}$ when we sample from a normal population, but these results hold approximately even when we are not sampling from a normal population.

## 4.5 the chi-square distribution

Given that the population variance is one of the two parameters necessary to define the normal distribution, it is of interest to investigate the sampling distribution of the sample variance, when we draw random samples from a normal distribution. This turns out to present a more difficult problem than the sampling distribution for $\overline{X}$, since the distribution of the sample variance is non-normal. The intuitive explanation for this is that the sample variance involves the calculation of

$$\frac{1}{n} \sum_{i=1}^{n} (X_i - \overline{X})^2 = \frac{1}{n}\left( \sum X_i^2 - n\overline{X}^2 \right) = \frac{1}{n}\sum X_i^2 - \overline{X}^2$$

and hence we are dealing with the distribution of squared normal variables. Thus while $X$ lies in the range $-\infty \leq X \leq \infty$, $X^2$, being positive, must lie in the range $0 \leq X^2 \leq \infty$.

To illustrate the behaviour of sampling distributions involving squared normal variables, Figure 4.2 presents the histograms of the frequency distributions obtained by drawing 100 random samples from the standard normal distribution $N(0, 1)$ and calculating (a) $Z_i^2$, (b) $\sum_{i=1}^{2} Z_i^2$ and (c) $\sum_{i=1}^{4} Z_i^2$. These histograms illustrate the skewness of distributions involving $\sum_{i=1}^{n} Z_i^2$, but show that the skewness decreases as $n$ increases. However, the distribution of $\sum_{i=1}^{n} Z_i^2$ does not tend to normality, even in the limit when $n$ tends to infinity. Since we shall need to explore the form of the distributions, we shall introduce a new symbol and call a variable which results from squaring normal variables a *chi-square* variable. We shall define

$$\chi^2 = \sum_{i=1}^{n} \left( \frac{X_i - \mu}{\sigma} \right)^2 , \tag{4.17}$$

where the $X_i$ are drawn at random (and hence are independent) from $N(\mu, \sigma^2)$. We note that $\chi^2$ is a continuous variable in the range $0 \leq \chi^2 \leq \infty$. The $\sum_{i=1}^{n} Z_i^2$ variables presented in the histograms are chi-square variables and serve to illustrate that we are dealing with a family of distributions rather than a single distribution. Thus the chi-square distribution is like the binomial distribution, where for a given value of $p$ the shape depends on $n$, rather than the normal distribution, $N(\mu, \sigma^2)$, which may be transformed to the standard form $N(0, 1)$ regardless of $\mu$ and $\sigma^2$.

We shall not explore the mathematics of 4.17, but were we to do so we should find that the formula for the distribution of $\chi^2$ contains a parameter which is a

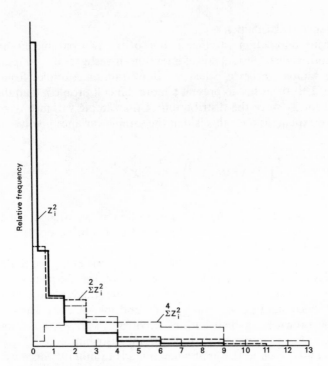

Figure 4.2 Empirical sampling distributions for chi-square variables

function of $n$. We shall call this parameter the *degrees of freedom* of the chi-square distribution and denote it by $v$. The variable defined in 4.17 follows a chi-square distribution with $n$ degrees of freedom. That is, since $\mu$ and $\sigma^2$ are constant from sample to sample, there are $n$ degrees of freedom for random variation in $\chi^2$. Clearly, the magnitude of $\chi^2$ depends on the dispersion of the $X$s about $\mu$.

We now consider the distribution of

$$\sum_{i=1}^{n} \left( \frac{X_i - \overline{X}}{\sigma} \right)^2 \tag{4.18}$$

If the repeated samples of the $X$s are chosen at random from $N(\mu, \sigma^2)$, it turns out that this variable also follows a chi-square distribution, but with $(n-1)$ degrees of freedom in this case. Why $(n-1)$ degrees of freedom? We have noted that the magnitude of the chi-square variable depends on the dispersion of the $X$s about the mean. To concentrate on this dispersion we may imagine the hypothetical experiment of drawing repeated random samples of $X$s from $N(\mu, \sigma^2)$

with $\overline{X}$ held *constant from sample to sample* and calculating the variable defined in 4.18. However, if we impose the constraint that $\overline{X}$ shall be the same for all samples, we are no longer able to draw all the $n$ $X$s in a sample at random: once we have drawn $X_1, ..., X_{n-1}$ at random, $X_n$ is determined by the values of the $X$s we have drawn and our constraint on $\overline{X}$. Thus

$$\frac{1}{n} \sum_{i=1}^{n} X_i = \overline{X}$$

$$\sum_{i=1}^{n} X_i = n\overline{X},$$

so that

$$X_n = n\overline{X} - \sum_{i=1}^{n-1} X_i$$

and we only have $(n-1)$ degrees of freedom for random variation in the chi-square variable 4.18.

Since the chi-square variable is continuous, it is convenient to present probabilities as areas under the curve of the frequency distribution, as was done for the normal distribution. However, there is a problem here since we have not a single curve but a family of curves, depending on the degrees of freedom, $v$. It would obviously require a voluminous set of tables to provide full information on the areas under each chi-square distribution, varying both $\chi^2$ and $v$. The solution is to condense the amount of information presented: instead of tabulating the area under the frequency distribution between specified values of the variable, as was done for the standard normal distribution $N(0, 1)$, the table for the chi-square distribution presents, for different numbers of degrees of freedom, values of $\chi^2$ which divide the area under the curve of the frequency distribution into specified proportions.

These probabilities are presented in the appendix, Table A3. As an example, Figure 4.3 illustrates the chi-square distribution with $v = 4$ degrees of freedom. From Table A3 we find that with $v = 4$, 95 per cent of the area under the curve lies to the right of $\chi^2 = 0.711$ and 5 per cent of the area under the curve lies to the right of $\chi^2 = 9.49$. By subtraction, 90 per cent of the area lies in the range $0.711 \leq \chi^2 \leq 9.49$.

To conclude our present discussion of the chi-square distribution, we note that for this distribution it may be proved that

$$\mu = n, \quad \text{and} \quad \sigma^2 = 2n.$$

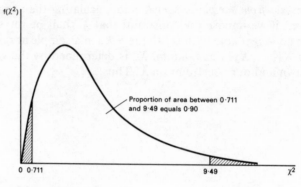

Figure 4.3

Thus as $n$ increases, the centre of the chi-square distribution shifts to the right and the dispersion increases, theoretical results which confirm the visual impression provided by the histograms in Figure 4.2. We shall return to the chi-square distribution and the use of 4.18 in estimation problems in the next chapter.

# problems

**4.1** Discuss any problems that may arise in the following sampling situations:

(a) In order to predict the result of a city election involving two candidates, a research organization telephones a sample of people randomly selected from the city's telephone directory and bases its prediction on the number of respondents in favour of each of the candidates.

(b) One question put to a random sample of households in a survey on water usage is whether the respondent takes a bath each day.

(c) To discover the attitude of passengers towards the new colour scheme inside its planes, an airline puts questionnaires in the seat pockets of the planes on a given route and analyses those completed.

(d) To estimate the annual average income of students ten years after they graduated from a particular university, questionnaires are sent in 1972 to all those who graduated in 1962 and the estimate is based on the questionnaires returned.

**4.2** How would you use a table of random numbers to draw samples in the following situations:

(a) a sample of 50 food shops in a given large city;

(b) a sample of 100 workers in a large factory;

(c) a sample of 1000 voters in a city;

(d) a sample of 250 social science students in a large university;

(e) a sample of 100 prices quoted on a particular stock market.

**4.3** A random sample of size $n$ is to be drawn from the standard normal population $N(0, 1)$ and the sample mean, $\bar{Z}$, calculated. Find $p(0 \cdot 10 \leq \bar{Z} \leq 0 \cdot 25)$ and $p(\bar{Z} < -0 \cdot 20)$ for the following values of $n$:

(a) 1,

(b) 2,

(c) 4,

(d) 16,

(e) 64,

(f) 100.

**4.4** For a given normal distribution $N(\mu, \sigma^2)$ and sample size $n$, calculate the

probabilities concerning the sample mean, $\bar{X}$, $p(\bar{X} < 23)$, $p(24 < \bar{X} < 26)$:
(a) $N(25, 25)$, $n = 25$;
(b) $N(25, 25)$, $n = 100$;
(c) $N(25, 100)$, $n = 25$;
(d) $N(25, 100)$, $n = 100$.

**4.5** A firm producing colour television sets wishes to estimate the average time before a set breaks down. It is required that the probability that the estimate differs from the true value by more than ten hours should be 0·05. It is believed that the standard deviation of the time to failure is 100 hours. How large a sample of sets should be tested?

**4.6** Given the 100 sample means presented in Table 4.1, use the data to calculate the mean of the means and the variance. Compare the calculated values with the theoretical values you evaluated in problem 3.6.

**4.7** Table 4.4 contains 100 values of $Z$ drawn at random from $N(0, 1)$, together with the values of $Z^2$. (These are the data presented in the histogram in Figure 4.2.)

   (i) Calculate the frequency distribution for $Z$ and plot the corresponding histogram.

   (ii) Calculate the mean and variance of the $Z$s and compare them with the theoretical values.

   (iii) Calculate the mean and variance for the values of $Z^2$ and compare them with the theoretical values.

Table 4.4

| $Z$ | $Z^2$ | $Z$ | $Z^2$ | $Z$ | $Z^2$ | $Z$ | $Z^2$ | $Z$ | $Z^2$ |
|---|---|---|---|---|---|---|---|---|---|
| −0·709 | 0·503 | 0·200 | 0·040 | −1·131 | 1·279 | 0·490 | 0·240 | −0·016 | 0·00026 |
| 0·710 | 0·504 | 0·041 | 0·002 | −0·650 | 0·422 | 0·806 | 0·650 | −2·153 | 4·635 |
| 0·427 | 0·182 | −0·038 | 0·001 | 1·419 | 2·014 | 0·107 | 0·011 | −0·103 | 0·011 |
| 0·016 | 0·00026 | 0·748 | 0·560 | 1·011 | 1·022 | 1·780 | 3.168 | 0·972 | 0·945 |
| −2·461 | 6·057 | 0·816 | 0·666 | −0·001 | 0·000001 | −1·445 | 2·088 | −1·456 | 2·120 |
| −1·283 | 1·646 | −0·380 | 0·144 | 1·938 | 3·756 | −0·685 | 0·469 | −0·075 | 0·006 |
| −0·383 | 0·147 | 0·967 | 0·935 | 0·896 | 0·803 | −1·853 | 3·434 | −0·289 | 0·084 |
| 0·172 | 0·030 | 1·521 | 2·313 | −0·147 | 0·022 | −0·945 | 0·893 | −0·048 | 0·002 |
| −0·331 | 0·110 | 1·304 | 0·700 | 1·200 | 1·440 | −0·260 | 0·068 | 0·383 | 0·147 |
| −0·870 | 0·757 | −2·421 | 0·861 | −0·169 | 0·029 | −0·803 | 0·645 | −0·123 | 0·015 |
| −0·728 | 0·530 | 0·538 | 0·289 | 0·418 | 0·174 | 1·116 | 1·245 | −0·884 | 0·781 |
| 0·465 | 0·216 | −0·357 | 0·127 | 2·119 | 4·490 | 0·339 | 0·115 | 0·001 | 0·000 |
| 0·299 | 0·089 | 1·737 | 3·017 | 2·012 | 4·048 | −1·436 | 2·062 | −0·529 | 0.280 |
| 0·742 | 0·551 | 1·250 | 1·562 | 0·518 | 0·268 | 2·199 | 4·836 | 0·695 | 0·483 |
| 1·778 | 3·161 | 0·618 | 0·382 | 0·819 | 0·671 | −0·712 | 0·507 | 0·115 | 0·013 |
| −1·092 | 1·192 | −1·103 | 1·217 | −0·298 | 0·089 | 0·631 | 0·398 | −0·082 | 0·007 |
| 1·420 | 2·016 | −1·147 | 1·316 | 0·673 | 0·453 | 0·294 | 0·086 | 0·477 | 0·228 |
| 1·634 | 2·670 | 1·624 | 2·637 | −1·453 | 2·111 | −1·685 | 2·839 | 1·453 | 2·111 |
| −0·823 | 0·677 | 1·157 | 1·339 | −0·884 | 0·781 | −0·188 | 0·935 | −1·178 | 1·388 |
| −0·030 | 0·0009 | 1·583 | 2·506 | −0·667 | 0·458 | −0·418 | 0·175 | 1·165 | 1·357 |

**4.8** Use Table A3 to calculate the values of $\chi_1^2$ and $\chi_2^2$ such that $p(\chi^2 > \chi_1^2) = 0.95$ and $p(\chi^2 > \chi_2^2) = 0.05$ for the following values of $v$:
  (a) 5,
  (b) 10,
  (c) 20,
  (d) 40,
  (e) 60,
  (f) 100.
Plot these values of $\chi^2$ against $v$ and comment.

# chapter 5

# estimation

## 5.1 introduction
With the results concerning the sampling distributions of random variables which we developed in the last chapter, we are now in a position to consider problems of estimation. We have shown that the main characteristics of a distribution depend on the parameters of the distribution, such as the mean and variance, and we shall consider the estimation of these parameters from sample data. Given a random sample of data, we may calculate sample statistics, such as the sample mean or median, standard deviation or quartile deviation, and consider these as potential estimators of the *population* parameters. For example, if we choose the sample mean as our estimator of the population mean, then the value of $\overline{X}$ calculated from our random sample is our *estimate* of $\mu$.

It is possible to calculate a large number of statistics from a single random sample, and we shall often be faced with a choice between alternative estimators of a given population parameter. One criterion which appeals to the intuition is to choose the sample statistic which corresponds to the population parameter, thus estimating the population mean and variance by the sample mean and variance, but this does not entirely solve the problem. For example, if we are sampling from a symmetrical distribution, the population mean, median and mode all equal the same value and we might choose either the mean or the median of the sample as our estimate of this population value. Since it is very unlikely that the mean and median of a given random sample would be equal; we are still left with a problem of choice. We shall now discuss some criteria for choosing between alternative estimators.[1]

## 5.2 the choice between alternative estimators
With a single random sample, each estimator produces one estimate of the parameter being considered and there is very little that we can say about these individual estimates. In the first place we do not know the value of the parameter, so that we have nothing with which to compare our estimates. Even if we do know the parameter value and find that one estimator gives an estimate which is closer to the true value than the others, can we assume that this will generally

happen in other samples, or is it a chance result produced by random sampling? To answer these questions we need to investigate the sampling properties of the estimators, which we might do empirically by drawing repeated random samples of a given size, calculating the value of the different estimates for each sample and building up the frequency distributions of the various estimators. There is no need to go through this sampling experiment in practice, as we may use the theoretical results which were obtained in the previous chapter.

Suppose we draw a random sample of $n$ observations $X_1, \ldots, X_n$ from some population and wish to estimate some parameter $\theta$. We assume that there are $k$ estimators of $\theta$, say $t_1, \ldots, t_k$, we could calculate from our sample data. To choose between the alternative $t$s, we shall consider the following criteria:

1. *Unbiasedness*. While $t_i$ may not (indeed most probably will not) equal $\theta$ in any one random sample, we shall where possible choose estimators which at least on average equal $\theta$. Thus if $f(t_i)$ is the sampling distribution of $t_i$ (for samples of a given size $n$), we require the mean of this distribution to equal $\theta$, that is,

$$E(t_i) = \theta. \tag{5.1}$$

If 5.1 holds, $t_i$ is an *unbiased* estimator of $\theta$, while if $E(t_i) \neq \theta$, $t_i$ is said to be a *biased* estimator. This is an important property, since 5.1 is independent of sample size and holds for all values of $n$. Figure 5.1 illustrates a case in which one estimator, say $t_1$, is unbiased while another, $t_2$, is biased with $E(t_2) = \theta^* \neq \theta$.

To consider some actual estimators, we have shown that if we sample from a population with mean $\mu$, and variance $\sigma^2$, then for the sample mean, $\overline{X}$, $E(\overline{X}) = \mu$, so that $\overline{X}$ is an unbiased estimator of $\mu$. What about the sample median? If we order our sample of observations from the smallest, $X_{(1)}$, to the largest, $X_{(n)}$,

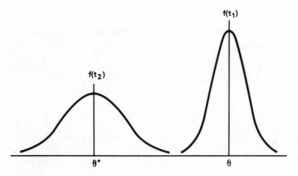

Figure 5.1

then the sample median is $M = X_{(m+1)}$ if $n$ is odd $(n = 2m + 1)$, and when $n$ is even $(n = 2m)$, $M = \frac{1}{2}(X_{(m)} + X_{(m+1)})$. If we are sampling from a symmetrical distribution, $E(X_{(i)}) = \mu, i = 1, \ldots, n$,

$$E(M) = \begin{cases} E(X_{(m+1)}) = \mu & n \text{ odd} \\ E[\frac{1}{2}(X_{(m)} + X_{(m+1)})] = \frac{1}{2}(2\mu) = \mu & n \text{ even,} \end{cases}$$

that is, the sample median is also an unbiased estimator of the population mean. As a final example here, consider the mean square deviation which we defined in 1.4 as

$$\text{MSD} = \frac{1}{n} \sum_{i=1}^{n} (X_i - \bar{X})^2.$$

Now

$$E\left[\frac{1}{n}\sum (X_i - \bar{X})^2\right] = E\left\{\frac{1}{n}\sum [(X_i - \mu) - (\bar{X} - \mu)]^2\right\}$$

$$= E\left\{\frac{1}{n}\left[\sum (X_i - \mu)^2 - 2(\bar{X} - \mu)\sum (X_i - \mu) + \sum (\bar{X} - \mu)^2\right]\right\}$$

$$= \frac{1}{n}\left[\sum E(X_i - \mu)^2 - 2E(\bar{X} - \mu)\sum (X_i - \mu) + \sum E(\bar{X} - \mu)^2\right]$$

$$= \frac{1}{n}\left[n\sigma^2 - 2\sigma^2 + n\left(\frac{\sigma^2}{n}\right)\right]$$

$$= \frac{n-1}{n}\sigma^2 \neq \sigma^2. \tag{5.2}$$

Thus the MSD is a biased estimator of the population variance. Suppose we now define the sample variance as

$$s^2 = \left(\frac{n}{n-1}\right) \cdot \left(\frac{1}{n}\right) \sum_{i=1}^{n} (X_i - \bar{X})^2 = \frac{1}{n-1} \sum_{i=1}^{n} (X_i - \bar{X})^2. \tag{5.3}$$

Then

$$E(s^2) = \left(\frac{n}{n-1}\right) E\left[\frac{1}{n}\sum (X_i - \bar{X})^2\right] = \left(\frac{n}{n-1}\right)\left(\frac{n-1}{n}\right)\sigma^2 = \sigma^2.$$

Hence $s^2$ is an unbiased estimator of $\sigma^2$.

2. *Minimum variance property.* If we have several unbiased estimators of $\theta$, say $t_1$, $t_3$ and $t_5$, we know that on average each of them is equal to $\theta$, but need a further criterion to choose between them. This criterion is based on the relative dispersion of their sampling distribution, and we choose from among the unbiased estimators of $\theta$ the one with the smallest sampling variance. This is the *minimum variance property*, and the estimator which has this property is often said to be the *best* estimator of $\theta$. The standard deviation of an estimator is often called its *standard error*.

Figure 5.2 illustrates the sampling distributions for a hypothetical case based on random samples of size $n$ in which the unbiased estimators of $\theta$ are $t_1$, $t_3$ and $t_5$, with $t_1$ being the best estimator of $\theta$. The minimum variance property is important since, as the figure makes clear, there is a higher probability that, given a single random sample, an estimate of $\theta$ based on $t_1$ will be close to $\theta$ (say in the interval $\theta \pm \varepsilon$) than there is for estimates based on $t_3$ or $t_5$.

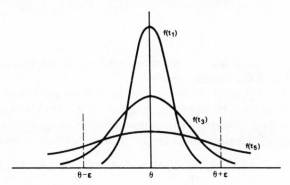

Figure 5.2

The variances of the sampling distributions are functions of the sample size $n$ and decrease with increases in the size of $n$. The distribution $f(t_1)$ is drawn for some given sample size $n$, and by increasing the sample size above $n$ we may eventually be able to achieve the same area under $f(t_2)$ in the range $\theta \pm \varepsilon$. Thus the minimum variance estimator $t_1$ achieves a given precision at the lowest cost in terms of sample size, and for this reason is often referred to as an *efficient* estimator of $\theta$. The relative efficiency of estimators may be considered by comparing their sampling variances, and the relative efficiency of $t_1$ compared with $t_3$ is defined as

$$\frac{\text{var}(t_3)}{\text{var}(t_1)}.$$

(5.4)

For the mean, $\overline{X}$, of a random sample of size $n$, we showed in the previous chapter that $\text{var}(\overline{X}) = \sigma^2/n$. In addition, if we are sampling from a normal population, $N(\mu, \sigma^2)$, it may be proved that $\overline{X}$ is an efficient estimator of $\mu$. Hence if we wish to estimate $\mu$, our best estimator is $\overline{X}$. We have shown that in random samples from a symmetrical distribution the median is an unbiased estimator of the population mean $\mu$. How do the median and the mean compare for samples from a normal population? It may be shown[2] that the variance of the sampling distribution is $k_n(\sigma^2/n)$, where $k_n$ depends on $n$. From 5.4, the relative efficiency of the mean compared to the median is

$$\frac{\text{var}(M)}{\text{var}(\overline{X})} = \frac{k_n\left(\dfrac{\sigma^2}{n}\right)}{\sigma^2/n} = k_n$$

and some representative values of $k_n$ are

| $n$   | 2     | 6     | 10    | 20    | $\infty$ |
|-------|-------|-------|-------|-------|----------|
| $k_n$ | 1·000 | 1·288 | 1·385 | 1·474 | 3·015    |

Thus except in very small samples, the sample mean is a very much more efficient estimator of $\mu$ than is the sample median.

In general, where unbiased estimators can be defined, we shall choose to work with efficient estimators. However, in chapter 11 we shall meet situations in which we cannot find unbiased estimators and have to choose between biased estimators. We shall therefore consider the problem of bias in more detail.

3. *Consistency.* In Figure 5.1 we illustrated a biased estimator ($t_2$) of $\theta$ with $E(t_2) = \theta^* \neq \theta$. We may define the bias as

$$B(t_2) = E(t_2) - \theta$$
$$= (\theta^* - \theta). \tag{5.5}$$

One measure of the magnitude of the bias is the *mean square error* (MSE), which may be defined (similarly to a variance) as

$$\text{MSE} = E(t_2 - \theta)^2. \tag{5.6}$$

Now introducing $\theta^*$,

$$\text{MSE} = E[(t_2 - \theta^*) + (\theta^* - \theta)]^2$$
$$= E[(t_2 - \theta^*)^2 + 2(\theta^* - \theta)(t_2 - \theta^*) + (\theta^* - \theta)^2]$$
$$= \text{var}(t_2) + 2(\theta^* - \theta)E(t_2 - \theta^*) + (\theta^* - \theta)^2$$
$$= \text{var}(t_2) + 0 + B(t_2)^2, \tag{5.7}$$

since $E(t_2 - \theta^*) = 0$. Thus the mean square error is equal to the variance of $t_2$ plus the square of the bias. If, as the sample size tends to infinity, both the variance of $t_2$ and the bias tend to zero (so that MSE tends to zero), then $t_2$ is said to be a *consistent* estimator of $\theta$. If as the sample size tends to infinity $B(t_2)$ tends to zero, but $\text{var}(t_2)$ does not, $t_2$ is called an *asymptotically* unbiased estimator of $\theta$. These two cases are illustrated in Figure 5.3.

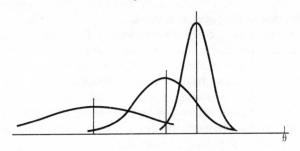

(a) *Consistency.* Distributions move to the right as $n$ increases and sampling variance tends to zero

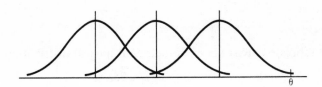

(b) *Asymptotic unbiasedness.* Distributions move to the right as $n$ increases, but sampling variance does not tend to zero

Figure 5.3

As examples of consistent estimators we have the sample mean, $\bar{X}$, which is unbiased and whose variance $(\sigma^2/n)$ tends to zero as $n$ tends to infinity. The same is true for the median, since its variance $[k_n(\sigma^2/n)]$ also tends to zero as $n$ tends to infinity. We have shown in 5.2 that the MSD is a biased estimate of $\sigma^2$, with

$$B(\text{MSD}) = E(\text{MSD}) - \sigma^2$$

$$= \frac{n-1}{n}\sigma^2 - \sigma^2 = -\frac{1}{n}\sigma^2.$$

It may be shown that the variance of the MSD tends to zero as $n$ increases,[3] and as both $B(\text{MSD})$ and $\text{var}(\text{MSD})$ tend to zero as $n$ tends to infinity, the MSD is a consistent estimate of $\sigma^2$. By a similar argument it may be shown that $s^2$ is a consistent and unbiased estimate of $\sigma^2$.

Having established these criteria for choosing between alternative estimators, we shall now consider the estimation of some population parameters.

### 5.3 the estimation of the population mean

If we have a random sample of $n$ observations and we wish to estimate the population mean, $\mu$, we have shown that we should take the mean of the sample as our estimate. This is our best estimate, although we cannot say how close it is to the unknown parameter. If we have to make a single estimate of $\mu$, this is the best we can do. Such a single estimate is often called a *point* estimate. We shall now consider how, by making use of our knowledge of the sampling distributions of estimators, we may go beyond the point estimate and make probability statements concerning the population parameter.

To develop these ideas, we shall first consider the case in which we are to draw a random sample of size $n$ from a normal population $N(\mu, \sigma^2)$. We know that the sampling distribution in this case is $N[\mu, (\sigma^2/n)]$, which may be transformed to the standard normal distribution $N(0, 1)$ if we write

$$Z = \left( \frac{\overline{X} - \mu}{\sigma/\sqrt{n}} \right).$$

From the tabulated areas under the standard normal distribution we know that

$$P(-1 \cdot 96 \leq Z \leq 1 \cdot 96) = 0 \cdot 95,$$

and substituting for $Z$,

$$P\left( -1 \cdot 96 \leq \frac{\overline{X} - \mu}{\sigma/\sqrt{n}} \leq 1 \cdot 96 \right) = 0 \cdot 95$$

$$P\left( -1 \cdot 96 \frac{\sigma}{\sqrt{n}} \leq \overline{X} - \mu \leq 1 \cdot 96 \frac{\sigma}{\sqrt{n}} \right) = 0 \cdot 95$$

$$P\left( 1 \cdot 96 \frac{\sigma}{\sqrt{n}} \geq -\overline{X} + \mu \geq -1 \cdot 96 \frac{\sigma}{\sqrt{n}} \right) = 0 \cdot 95$$

$$P\left( \overline{X} - 1 \cdot 96 \frac{\sigma}{\sqrt{n}} \leq \mu \leq \overline{X} + 1 \cdot 96 \frac{\sigma}{\sqrt{n}} \right) = 0 \cdot 95. \tag{5.8}$$

That is, if we calculate $\overline{X}$ from a random sample of size $n$ and calculate the range $\overline{X} \pm 1.96(\sigma/\sqrt{n})$, then the probability is 0·95 that this range will include the value of $\mu$. The range we have just calculated is usually called a *confidence interval* and enables us to say something about the precision of our estimate of $\mu$.

For a given value of $\sigma$, the width of the confidence interval depends on two things: the level of probability we choose and the sample size. For example, suppose we wish to increase the probability that our confidence interval includes $\mu$ from 0·95 to 0·99. From the tabulated areas for $N(0, 1)$ we have

$$P(-2.58 < Z < 2.58) = 0.99,$$

so that

$$P\left( \overline{X} - 2.58 \frac{\sigma}{\sqrt{n}} < \mu < \overline{X} + 2.58 \frac{\sigma}{\sqrt{n}} \right) = 0.99.$$

Thus we get the common-sense result that to obtain the higher probability of being correct we need to increase the width of the confidence interval. Suppose we leave the probability level at 0·95, but consider drawing a sample of size $4n$ instead of $n$. In that case the confidence interval will be

$$\overline{X} \pm 1.96 \frac{\sigma}{2\sqrt{n}}.$$

That is, by quadrupling the sample size we have halved the width of the confidence interval.

As a numerical example, consider sampling from $N(0, 1)$, in which case with the probability level $p = 0.95$ the confidence interval is $\overline{Z} \pm (1.96/\sqrt{n})$. Some values of this range for different values of $n$ are given below:

| $n$ | 5 | 10 | 20 | 25 | 1000 |
|---|---|---|---|---|---|
| $\pm (1.96/\sqrt{n})$ | 0·877 | 0·620 | 0·438 | 0·392 | 0·196 |

Table 4.4 on page 72 contains 100 random observations drawn from $N(0, 1)$, and we have taken these (down the columns) in groups of 5 to give twenty random samples, for each of which we have calculated $\overline{Z}$. These twenty values of $\overline{Z}$ are presented in Table 5.1. The first sample mean is $-0.403$, giving a $p = 0.95$ confidence interval of $-0.403 \pm 0.877$, which does include the true value of $\mu = 0$. Forming the $p = 0.95$ confidence intervals for the remaining values of $\overline{Z}$, we see that only in the case of the eleventh sample, where the confidence interval is $0.240 < \mu < 1.994$, and the fourteenth, where the confidence interval is $-1.786 < \mu < -0.032$, does our confidence interval not include the true value of $\mu = 0$.

We have shown how a confidence interval may be set up around our estimate

*Table 5.1.*

| Sample number | z | Sample number | z | Sample number | z | Sample number | z |
|---|---|---|---|---|---|---|---|
| 1 | $-0.430$ | 6 | $0.198$ | 11 | $1.177$ | 16 | $-0.273$ |
| 2 | $-0.539$ | 7 | $0.757$ | 12 | $-0.528$ | 17 | $0.031$ |
| 3 | $0.511$ | 8 | $0.423$ | 13 | $0.348$ | 18 | $0.085$ |
| 4 | $0.222$ | 9 | $0.130$ | 14 | $-0.909$ | 19 | $-0.120$ |
| 5 | $0.353$ | 10 | $0.744$ | 15 | $0.301$ | 20 | $0.367$ |

of the population mean when we are sampling from a normal population and the population variance is known. These may seem to be somewhat restrictive assumptions, particularly the second, so we shall explore what happens if we relax them. First assume the variance is known, but that the population is not normal. In this case we may recall that the Central Limit Theorem states that the distribution of

$$\frac{\overline{X} - \mu}{\sigma/\sqrt{n}}$$

tends to the standard normal distribution as the sample size increases. Thus the probability statement given in 5.8, which holds exactly when the population is normal, will hold *approximately*, even when the population is not normal. In practice the approximation is good for quite small sample sizes, so that relaxing the normality assumption is not crucial.

When the population variance is not known, which is generally the case, we have a more serious problem to deal with, since the formula for the confidence interval requires the value of $\sigma$. In attempting to solve this problem we shall return to the case of sampling from a normal population and introduce a new distribution which is derived from the normal distribution.

### 5.4 the *t*-distribution
The confidence intervals we have discussed have been based on the standard normal variable

$$Z = \frac{\overline{X} - \mu}{\sigma/\sqrt{n}}$$

and one possible solution to the problem is to substitute an estimate of $\sigma$ into the expression and to assume that the resulting variable is approximately standard normal. The obvious candidate to estimate $\sigma^2$ is $s^2$, which is unbiased and consistent, so that (we hope)

$$Z \approx \frac{\overline{X} - \mu}{s/\sqrt{n}}.$$

In fact, provided the sample size is reasonably large, say greater than 40, this approximation is good and may safely be used to calculate confidence intervals. As we see below, however, the approximation is poor for small sample sizes and the approximation needs to be replaced by an investigation of the exact distribution of

$$t = \frac{\overline{X} - \mu}{s/\sqrt{n}}. \tag{5.9}$$

The mathematical properties of the distribution of the variable, $t$, we have defined in 5.9, have been evaluated and the results are presented in the Appendix, Table A2. The mathematical form of the distribution is too complicated to be worth exploring here, but we may note some of its features. Firstly, without changing the numerical value of 5.9, we may write the ratio as

$$t = \frac{\overline{X} - \mu}{s/\sqrt{n}} = \frac{\overline{X} - \mu}{(s/\sigma)(\sigma/\sqrt{n})}$$

$$= \frac{(\overline{X} - \mu)/(\sigma/\sqrt{n})}{(s/\sigma)}. \tag{5.10}$$

We note that the numerator is the standard normal variable we would be considering if $\sigma^2$ were known, while the denominator introduces the additional variability arising from the fact that we do not know $\sigma^2$ and have to calculate $s^2$. We have seen that $s^2$ is a consistent estimator of $\sigma^2$, hence $(s/\sigma)$ will tend to unity as the sample size increases, so that the $t$-distribution will tend to the standard normal distribution as the sample size increases. Given the numerator, the $t$-distribution is symmetrical regardless of sample size. We may also note that substituting for $s^2$ gives

$$\frac{s^2}{\sigma^2} = \frac{\sum(X_i - \overline{X})^2}{(n-1)\sigma^2} = \frac{1}{n-1}\left(\frac{\sum(X_i - \overline{X})^2}{\sigma^2}\right) = \frac{1}{n-1}(\chi^2_{n-1}).$$

Thus the random variable which appears in the denominator involves the square root of the chi-square variable with $(n-1)$ degrees of freedom which we discussed in the last chapter, section 4.5.[4] We saw there that the chi-square distribution varies with the degrees of freedom parameter, and the same is true of the $t$-distribution. The expression in 5.9 follows a $t$-distribution with $(n-1)$ degrees of freedom. Hence Table A2 presents the values of $t$ which divide the area under the curve into specified proportions against the degrees of freedom. The $t$-distribution with 4 degrees of freedom (df) is compared to the normal distribution in Figure 5.4.

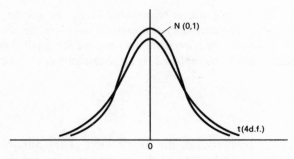

Figure 5.4

As an example, we note that with 9 degrees of freedom 95 per cent of the area of the corresponding *t*-distribution lies within the range $\pm 2\cdot26$. That is,

$$P(-2\cdot26 < t < 2\cdot26) \qquad\qquad = 0\cdot95,$$

and substituting for *t* from 5.9,

$$P\left(-2\cdot26 < \frac{\overline{X} - \mu}{s/\sqrt{10}} < 2\cdot26\right) \qquad = 0\cdot95$$

$$P\left(-2\cdot26\,\frac{s}{\sqrt{10}} < \overline{X} - \mu < 2\cdot26\,\frac{s}{\sqrt{10}}\right) \qquad = 0\cdot95$$

$$P\left(\overline{X} - 2\cdot26\,\frac{s}{\sqrt{10}} < \mu < \overline{X} + 2\cdot26\,\frac{s}{\sqrt{10}}\right) = 0\cdot95. \qquad (5.11)$$

Here we may compare the confidence interval of $\overline{X} \pm 2\cdot26\,(s/\sqrt{10})$ with the corresponding confidence interval based on the standard normal distribution, $\overline{X} \pm 1\cdot96(s/\sqrt{10})$, which understates by a factor of $13\cdot3$ per cent. Hence for small samples it is important to base confidence intervals on the *t*-distribution.

We have already argued that the confidence interval defined in 5.8 could be applied as an approximation to samples from a non-normal population when the variance was known. By a similar argument the confidence interval derived from 5.9, which is an exact result for random samples from a normal population, may be used as a good approximation for samples from non-normal populations.

For example, suppose we had drawn the first five of the numbers reported in Table 4.4 without knowing that it had been drawn from $N(0, 1)$. In that case we should have to calculate an estimate of $\sigma^2$ from our sample observations $-0\cdot709, 0\cdot710, 0\cdot427, 0\cdot016, -2\cdot461$. Here $s^2 = \frac{1}{4}\sum(X_i - \overline{X})^2 = \frac{1}{4}(6\cdot434) = 1\cdot608$, so that $s = 1\cdot268$. With $p = 0\cdot95$ and 4 degrees of freedom,

$p(-2\cdot78 < t < 2\cdot78) = 0\cdot95$

and our confidence interval is

$$p\left(-0\cdot403 - 2\cdot78\left(\frac{1\cdot268}{\sqrt{5}}\right) < \mu < -0\cdot403 + 2\cdot78\left(\frac{1\cdot268}{\sqrt{5}}\right)\right) = 0\cdot95$$

$$= p(-1\cdot979 < \mu < 1\cdot173).$$

The reader may compare this confidence interval with that obtained when we know that $\sigma^2 = 1$, that is, $-1\cdot280 < \mu < 0\cdot477$.

### 5.5 the estimation of the population variance

Having discussed the estimation of the population mean, it is reasonable to proceed next to the estimation of the population variance. We have already shown that $s^2$, the estimator of $\sigma^2$ defined in 5.3, is an unbiased estimator. Hence if we want a point estimate of $\sigma^2$ we would choose $s^2$. To set up a confidence interval as opposed to giving a single estimate involves knowledge of the sampling distribution of $s^2$. This depends on the population we are sampling from, and to begin with we shall consider sampling from a normal population.

If we are sampling from a normal population we may utilize the results concerning the chi-square distribution developed in the last chapter, section 4.5, to set up a confidence interval. There it was argued that $\sum^n (X_i - \bar{X})^2/\sigma^2$ followed the chi-square distribution with $(n-1)$ degrees of freedom. Given the value of $(n-1)$, we may use Table A3 to find two values of $\chi^2$ such that

$$p(\chi^2 \geq \chi_1^2) = 0\cdot975 \quad \text{and} \quad p(\chi^2 \geq \chi_2^2) = 0\cdot025.$$

Then by subtraction

$$p(\chi_1^2 \leq \chi^2 \leq \chi_2^2) \qquad\qquad = 0\cdot95,$$

and substituting for $\chi^2$,

$$p\left(\chi_1^2 \leq \frac{\sum(X_i - \bar{X})^2}{\sigma^2} \leq \chi_2^2\right) \qquad = 0\cdot95$$

and by rearrangement

$$p\left(\frac{1}{\chi_1^2} \geq \frac{\sigma^2}{\sum(X_i - \bar{X})^2} \geq \frac{1}{\chi_2^2}\right) \qquad = 0\cdot95,$$

or

$$p\left(\frac{\sum(X_i - \bar{X})^2}{\chi_2^2} \leq \sigma^2 \leq \frac{\sum(X_i - \bar{X})^2}{\chi_1^2}\right) = 0\cdot95. \qquad (5.12)$$

As an example, for the random sample of 5 observations from a normal distribution we calculated that $\sum (X_i - \overline{X})^2 = 6\cdot434$. In this case $(n - 1) = 4$, and with 4 degrees of freedom $p(\chi^2 \geq 0\cdot484) = 0\cdot975$ and $p(\chi^2 \geq 11\cdot14) = 0\cdot025$. Hence

$$p\left( \frac{6\cdot434}{11\cdot14} \leqq \sigma^2 \leqq \frac{6\cdot434}{0\cdot484} \right) = 0\cdot95,$$

that is,

$$p(0\cdot578 \leq \sigma^2 \leq 13\cdot29) = 0\cdot95.$$

In this case we know that the true value is $\sigma^2 = 1$ and we see that the confidence interval does include the true value of $\sigma^2$. We may also note that the very small sample size is reflected in the very wide confidence interval we have obtained.

The confidence interval derived in equation 5.12 holds as an exact result if the sample is drawn from a normal distribution. What happens if the sample is drawn from a non-normal distribution: does 5.12 hold as an approximate result? The answer here is that, unfortunately, this result is very sensitive to the normality assumption, since we do not have any result like the Central Limit Theorem that can be proved for variances. This may be explained by noting that whereas the calculation of the sample mean tends to average out deviations from normality, the calculation of the variance involves squaring deviations from the mean and hence tends to emphasize extreme observations. Thus if the distribution being sampled is skewed, or has long tails compared with the normal distribution, 5.12 will not yield a good approximation, particularly with small sample sizes. Since its properties are not greatly affected by relaxing the assumption of normality, the sample mean is often referred to as a *robust* estimator. The sample variance is a considerably less robust estimator.[5]

### 5.6 the estimation of population proportions
The cases we have considered concerning the mean and variance generally involve measurement rather than counting, but there are many situations in which counting is very important. For example, in attempting to forecast the outcome of an election the proportion of the relevant population which will vote for a given candidate is of interest. Thus we may be interested in estimating the proportion of members of a population having a given characteristic from a random sample of observations drawn from the population. We shall denote the proportion in the population having the characteristic by $\pi$ and the sample estimate by $\hat{\pi}$.

Suppose we have drawn a random sample of size $n$ from some population and wish to estimate the proportion of members of the population having a particular characteristic. If the number of members of the sample having the characteristic is *m*, then the proportion in the sample is[6]

$$\hat{\pi} = \frac{m}{n}. \tag{5.13}$$

It may be shown that this is the best single estimate we may obtain of $\pi$, since $\hat{\pi}$ is unbiased and consistent.

To go beyond this point estimate we need to develop a confidence interval, which depends on a knowledge of the sampling distribution of $\hat{\pi}$. In developing the relevant sampling theory here the normal distribution is once more important, in this case the results developed for the normal approximation to the binomial distribution. In that context it was shown that the probability of *m* successes out of *n* trials was given by

$$Z \approx \frac{m - \mu}{\sigma} = \frac{m - np}{\sqrt{np(1 - p)}}. \tag{5.14}$$

If the probability that any member of the population has the characteristic is *p*, then the proportion of members we would expect to have the characteristic is also *p*. In what follows we shall denote the population proportion by $\pi$. Suppose we now divide through the numerator and denominator of 5.14 by *n*, obtaining

$$Z \approx \frac{\dfrac{m}{n} - \pi}{\sqrt{\dfrac{\pi(1 - \pi)}{n}}} = \frac{\hat{\pi} - \pi}{\sqrt{\dfrac{\pi(1 - \pi)}{n}}}. \tag{5.15}$$

Now *Z* is a standard normal variable, and hence the approximation in 5.15 may be used to set up a confidence interval for $\pi$. Thus

$$p\left( -1 \cdot 96 < \frac{\hat{\pi} - \pi}{\sqrt{\dfrac{\pi(1 - \pi)}{n}}} < 1 \cdot 96 \right) \approx 0 \cdot 95$$

$$p\left( -1 \cdot 96 \sqrt{\frac{\pi(1 - \pi)}{n}} < \hat{\pi} - \pi < 1 \cdot 96 \sqrt{\frac{\pi(1 - \pi)}{n}} \right) \approx 0 \cdot 95$$

$$p\left( \hat{\pi} - 1\cdot96 \sqrt{\frac{\pi(1-\pi)}{n}} < \pi < \hat{\pi} + 1\cdot96 \sqrt{\frac{\pi(1-\pi)}{n}} \right) \approx 0\cdot95. \tag{5.16}$$

There is one problem that arises in applying this approximation. The standard error of $\hat{\pi}$,

$$\text{s.e.}(\hat{\pi}) = \sqrt{\frac{\pi(1-\pi)}{n}},$$

is a function of $\pi$, so that to use 5.16 to set up a confidence interval for $\pi$, we need to know $\pi$. The solution to this problem is to substitute an estimate for the unknown value of $\pi$, the obvious candidate being $\hat{\pi}$. Thus we rewrite 5.16 as

$$p\left( \hat{\pi} - 1\cdot96 \sqrt{\frac{\hat{\pi}(1-\hat{\pi})}{n}} < \pi < \hat{\pi} + 1\cdot96 \sqrt{\frac{\hat{\pi}(1-\hat{\pi})}{n}} \right) \approx 0\cdot95 \tag{5.17}$$

and use 5.17 as our confidence interval.

We shall calculate a confidence interval for the following case. Table 4.4 contains a random sample of observations on $Z$ and we shall estimate the proportion of negative numbers in the population from which the sample was drawn. If we examine the signs of the $Z$s we find that 46 are negative, so that our best estimate of $\pi$ is $0\cdot46$. To set up a confidence interval with a probability of $0\cdot95$ that it will include the true value of $\pi$, we may substitute $\hat{\pi} = 0\cdot46$ and $n = 100$ in 5.17. Then

$$p\left( 0\cdot46 - 1\cdot96 \sqrt{\frac{(0\cdot46)(0\cdot54)}{100}} < \pi < 0\cdot46 + 1\cdot96 \sqrt{\frac{(0\cdot46)(0\cdot54)}{100}} \right) \approx 0\cdot95$$

$$p(0\cdot46 - 0\cdot098 < \pi < 0\cdot46 + 0\cdot098) \qquad\qquad \approx 0\cdot95$$

$$p(0\cdot362 < \pi < 0\cdot558) \qquad\qquad \approx 0\cdot95.$$

In this particular case we see that the confidence interval does include the true value of $\pi = 0\cdot5$.

For small values of $n$, the normal approximation will not hold very closely and, in addition, the confidence interval tends to become too wide to be of much interest. For example, suppose our random sample had consisted of only the first five observations on $Z$ from Table 4.4. In that case we observe two negative numbers, so that $\hat{\pi} = \frac{2}{5} = 0\cdot4$. Our $p = 0\cdot95$ confidence interval in this case with $\hat{\pi} = 0\cdot4$ and $n = 5$ is $0\cdot4 \pm 0\cdot430$ or $-0\cdot030 < \pi < 0\cdot830$, a rather wide range.

### 5.7 conclusion

We have now considered the problem of estimating means, variances and population proportions. We shall consider some further estimation problems below, particularly in the second part of the text in our discussion of correlation and regression analysis.

# problems

**5.1** Suppose we draw two observations $X_1$ and $X_2$ at random from $N(\mu, \sigma^2)$ and wish to estimate $\mu$. We define an estimate of $\mu$ as

$$\text{est}(\mu) = \lambda X_1 + \theta X_2.$$

What values should we give to $\lambda$ and $\theta$ so that $\text{est}(\mu)$ will be
   (a) an unbiased estimator of $\mu$,
   (b) the minimum variance estimator of $\mu$?

**5.2** A sample of 25 is drawn from a normal population with unknown mean and known variance 64. If the sample mean is $21 \cdot 2$, estimate the population mean and set up a confidence interval (with $p = 0 \cdot 95$) for this estimate.

**5.3** A random sample of 36 observations is drawn from a normal population with unknown mean and variance. The sample mean is $114 \cdot 2$ and variance $31 \cdot 28$. Set up a confidence interval (with $p = 0 \cdot 95$) for the population mean. How would your procedure be modified if the population variance was known to be $32 \cdot 8$?

**5.4** Given that $\overline{X} = 20$, $s = 4$ and $n = 10$ and $X$ is normally distributed, use the $t$-distribution to set up confidence intervals of $\mu$ with
   (a) $p = 0 \cdot 95$,
   (b) $p = 0 \cdot 99$.

**5.5** A random sample of 12 wine stores is selected in a given town and the following prices are recorded (in new pence) for a half-bottle of Chateau d'Ipswich 1970:

30, 22, 32, 26, 24, 40, 34, 36, 32, 33, 28, 30.

Estimate $\mu$ and $\sigma^2$ and set up confidence intervals, with $p = 0 \cdot 95$, for $\mu$ and $\sigma^2$.

**5.6** A random sample of 400 voters in a city is polled and 240 say that they intend to vote in a coming election. Set up a confidence level (with $p = 0 \cdot 95$) for the proportion of the voters in the city who will take part in the election.

**5.7** In order to estimate the proportion of housewives who have tried a new detergent, a random sample of 196 housewives is interviewed. If 108 members of the sample have tried the detergent, construct a confidence interval (with $p = 0 \cdot 99$) for the proportion of the population who have tried the new detergent.

**5.8** For a random sample of 20 observations the sample variance is calculated as $14 \cdot 5$. Calculate a confidence interval (with $p = 0 \cdot 95$) for the population variance.

# chapter 6
# hypothesis testing

## 6.1 introduction

The word 'hypothesis' is defined by *Chambers Twentieth Century Dictionary* as 'a supposition: a proposition assumed for the sake of argument: a theory to be proved or disproved by reference to the facts: a provisional explanation of anything'. At this stage we shall be much less general than this definition and restrict our interpretation of the word 'hypothesis' to imply a theory concerning the value of a population parameter such as the mean, $\mu$, or the values of several population parameters, such as the mean, $\mu$, and variance, $\sigma^2$. This may seem to be a somewhat restrictive approach, but in the second part of this text, when we explore the relationship between economic variables, we shall show that the methodology developed here can be applied to the testing of an interesting range of hypotheses.

To begin with, we shall consider a problem involving a population mean. For example, suppose the length of a metal bar produced by a given process is a random variable which follows a normal distribution with a standard deviation of 15 cm. The bars have two uses, for one of which the process is set to produce bars with a mean length $\mu = 150$ cm, while for the other the process is set to produce bars with a mean length of $\mu = 160$ cm. The bars produced by the machines set for a given $\mu$ are collected, measured, packed in boxes of 25 and sent to a dispatch department, whence they are sent to their end uses. Each box has on it a label listing the lengths of the 25 bars and the level of $\mu$ at which the process has been set. However, the dispatcher finds that the label on one box has been smudged, so that he cannot read the entry for $\mu$, although he thinks it looks like $\mu = 150$ cm. What should he do, given that he knows the process is normal and the standard deviation is 15 cm?

He might proceed to argue as follows. If $\mu$ does equal 150 cm, then the distribution of the means of random samples of 25 bars drawn from $N(150, 225)$ is $N(150, 9)$. On the basis of this information he knows that there is a probability of 0·95 that the mean of a random sample of 25 observations will lie in the range $\mu \pm 1.96\sigma/\sqrt{n}$, that is, $150 \pm 1.96(3)$, or 144·12 to 155·88. There is a probability of 0·99 that $\bar{X}$ lies in the range $150 \pm 2.58(3)$, or 142·26 to 157·74, while the probability is

0·999 that $\overline{X}$ lies in the range $150 \pm 3\cdot32(3)$, or 140·04 to 159·96. From this last calculation it follows that if $\mu = 150$ cm, it is unlikely that we would observe a value $\overline{X}$ which is less than 140·04 or greater than 159·96, since the probability of the event is only 0·001. Hence if our dispatcher adopts the rule that he will calculate the mean length of the rods in the box and accept the hypothesis that $\mu = 150$ cm if $\overline{X}$ falls in the range 140·04 cm to 159·96 cm, there is only a small probability of rejecting the hypothesis if it is true.

While this rule means that our dispatcher has a high probability of accepting the hypothesis $\mu = 150$ cm when it is correct, there is some probability that he will reject the hypothesis that $\mu = 150$ cm when it is true. For example, if he chooses a probability level of $p = 0\cdot95$, he will accept $\mu = 150$ cm if $\overline{X}$ falls inside the range 144·12 cm to 155·88 cm and reject the hypothesis if $\overline{X} < 144\cdot12$ cm or $\overline{X} > 155\cdot88$ cm. Since

$$p(\overline{X} < 144\cdot12 | \mu = 150) + p(\overline{X} > 155\cdot88 | \mu = 150) = 0\cdot05,$$

the decision rule implies that the probability that he will reject the hypothesis when it is true is 0·05. In general, if he chooses a probability level of $p = 1 - \alpha$, then the probability that he will reject the hypothesis when in fact it is true is $\alpha$.

If he dislikes the probability of rejecting the hypothesis when it is true, he may decide to choose a very high value of $1 - \alpha$. For example, if he adopts the rule that he will accept $\mu = 150$ cm if $\overline{X}$ falls in the range 138 cm to 162 cm (that is, four standard deviations on either side of the hypothetical mean), then $\alpha = 0\cdot00003$. At first sight this might appear to be a highly satisfactory state of affairs, but this is only so if he ignores the fact that there is an *alternative* hypothesis, $\mu = 160$ cm. Suppose the hypothesis that $\mu = 150$ cm is false and in fact $\mu = 160$ cm is true. In that case the correct sampling distribution for values of $\overline{X}$ is $N(160, 9)$, so that $p(\overline{X} < 162) = 0\cdot7454$. In other words, while the rule 'accept $\mu = 150$ cm if $138 < \overline{X} < 162$' produces a very low value for $\alpha$, it produces the very high probability of 0·7454 that the hypothesis $\mu = 150$ cm will be accepted when in fact $\mu = 160$ cm.

As this example makes clear, there are two kinds of error that we must consider when testing statistical hypotheses: the possibility that we reject a hypothesis when in fact it is true (the Type I error) and the possibility that we accept the hypothesis as being true when in fact it is false (the Type II error). The Type I error is the probability $\alpha$ considered above, the Type II error we shall denote by $\beta$. The first point to notice is that the test procedure involves a trade-off between the two kinds of error. For example, suppose our dispatcher feels that a value of $\beta = 0\cdot7454$ is too high a price to pay to obtain $\alpha = 0\cdot00003$ and therefore changes his rule so that he now only accepts $\mu = 150$ cm if $\overline{X}$ falls in the range $144\cdot12 < \overline{X} < 155\cdot88$, giving $\alpha = 0\cdot05$. Now if $\mu = 160$ cm,

$$p(\overline{X} < 155\cdot88) = p(Z < 160/3) = p(Z < -1\cdot37) = 0\cdot0853,$$

so that $\beta = 0\cdot0853$.

We may summarize these conclusions in a general form if we denote the hypothesis concerning some population parameter which is to be tested by $H_0$, and the alternative hypothesis by $H_1$. Then the probabilities concerning the errors may be presented in the following table:

| If in fact | Decision concerning $H_0$ | |
|---|---|---|
| | Accept | Reject |
| $H_0$ true | $\checkmark$ <br> $(1 - \alpha)$ | Type I error <br> $\alpha$ |
| $H_0$ false <br> and $H_1$ true | Type II error <br> $\beta$ | $\checkmark$ <br> $(1 - \beta)$ |

$\checkmark$ denotes a correct decision

Since $(1 - \beta)$ represents the probability of correctly rejecting the hypothesis $H_0$, it is often referred to as the *power* of the test to discriminate between the two hypotheses.

We have shown that there is a trade-off between $\alpha$ and $\beta$, so that decreasing $\alpha$ involves an increase in $\beta$. However, it is possible to reduce $\beta$ without increasing $\alpha$. Consider our example in which the dispatcher is attempting to decide between $H_0: \mu = 150$ and $H_1: \mu = 160$. Assuming $H_0$ is true, the sampling distribution for $\overline{X}$ is $N(150, 9)$, and if he chooses a probability level of $(1 - \alpha)$, say $p = 0\cdot95$, he rejects the hypothesis $H_0$ if $\overline{X}$ falls outside the range $144\cdot12 < \overline{X} < 155\cdot88$. However, from the point of view of *discriminating* between $H_0$ and $H_1$, the left-hand tail of $N(150, 9)$ is not of as much interest as the right-hand tail, since if $\overline{X} < 144\cdot12$ we must reject $H_1$ as well as $H_0$. This suggests the following modification in the test procedure we have been considering: given we have chosen $p = (1 - \alpha)$, instead of using two values which cut off $\alpha/2$ of the area in either tail of $N(150, 9)$, we now choose a value of $\overline{X}$ which cuts off $\alpha$ in the right-hand tail. For example, if $(1 - \alpha) = 0\cdot95$, then

$$\overline{X} = 150 + 1\cdot64(3) = 154\cdot92$$

cuts off an area of $0\cdot05$ in the right-hand tail of $N(150, 9)$. Our test procedure is now to reject $H_0$ in favour of $H_1$ if $\overline{X} > 154\cdot92$. Our Type I error remains at $\alpha = 0\cdot05$, but in this case the value of $\beta$ is

$$p(\overline{X} < 154\cdot92) = p(Z < (154\cdot92 - 160)/3) = p(Z < -1\cdot69) = 0\cdot0455,$$

as compared with the value $\beta = 0\cdot0853$ obtained for $\alpha = 0\cdot05$ when both tails of $N(150, 9)$ were used in the test.

We shall call a test based on $\alpha/2$ in each tail a *two-tail* test (2TT), while if the area $\alpha$ is concentrated in either the left-hand or the right-hand tail we shall talk of a *one-tail* test (1TT).

In the example we have been discussing we had a specific alternative hypothesis, $H_1:\mu = 160$. This will not be true in all situations, but we should always remember that *some* alternative hypothesis must always exist. The sobering implication of this should be that whenever we test a statistical hypothesis we are making some Type II error. At a more practical level, even if we do not know the precise form of the relevant alternative hypothesis, we should try to determine whether a one-tail or two-tail test is appropriate. For example, if the hypothesis being tested is $H_0:\mu = \mu_0$ and we have an idea that the alternative hypothesis is $H_1:\mu > \mu$, then we shall commit a smaller Type II error by carrying out a one-tail test. If, however, our alternative hypothesis is $H_1:\mu \neq \mu_0$, then a two-tail test is appropriate.

In the light of our discussion of Type I and Type II errors, it is clear that one of the dictionary definitions of the word 'hypothesis', namely 'a theory to be proved or disproved by reference to the facts', does not fit too happily into the context of statistical theory. Given $\alpha$ and $\beta$ are never zero, we can never 'prove' or 'disprove' any hypothesis with complete certainty. All we can hope to do is to establish high probabilities for our decisions. Given this constraint, it is obvious that it is easier to disprove a particular hypothesis (that is, achieve a low value for $\alpha$) than it is to prove a particular hypothesis is true (that is, to achieve a low value for $\beta$). For this reason when we speak of 'accepting' a hypothesis we mean that the available evidence does not lead us to reject the hypothesis and not that the hypothesis has been proved true.

### 6.2 testing hypotheses concerning the population mean
Suppose the box examined by the dispatcher contains 25 bars with lengths

| | | | | |
|---|---|---|---|---|
| 120·15 | 179·49 | 145·34 | 180·84 | 133·11 |
| 149·84 | 151·53 | 165·63 | 174·60 | 163·62 |
| 140·52 | 139·32 | 165·84 | 162·03 | 126·64 |
| 148·95 | 175·35 | 127·28 | 142·35 | 164·98 |
| 144·34 | 157·46 | 145·78 | 147·75 | 148·44 |

From these data he calculates that $\overline{X} = 152\cdot04$ cm. Having already chosen the probability level $p = 0\cdot95$ and, since $H_1:\mu = 160$, opted for a one-tail test, he will reject $H_0:\mu = 150$ if $\overline{X} > 150 + 1\cdot64(3)$, that is, if

$$\frac{\overline{X} - 150}{3} > 1\cdot64.$$

In this case

$$\frac{152\cdot04 - 150}{3} = 0\cdot68$$

so that we would not reject the hypothesis that $\mu = 150$ cm.

We may now restate this test procedure in general terms as follows. Let the hypothesis to be tested be $H_0:\mu = \mu_0$ and the probability level be $(1 - \alpha)$. Suppose first that the alternative hypothesis is $H_1:\mu > \mu_0$, so that a one-tail test is appropriate. If $\bar{X}$ is calculated from a sample of size $n$, the hypothesis is rejected if

$$\frac{\bar{X} - \mu_0}{\sigma/\sqrt{n}} > Z_\alpha, \tag{6.1}$$

where $Z_\alpha$ is the standard normal deviate which cuts off $\alpha$ in the right-hand tail of $N(0, 1)$. On the other hand, if the alternative hypothesis had been $H_1:\mu \neq \mu_0$, the hypothesis $H_0$ is rejected if either

$$\frac{\bar{X} - \mu_0}{\sigma/\sqrt{n}} < -Z_{\alpha/2} \quad \text{or} \quad \frac{\bar{X} - \mu_0}{\sigma/\sqrt{n}} > Z_{\alpha/2}, \tag{6.2}$$

where $-Z_{\alpha/2}$ and $Z_{\alpha/2}$ cut off proportions of $\alpha/2$ in the left-hand and right-hand tails of $N(0, 1)$ respectively.

The procedure we have developed for testing a hypothesis concerning the mean of a population is based upon two assumptions: (1) that the population being sampled is normal and (2) that the variance $\sigma^2$ is known. Relaxing the first assumption does not cause any real problems since, as was suggested by Figure 4.1, $N(0, 1)$ provides a good approximation for the distribution of $(\bar{X} - \mu)/(\sigma/\sqrt{n})$ even if the population being sampled is non-normal.

If the variance, $\sigma^2$, is unknown we may proceed as follows: calculate an unbiased estimate of $\sigma^2$ from the sample as

$$s^2 = \frac{1}{n-1} \sum_{i=1}^{n} (X_i - \bar{X})^2,$$

then we may use the result developed in the previous chapter, section 5.4, that

$$t = \frac{\bar{X} - \mu}{s/\sqrt{n}}$$

follows the $t$-distribution with $(n - 1)$ degrees of freedom. Thus we substitute

$$\frac{\bar{X} - \mu_0}{s/\sqrt{n}} > t_\alpha \tag{6.3}$$

for 6.1 and

$$\frac{\overline{X} - \mu_0}{s/\sqrt{n}} < -t_{\alpha/2} \quad \text{or} \quad \frac{\overline{X} - \mu_0}{s/\sqrt{n}} > t_{\alpha/2} \tag{6.4}$$

for 6.2.

For example, if our dispatcher had not known that the population variance of the length of bars was 225, he could have calculated $s^2 = 277 \cdot 18$ from the sample data contained in the table on page 94, and then $s/\sqrt{n} = 16 \cdot 65/5 = 3 \cdot 33$. Having chosen a probability level of $p = 0 \cdot 95$, then with 24 degrees of freedom the appropriate $t$-value for a one-tail test $t = 1 \cdot 71$ and he will reject $H_0 : \mu = 150$ if

$$t = \frac{\overline{X} - 150}{3 \cdot 33} > 1 \cdot 71$$

since $\overline{X} = 152 \cdot 04$, $t = 2 \cdot 04/3 \cdot 33 = 0 \cdot 61$, a value which falls in the acceptance region, so we do not reject $H_0 : \mu = 150$.

### 6.3 testing hypotheses concerning the population variance
In section 5.5 we showed that the chi-square distribution could be used to set up a confidence interval in estimating the variance of a normal distribution. It is easy to show that this distribution may also be used to test hypotheses concerning the variance of a normal population. Suppose our hypothesis is $H_0 : \sigma^2 = \sigma_0^2$ and we have a random sample, $X_1, \ldots, X_n$. Then, if our hypothesis is true,

$$\chi^2 = \frac{\sum (X_i - \overline{X})^2}{\sigma^2} \tag{6.5}$$

follows a chi-square distribution with $(n - 1)$ degrees of freedom. If we choose a probability level of $p = (1 - \alpha)$ and our alternative hypothesis is $H_1 : \sigma^2 \neq \sigma_0^2$, then we may use Table 4.3 to find two values of $\chi^2$ such that

$$p(\chi^2 \geq \chi_1^2) = 1 - \alpha/2 \quad \text{and} \quad p(\chi^2 \geq \chi_2^2) = \alpha/2$$

and hence

$$p\left( \chi_1^2 \leq \frac{\sum (X_i - \overline{X})^2}{\sigma_0^2} \leq \chi_2^2 \right) = 1 - \alpha.$$

Our test procedure then is to calculate the ratio defined in 6.5 and reject the hypothesis $H_0 : \sigma^2 = \sigma_0^2$ if this ratio is either less than $\chi_1^2$ or greater than $\chi_2^2$. The Type I error in this case is $\alpha$. Had the alternative hypothesis been $H_1 : \sigma^2 > \sigma_0^2$, then a one-tail test would minimize the Type II error for a given $\alpha$. In this case we would use Table A3 to find a value of $\chi^2$, say $\chi_3^2$, such that $p(\chi^2 \geq \chi_3^2) = \alpha$, and reject $H_0$ if the ratio defined in 6.5 is greater than $\chi_3^2$.

As an example, let us suppose that the dispatcher had not been certain about the variance of the length of rods, but that he thought it was 225. He may then have estimated $\sigma^2$ from the sample and used the $t$-ratio in 6.2 to test $H_0:\mu = 150$. He may now decide to test his hypothesis $H_0:\sigma^2 = 225$ at a probability level $p = 0.95$, and we shall assume that he has no reason to expect the variance to be greater than or less than 225 if $H_0$ is false, that is, the alternative hypothesis is $H_1:\sigma^2 \neq 225$, implying a two-tail test. With 24 degrees of freedom

$$p(12.40 \leq \chi^2 \leq 39.36) = 0.95$$

and for the sample data presented in Table 6.1,

$$\frac{\sum (X_i - \overline{X})^2}{\sigma_0^2} = \frac{6652.42}{225} = 29.57.$$

On the basis of this $\chi^2$ value we would not reject the hypothesis that $\sigma^2$ is equal to 225.

### 6.4 testing hypotheses concerning population proportions

The discussion of hypothesis testing for the population mean and variance will have made clear that the results we obtained earlier in the context of estimation are relevant here. This is also the case if we are interested in testing hypotheses concerning population proportions. In section 5.6 we showed that if $\hat{\pi}$ was the proportion in a sample of $n$ and $\pi$ the population proportion, then as an approximation

$$Z \approx \frac{\hat{\pi} - \pi}{\sqrt{\dfrac{\pi(1 - \pi)}{n}}}, \tag{5.15}$$

where $Z$ is $N(0, 1)$. Suppose we have a hypothesis to test that $H_0:\pi = \pi_0$ against an alternative hypothesis that $H_1:\pi \neq \pi_0$. Given a random sample of $n$, we first calculate the sample proportion $\hat{\pi}$ and then the ratio

$$Z \approx \frac{\hat{\pi} - \pi_0}{\sqrt{\dfrac{\pi_0(1 - \pi_0)}{n}}}. \tag{6.6}$$

For a given probability level, $p = (1 - \alpha)$, we use Table A1 to find $\pm Z_{\alpha/2}$, such that $p(-Z_{\alpha/2} \leqq Z \leqq Z_{\alpha/2}) = 1 - \alpha$, and accept the hypothesis $H_0$ if the value of the ratio defined in 6.6 falls in the range $-Z_{\alpha/2} < Z < Z_{\alpha/2}$.

As an example, suppose a politician claims that 75 per cent of the voters in a country are in favour of a particular piece of legislation. An independent research organization carries out a poll and finds that, out of a random sample of 400 voters, 280 support the legislation and 120 oppose it. The null hypothesis here is $H_0 : \pi = 0.75$ and, if we suspect politicians tend to exaggerate, we may choose as our alternative hypothesis $H_1 : \pi < 0.75$, which would imply that a one-tail test was appropriate. We shall choose a probability level of $p = 0.95$. From our sample data $\hat{\pi} = 280/400 = 0.7$, so that

$$Z \approx \frac{0.7 - 0.75}{\sqrt{\dfrac{(0.75)(0.25)}{400}}} = -2.31.$$

With a one-tail test, we reject $H_0$ if $Z < -1.64$, so that in this case the sample evidence suggests that we should reject the politician's claim that 75 per cent of voters support this piece of legislation.

### 6.5 testing the difference between two population means

So far in our discussion of hypothesis testing we have considered cases in which we are sampling from a single population. However, in some economic contexts it may be interesting to consider the possibility that when we draw random samples they may come from different populations. For example, suppose we have drawn two random samples, one of size $n_1$ of urban families and the other of size $n_2$ of rural families. We calculate the mean family expenditure on coffee for each sample and find, not surprisingly, that the sample means are different. The difference between the two sample means may be explained by the fluctuations involved in drawing the random samples from a single population, but it may be due to the fact that the samples have been drawn from two different populations. In other words, we need to test the hypothesis that the mean value of coffee expenditure is different for urban and rural families.

We may develop the answer to this question in stages. Suppose we have drawn random samples from two normal populations, so that we have $n_1$ observations from $N(\mu_1, \sigma_1^2)$ in the first sample and $n_2$ observations from $N(\mu_2, \sigma_2^2)$ in the second sample. If the mean of the first sample is $\overline{X}_1$ and that of the second is $\overline{X}_2$, we shall first consider the expected value of $\overline{X}_1 - \overline{X}_2$:

$$E(\overline{X}_1 - \overline{X}_2) = E\left[ \frac{1}{n_1} \sum_{i=1}^{n_1} X_{1i} - \frac{1}{n_2} \sum_{j=1}^{n_2} X_{2j} \right]$$

$$= \frac{1}{n_1} \sum_{i=1}^{n_1} E(X_{1i}) - \frac{1}{n_2} \sum_{j=1}^{n_2} E(X_{2j})$$

$$= \frac{1}{n_1}(n_1\mu_1) - \frac{1}{n_2}(n_2\mu_2) = (\mu_1 - \mu_2). \tag{6.7}$$

That is, the expected value of the difference between the two sample means equals the difference between the two population means.

The variance of the difference between the two sample means is

$$\mathrm{var}(\overline{X}_1 - \overline{X}_2) = E[(\overline{X}_1 - \overline{X}_2) - E(\overline{X}_1 - \overline{X}_2)]^2 = E[(\overline{X}_1 - \overline{X}_2) - (\mu_1 - \mu_2)]^2$$

$$= E[(\overline{X}_1 - \mu_1) - (\overline{X}_2 - \mu_2)]^2$$

$$= E[(\overline{X}_1 - \mu_1)^2 - 2(\overline{X}_1 - \mu_1)(\overline{X}_2 - \mu_2) + (\overline{X}_2 - \mu_2)^2]$$

$$= \mathrm{var}(\overline{X}_1) - 2\,\mathrm{cov}(\overline{X}_1, \overline{X}_2) + \mathrm{var}(\overline{X}_2).$$

Now $\mathrm{var}(\overline{X}_1) = \sigma_1^2/n_1$ and $\mathrm{var}(\overline{X}_2) = \sigma_2^2/n_2$. Also, if the samples are drawn by a random process we may assume that the expected covariance between $\overline{X}_1$ and $\overline{X}_2$ is zero. Hence

$$\mathrm{var}(\overline{X}_1 - \overline{X}_2) = \mathrm{var}(\overline{X}_1) + \mathrm{var}(\overline{X}_2) = \frac{\sigma_1^2}{n_1} + \frac{\sigma_2^2}{n_2}. \tag{6.8}$$

The results derived in 6.7 and 6.8 are quite general, since we have not used any assumptions concerning the distributions being sampled. However, if we assume that the two populations are normal, then the random variable $(\overline{X}_1 - \overline{X}_2)$ involves *linear* combinations of random normal variables and so $(\overline{X}_1 - \overline{X}_2)$ will also follow a normal distribution. We may summarize these results by stating that the distribution of $(\overline{X}_1 - \overline{X}_2)$ is

$$N\left[ (\mu_1 - \mu_2), \left( \frac{\sigma_1^2}{n_1} + \frac{\sigma_2^2}{n_2} \right) \right]. \tag{6.9}$$

The normal distribution expressed in 6.9 may then be transformed to the standard normal distribution, $N(0, 1)$, by writing

$$Z = \frac{(\overline{X}_1 - \overline{X}_2) - (\mu_1 - \mu_2)}{\sqrt{\dfrac{\sigma_1^2}{n_1} + \dfrac{\sigma_2^2}{n_2}}}. \tag{6.10}$$

The result expressed in 6.10 provides us with a starting-point for our exploration of the difference $(\bar{X}_1 - \bar{X}_2)$. First, let us make the rather strong assumption that $\sigma_1^2$ and $\sigma_2^2$ are known. This enables us to obtain some rather simple results, and we shall explore the effect of relaxing this assumption below.

As an example, suppose it is known that for urban families the standard deviation of monthly expenditure on coffee is 5p, while for rural families the standard deviation is 4·472p. Random samples, each of size 100, are drawn from the urban and rural populations and it is calculated that mean expenditure on coffee in the sample of urban families is 125p, while in the sample of rural families the mean expenditure on coffee is 100p. If the population mean values of coffee expenditure are unknown, the hypothesis that they are the same may be tested by means of equation 6.10. Our null hypothesis in this case is $H_0 : \mu_1 - \mu_2 = 0$, a convenient way of saying that $\mu_1$ equals $\mu_2$ in a form which may be fitted into equation 6.10. We shall assume that there is no *a priori* evidence to suggest that we should expect mean expenditure by urban families to be either greater than or less than that of rural families, so that our alternative hypothesis is $H_1 : \mu_1 - \mu_2 \neq 0$, which suggests a two-tail test. Then we proceed by assuming that $H_0$ is true and calculating 6.10 under this assumption. In this case we obtain

$$Z = \frac{(\bar{X}_1 - \bar{X}_2)}{\sqrt{\dfrac{\sigma_1^2}{n_1} + \dfrac{\sigma_2^2}{n_2}}} = \frac{125 - 120}{\sqrt{\dfrac{25}{100} + \dfrac{20}{100}}} = \frac{5}{\sqrt{0.45}}$$

$$= \frac{5}{0.671} = 7.41.$$

Given that $Z$ is a standard normal variable, this value of $Z$ represents the occurrence of a very rare event. Hence we may conclude that average monthly expenditure on coffee does differ between the urban and rural populations.

In general we would not know the standard deviations of the two populations being sampled, and in practice the hypothesis test we have just discussed must be modified to allow for this additional ignorance on our part. The form of the modification is similar to that introduced in testing hypotheses concerning a single sample mean, but becomes more complicated in this new context. For example, we develop *two* modifications, the choice of which depends on whether the sample sizes are large or small.

### 1. *Large Sample Procedure*

Let our first random sample consist of the $n_1$ observations $X_{11}, \ldots, X_{1n_1}$, while the second random sample consists of the $n_2$ observations $X_{21}, \ldots, X_{2n_2}$. Denote the two sample means as $\bar{X}_1$ and $\bar{X}_2$. Then we may define unbiased estimators of the unknown variances, $\sigma_1^2$ and $\sigma_2^2$, as

$$s_1^2 = \frac{1}{n_1 - 1} \sum_{i=1}^{n_1} (X_{1i} - \bar{X}_1)^2 \quad \text{and} \quad s_2^2 = \frac{1}{n_2 - 1} \sum_{j=1}^{n_2} (X_{2j} - \bar{X}_2)^2. \tag{6.11}$$

If we now substitute $s_1^2$ for $\sigma_1^2$ and $s_2^2$ for $\sigma_2^2$ in 6.10, we obtain

$$\frac{(\bar{X}_1 - \bar{X}_2) - (\mu_1 - \mu_2)}{\sqrt{\dfrac{s_1^2}{n_1} + \dfrac{s_2^2}{n_2}}}, \tag{6.12}$$

and if our samples are large (both $n_1$ and $n_2$ greater than 30), then the variable defined in 6.12 will be approximately a standard normal variable and we may proceed to use $N(0, 1)$ to calculate probability levels as before.

For example, suppose we have $n_1 = 100, \bar{X}_1 = 125$ and $s_1^2 = 78$ for one random sample, and $n_2 = 100, \bar{X}_2 = 130$ and $s_2^2 = 96$ for a second random sample. We wish to test $H_0 : H_1 - H_2 = 0$ against $H_1 : \mu_1 - \mu_2 \neq 0$ at the probability level $p = 0.95$. Substituting these values into 6.12, we have

$$\frac{125 - 130}{\sqrt{\dfrac{78}{100} + \dfrac{96}{100}}} = -\frac{5}{\sqrt{1.74}} = -\frac{5}{1.32} = -3.8.$$

Since $-3.8$ lies outside the range $\pm 1.96$, we reject the hypothesis that $\mu_1 - \mu_2' = 0$. Thus the modification is relatively simple in the large sample case, so we now turn to the small sample case.

## 2. Small sample procedure

The reader with an extrapolative turn of mind may well expect at this point that when the sample sizes are small the distribution of the ratio defined in 6.12 will follow the *t*-distribution. *Unfortunately, this is not the case.* We can only make use of the *t*-distribution if we are prepared to confine ourselves to the rather more restrictive case in which the variances of the two populations being sampled are equal, that is, $\sigma_1^2 = \sigma_2^2 = \sigma^2$. Given this assumption, 6.10 may be written as

$$\frac{(\bar{X}_1 - \bar{X}_2) - (\mu_1 - \mu_2)}{\sqrt{\dfrac{\sigma^2}{n_1} + \dfrac{\sigma^2}{n_2}}} = \frac{(\bar{X}_1 - \bar{X}_2) - (\mu_1 - \mu_2)}{\sigma\sqrt{\dfrac{1}{n_1} + \dfrac{1}{n_2}}}, \tag{6.13}$$

and if the appropriate estimate is substituted for $\sigma$ in 6.13, the new ratio does follow a *t*-distribution. What is the appropriate estimate of $\sigma^2$ in this case? It is obtained by pooling the unbiased estimates of $\sigma^2$, $s_1^2$ and $s_2^2$, which were defined in 6.11, to give $s^2$, which is defined as

$$s^2 = \frac{(n_1 - 1)\,s_1^2 + (n_2 - 1)\,s_2^2}{n_1 + n_2 - 2}. \tag{6.14}$$

The weights here are the degrees of freedom for each sample, thus giving more weight in the pooling to the estimate based on the larger sample if $n_1 \neq n_2$. Substituting this estimate of $\sigma^2$ into 6.14, we obtain

$$t = \frac{(\overline{X}_1 - \overline{X}_2) - (\mu_1 - \mu_2)}{s\sqrt{\dfrac{1}{n_1} + \dfrac{1}{n_2}}} \tag{6.15}$$

which follows the $t$-distribution with $(n_1 + n_2 - 2)$ degrees of freedom.[1]

To illustrate the small sample procedure, suppose our data are $n_1 = 9$, $\overline{X}_1 = 165\cdot3$, $s_1^2 = 10\cdot65$; $n_2 = 16$, $\overline{X}_2 = 173\cdot5$, $s_2^2 = 9\cdot65$, and we wish to test the hypothesis that $H_0 : \mu_1 - \mu_2 = 0$ at the probability level $p = 0\cdot95$, with $H_1 : \mu_1 - \mu_2 \neq 0$. Substituting the data into 6.14, we obtain

$$s^2 = \frac{(8)(10\cdot65) + (15)(9\cdot65)}{23} = \frac{85\cdot20 + 144\cdot75}{23} = 10,$$

and substituting $s$ into 6.15 gives

$$t = \frac{165\cdot3 - 173\cdot5}{\sqrt{10\left(\dfrac{1}{9} + \dfrac{1}{16}\right)}} = -\frac{8\cdot2}{\sqrt{10\left(\dfrac{25}{9.16}\right)}} = -\frac{98\cdot4}{15\cdot81} = -6\cdot22.$$

Given our alternative hypothesis, a two-tail test is appropriate, and with $p = 0\cdot95$ and $(n_1 + n_2 - 2) = 23$ degrees of freedom, our acceptance region is $-2\cdot07 < t < 2\cdot07$. Since $-6\cdot22 < -2\cdot07$, we reject the hypothesis $H_0 : \mu_1 - \mu_2 = 0$.

This test procedure depends on the assumption that $\sigma_1^2 = \sigma_2^2 = \sigma^2$ and is sensitive to its violation. In practice, even if $\sigma_1^2 = \sigma_2^2$, it would be unlikely that in two random samples $s_1^2 = s_2^2$, and it would be useful to develop the theory to enable us to distinguish, with some probability of being right, between differences due to sampling fluctuations and those due to the fact that $\sigma_1^2 \neq \sigma_2^2$. We shall develop such theory below, but before we turn to the testing of hypotheses concerning two population variances, it is convenient to consider a method of testing hypotheses concerning the difference between proportions in two populations.

## 6.6 testing the difference between two population proportions

Suppose we are asked to assess whether two different machines are equally reliable in producing some product. From the first machine we have a random sample of 250 units of the product, of which 50 are found to be defective. From the second machine we have a random sample of 300 units of the product, of which 70 are found to be defective. We shall assume that there is no *a priori* information to suggest that either machine should be considered more efficient than the other. How might we attempt to interpret these data? One obvious step is to estimate the proportions of defectives in the two samples, in which case, if $m_1$ and $m_2$ denote the number of observations in each sample having the characteristic, we obtain $\hat{\pi}_1 = \frac{50}{250} = 0.200$ and $\hat{\pi}_2 = \frac{70}{300} = 0.233$. We note that the proportions differ in the two random samples, but do they differ sufficiently to suggest that the population proportions $\pi_1$ and $\pi_2$ differ?

To investigate this possibility, we extend our earlier use of the normal approximation in testing a hypothesis concerning a population proportion. If we draw random samples of size $n_1$ from a population with proportion $\pi_1$ having a given characteristic, we expect the sample proportion, $\hat{\pi}_1$, to follow (approximately) a normal distribution with mean $\pi_1$ and variance $\pi_1(1 - \pi_1)/n_1$. Similarly, if we have drawn random samples of size $n_2$ from a second population which has a proportion $\pi_2$ having this characteristic, the sample proportion, $\hat{\pi}_2$, follows (approximately) a normal distribution with mean $\pi_2$ and variance $\pi_2(1 - \pi_2)/n_2$. Given the random sampling process ensures that $\hat{\pi}_1$ and $\hat{\pi}_2$ are independent, we may modify the result obtained in the previous section on the distribution of the difference of two sample means to obtain the result that the sampling distribution of $(\hat{\pi}_1 - \hat{\pi}_2)$ is (approximately) normal with a mean of $(\pi_1 - \pi_2)$ and a variance of

$$\frac{\pi_1(1 - \pi_1)}{n_1} + \frac{\pi_2(1 - \pi_2)}{n_2}. \tag{6.16}$$

Applying the usual transformation yields

$$\frac{(\hat{\pi}_1 - \hat{\pi}_2) - (\pi_1 - \pi_2)}{\sqrt{\dfrac{\pi_1(1 - \pi_1)}{n_1} + \dfrac{\pi_2(1 - \pi_2)}{n_2}}}, \tag{6.17}$$

a random variable which is approximately $N(0, 1)$. To test if the two population proportions are equal, we may formulate the hypothesis $H_0 : \pi_1 - \pi_2 = 0$. The implication for the ratio in 6.17 is that the term $(\pi_1 - \pi_2)$ is set equal to zero in the numerator. However, there is a second implication for the determinator of 6.17, since if the hypothesis is true and $\pi_1 = \pi_2 = \pi$, the variance becomes

$$\frac{\pi_1(1-\pi_1)}{n_1} + \frac{\pi_2(1-\pi_2)}{n_2} = \frac{\pi(1-\pi)}{n_1} + \frac{\pi(1-\pi)}{n_2} = \pi(1-\pi)\left(\frac{1}{n_1} + \frac{1}{n_2}\right).$$

$$(6.18)$$

This modification requires the pooling of the data to obtain an estimate of the common proportion $\pi$. The efficient estimate of $\pi$, denoted by $\hat{\pi}$, is a weighted average of $\hat{\pi}_1$ and $\hat{\pi}_2$ which allows for sample size,

$$\hat{\pi} = \frac{n_1\hat{\pi}_1 + n_2\hat{\pi}_2}{n_1 + n_2} = \frac{n_1(m_1/n_1) + n_2(m_2/n_2)}{n_1 + n_2}$$

$$= \frac{m_1 + m_2}{n_1 + n_2}.$$

$$(6.19)$$

Combining 6.18 and 6.19, we may rewrite 6.17 when $H_0: \pi_1 - \pi_2 = 0$ as

$$\frac{\hat{\pi}_1 - \hat{\pi}_2}{\sqrt{\hat{\pi}(1-\hat{\pi})\left(\frac{1}{n_1} + \frac{1}{n_2}\right)}},$$

$$(6.20)$$

a random variable which is approximately $N(0, 1)$ and which serves as the basis for our test.

In our example, we shall choose a probability level of $p = 0\cdot95$ and an alternative hypothesis $H_1: \pi_1 - \pi_2 \neq 0$, $\hat{\pi} = (50 + 70)/(250 + 300) = 0\cdot218$, so that when we substitute our data into 6.20 we obtain

$$\frac{0\cdot200 - 0\cdot233}{\sqrt{(0\cdot218)(0\cdot782)\left(\frac{1}{250} + \frac{1}{300}\right)}} = -\frac{0\cdot033}{\sqrt{0\cdot00124}} = -\frac{0\cdot033}{0\cdot0352}$$

$$= -0\cdot94.$$

This result is clearly not significant and hence we cannot reject the hypothesis $H_0: \pi_1 - \pi_2 = 0$.

Having discussed this area of hypothesis testing, we shall turn to the problem of testing hypotheses concerning more than one variance.

### 6.7 testing the difference between population variances. the $F$-distribution

The small sample procedure for testing the difference between two population means involved the assumption that the variances of the populations being sampled were equal. In this section we shall extend our discussion so that we may take the hypothesis $H_0: \sigma_1^2 = \sigma_2^2 = \sigma^2$ and test its statistical significance. In view of the previous tests of the difference between two population means and two population proportions, the reader might well expect the test for the difference between the variances of two populations to be formulated as $H_0: \sigma_1^2 - \sigma_2^2 = 0$. However, the mathematical solution to the problem is much simpler when the hypothesis is stated in a different form.

We have already discussed the chi-square distribution

$$\chi^2 = \frac{(n-1)s^2}{\sigma^2}$$

when dealing with the estimation of a single variance, $\sigma^2$, and in dealing with two variances we are able to build on this result. Suppose we have drawn two random samples, the first of size $n_1$ from a normal population with variance $\sigma_1^2$ and the second of size $n_2$ from a normal population with variance $\sigma_2^2$. Let the two sample variances be $s_1^2$ and $s_2^2$ and consider the ratio $s_1^2/s_2^2$. If the hypothesis $H_0: \sigma_1^2 = \sigma_2^2 = \sigma^2$ is true, then $s_1^2$ and $s_2^2$ are independent estimates of $\sigma^2$ and hence we would expect the ratio $s_1^2/s_2^2$ to be unity on average in repeated sampling and tend to unity as the sample sizes increase. Similarly, if $\sigma_1^2 = \sigma_2^2 = \sigma^2$,

$$\frac{s_1^2}{s_2^2} = \frac{s_1^2/\sigma_1^2}{s_2^2/\sigma_2^2}$$

$$= \frac{(n_1 - 1)\,s_1^2/\sigma_1^2(n_1 - 1)}{(n_2 - 1)\,s_2^2/\sigma_2^2(n_2 - 1)}$$

$$= \frac{\chi_1^2/(n_1 - 1)}{\chi_2^2/(n_2 - 1)}, \tag{6.21}$$

that is, under the hypothesis $H_0: \sigma_1^2 = \sigma_2^2 = \sigma^2$, $s_1^2/s_2^2$ involves the ratio of two chi-square variables, with $(n_1 - 1)$ and $(n_2 - 1)$ degrees of freedom respectively, where each chi-square variable is divided by its degrees of freedom.

It turns out that the ratio of chi-square variables in 6.21 is relatively easy to deal with mathematically, and its probability distribution has been investigated and tabulated. This distribution has been called the $F$-distribution[2] and is presented in the Appendix, Table A4. Since both $s_1^2$ and $s_2^2$ are positive, the range of $F$ is $0 \leq F \leq \infty$. Its tabulation is complicated by the fact that it involves a family of distributions, each of which depends on two degrees of freedom parameters, $(n_1 - 1)$ and $(n_2 - 1)$. In order to tabulate the areas under the curve of $F$-distribution in concise form, we choose various probability levels and tabulate the values of $F$ which divide the area under the curve appropriately, as was done with the $t$ and chi-square distributions. In addition here, since the $F$-distribution has to be tabulated by the two degrees of freedom of parameters, the volume is reduced by concentrating only on one tail of the distribution, the upper tail. Thus we form the $F$-ratio to be tested by taking the *larger* sample variance divided by the *smaller*, obtaining a ratio which must be greater than unity and which, if significant, will fall in the upper tail of the appropriate $F$-distribution. The distribution is tabulated so that the degrees of freedom of the numerator identifies the appropriate column and the denominator's degrees of freedom gives the appropriate row.

To illustrate the use of the $F$-distribution to test the hypothesis that $\sigma_1^2 = \sigma_2^2$, we shall return to the example which was analysed on p. 102, in which $n_1 = 9$, $s_1^2 = 10{\cdot}65$, $n_2 = 16$ and $s_2^2 = 9{\cdot}65$. Here the ratio is $F = 10{\cdot}65/9{\cdot}65 = 1{\cdot}10$. If we choose a probability level of $p = 0{\cdot}95$, then with 8 and 15 degrees of freedom we obtain the value $F = 2{\cdot}64$ from Table A4. Since our calculated $F$ ratio of $1{\cdot}10$ is less than $2{\cdot}64$, we cannot reject the hypothesis that $\sigma_1^2 = \sigma_2^2$.

In the example given to illustrate the large sample procedure on p. 101, we had $n_1 = 100$, $s_1^2 = 78$, $n_2 = 100$ and $s_2^2 = 96$. To test the hypotheses that $\sigma_1^2 = \sigma_2^2$ we calculate $F = 96/78 = 1{\cdot}22$. Choosing the probability level $p = 0{\cdot}95$ with $(n_1 - 1) = (n_2 - 1) = 99$ degrees of freedom, $F = 1{\cdot}39$, so we accept the hypothesis that $\sigma_1^2 = \sigma_2^2$.

Before passing on to consider further tests based on the $F$-distribution, the reader should recall the problem of relacing the normality assumption that was raised in the discussion of the chi-square distribution (p. 86). Since the $F$-ratio involves squared values of the variables, it also tends to be sensitive to the non-normality of the populations being sampled. To follow this problem further, the reader should consult the references given in note 5 to chapter 5.

## 6.8 elementary analysis of variance

In section 6.5 we developed a procedure for testing the hypothesis that two random samples might have been drawn from populations having different expected values. Having developed the $F$-distribution in the previous section, we are now in a position to extend our test of the difference between means to include more than two samples.

We shall develop the method by assuming that $k$ random samples, each of size $n$, have been drawn.[3] It is assumed that the variable being sampled is normally distributed, and what is of interest is to consider two alternative hypothesis as to whether the samples are drawn from normal populations having different means, so that

$$H_1 : \mu_1 \neq \mu_2 \neq \dots \neq \mu_k, \tag{6.22}$$

or whether the population means are the same, so that

$$H_0 : \mu_1 = \mu_2 = \dots = \mu_k = \mu. \tag{6.23}$$

In order to concentrate on the difference between means, it will be assumed that the population variances are all equal.

We shall introduce the following notation. Let $X_{ij}$ represent the $i$th observation of the $j$th sample, $(i = 1, \ldots, n; j = 1, \ldots, k)$, $\overline{X}_{\cdot j}$ the mean of the $j$th sample, and $\overline{X}_{\cdot\cdot}$ the mean of all the observations, that is, $\overline{X}_{\cdot\cdot} = \sum_{i=1}^{n} \sum_{j=1}^{k} X_{ij}/nk$. Now the variations of the observations about the overall mean may be represented as $\sum_{i}^{n} \sum_{j}^{k} (X_{ij} - \overline{X}_{\cdot\cdot})^2$, which without altering its value may be rewritten

$$\sum_{i}^{n} \sum_{j}^{k} (X_{ij} - \overline{X}_{\cdot\cdot})^2 = \sum_{i}^{n} \sum_{j}^{k} [(X_{ij} - \overline{X}_{\cdot j}) + (\overline{X}_{\cdot j} - \overline{X}_{\cdot\cdot})]^2$$

$$= \sum_{i} \sum_{j} (X_{ij} - \overline{X}_{\cdot j})^2 + 2\sum_{i} \sum_{j} (X_{ij} - \overline{X}_{\cdot j})(\overline{X}_{\cdot j} - \overline{X}_{\cdot\cdot}) + \sum_{i} \sum_{j} (\overline{X}_{\cdot j} - \overline{X}_{\cdot\cdot})^2$$

$$= \sum_{i} \sum_{j} (X_{ij} - \overline{X}_{\cdot j})^2 + 2\sum_{j} (\overline{X}_{\cdot j} - \overline{X}_{\cdot\cdot})\sum_{i}(X_{ij} - \overline{X}_{\cdot j}) + n\sum_{j} (\overline{X}_{\cdot j} - \overline{X}_{\cdot\cdot})^2$$

$$= \sum_{i} \sum_{j} (X_{ij} - \overline{X}_{\cdot j})^2 + n\sum_{j} (\overline{X}_{\cdot j} - \overline{X}_{\cdot\cdot})^2, \tag{6.24}$$

as $2\sum_{j}(\overline{X}_{\cdot j} - \overline{X}_{\cdot\cdot})\sum_{i}(X_{ij} - X_{\cdot j}) = 0$ by virtue of the fact that $\sum_{i}(X_{ij} - X_{\cdot j}) = 0$. Looking at the right-hand side of 6.24, we note that the original variation about the overall mean has been split into two components which we can identify and interpret. The first term, $\sum_{i} \sum_{j} (X_{ij} - \overline{X}_{\cdot j})^2$, measures the deviations of the observations from the respective sample means when summed across all samples, and since all the sample means are used as reference points in this calculation, this term provides a measure of the 'within sample' variation. The second term, $n\sum_{j} (\overline{X}_{\cdot j} - \overline{X}_{\cdot\cdot})^2$, measures the dispersion of the means of all the samples about the overall mean and provides a measure of the 'between sample' variation.

Consider first the 'within sample' variation. Given the assumption that all the populations being sampled have the same variance, $\sigma^2$, the expression $\sum_{i}^{n} (X_{ij} - \overline{X}_{\cdot j})^2$ (when divided by $n - 1$ degrees of freedom) represents an unbiased estimate of $\sigma^2$. Hence the expected value of $\sum_{i}(X_{ij} - \overline{X}_{\cdot j})^2$ is $(n - 1)\sigma^2$. This is true for each value of $j$, so that the summation across the $k$ samples is equivalent to pooling $k$ estimates of $(n - 1)\sigma^2$. In other words, if we divide $\sum_{i} \sum_{j} (X_{ij} - \overline{X}_{\cdot j})^2$ by $k(n - 1)$ degrees of freedom, we have an unbiased estimate of $\sigma^2$ *regardless of whether* $H_0 : \mu_1 = \mu_2 = \ldots = \mu_k = \mu$ *is true or not.*

The expected value of the 'between sample' variation, however, depends crucially on whether $H_0$ is true or not. If $H_0$ is true, $E(\overline{X}_{\cdot\cdot}) = \mu$, and since the $\overline{X}_{\cdot j}$ follow a distribution with variance $\sigma^2/n$, the expression $\sum_{j}^{n} (\overline{X}_{\cdot j} - \overline{X}_{\cdot\cdot})^2$ (when divided by $k - 1$ degrees of freedom) will be an unbiased estimate of $\sigma^2/n$. Hence the expected value of $n\sum_{j}^{k} (\overline{X}_{\cdot j} - \overline{X}_{\cdot\cdot})^2$ is $(k - 1)\sigma^2$ *if $H_0$ is true.* However, if $H_1 : \mu_1 \neq \ldots \neq \mu_k$ is true, then the expected value of $\overline{X}_{\cdot\cdot}$ is a weighted average of the $\mu_j$ and the expected deviations $(\overline{X}_{\cdot j} - \overline{X}_{\cdot \cdot})^2$ will be *larger* than the deviations $(\overline{X}_{\cdot j} - \mu_j)^2$. Hence we would expect $n \sum_j^k (\overline{X}_{\cdot j} - \overline{X}_{\cdot \cdot})^2/(k-1)$ to be larger than $\sigma^2$ *if $H_1$ is true.*

These arguments may now be combined to yield a procedure for testing $H_0$ against $H_1$. If $H_0$ is true, then both

$$s_1^2 = \frac{n \sum_j^k (\overline{X}_{\cdot j} - \overline{X}_{\cdot \cdot})^2}{(k-1)}$$

and

$$s_2^2 = \frac{\sum_i^n \sum_j^k (X_{ij} - \overline{X}_{\cdot j})^2}{k(n-1)}$$

are unbiased estimates of $\sigma^2$. Since we have assumed that the populations being sampled are normal, the ratio

$$F = \frac{s_1^2}{s_2^2}$$

will follow the $F$-distribution with $(k-1)$ and $k(n-1)$ degrees of freedom. We would expect the ratio to equal unity if $H_0$ is true, but in any actual set of sample data it will almost certainly differ from unity because of sampling fluctuations, the $F$-distribution providing us with information concerning the probability of different deviations from unity. If we obtain a large value for the ratio $s_1^2/s_2^2$, we may conclude either that $H_0$ is true and we have observed a rare event, or that $H_0$ is probably false and it is reasonable to reject $H_0$ in favour of $H_1$.

This technique of partitioning a sum of squared deviations from a mean into components which, under a null hypothesis, follow the $F$-distribution, is known as the *analysis of variance*. The partitioning is usually presented in a convenient tabular form, as in Table 6.1. We shall now illustrate the use of the analysis of variance with a numerical example. Suppose that students from four different colleges sit the same examination in economics. After the examination a random sample of five students is chosen from each college and their grades (out of a maximum of 200) are recorded. These data are presented below.

The average scores by college are $\overline{X}_{\cdot j} = 108, \overline{X}_{\cdot 2} = 100, \overline{X}_{\cdot 3} = 122$ and $\overline{X}_{\cdot 4} = 126$. They indicate some variation between colleges, which might have arisen because the colleges represent populations of grades which have different means, $H_1 : \mu_1 \neq \mu_2 \neq \mu_3 \neq \mu_4$. Alternatively, the means could all be equal, $H_0 : \mu_1 =$

| College number | | | |
|---|---|---|---|
| 1 | 2 | 3 | 4 |
| 84 | 88 | 114 | 140 |
| 124 | 76 | 124 | 116 |
| 112 | 116 | 120 | 120 |
| 96 | 116 | 136 | 124 |
| 124 | 104 | 116 | 130 |

$\mu_2 = \mu_3 = \mu_4 = \mu$, and the variation might be due to sampling fluctuations. To decide between $H_0$ and $H_1$ we may carry out an analysis of variance following the partitioning in Table 6.1. This analysis is presented in Table 6.2. The calculations may be kept to a minimum by computing the overall mean, $\overline{X}.. = 114$,

*Table 6.1.* Tabulation of analysis of variance

| Source of variation | Sum of squares | Degrees of freedom | Mean sum of squares | F |
|---|---|---|---|---|
| Between sample | $n \sum_{j}^{k} (\overline{X}._{j} - \overline{X}..)^2$ | $k - 1$ | $s_1^2 = \dfrac{n \sum (\overline{X}._{j} - \overline{X}..)^2}{k - 1}$ | $F = \dfrac{s_1^2}{s_2^2}$ |
| Within sample | $\sum_{i} \sum_{j} (X_{ij} - \overline{X}._{j})^2$ | $k(n - 1)$ | $s_2^2 = \dfrac{\sum \sum (X_{ij} - \overline{X}._{j})^2}{k(n - 1)}$ | |
| Total | $\sum_{i} \sum_{j} (X_{ij} - \overline{X}..)^2$ | $nk - 1$ | — | |

*Table 6.2.* Analysis of variance for college example

| Source of variation | Sum of squares | Degrees of freedom | Mean sum of squares | F |
|---|---|---|---|---|
| Between colleges | 2200 | 3 | 733·3 | 3·72 |
| Within colleges | 3152 | 16 | 197·0 | |
| Total | 5352 | 19 | — | |

and the 'between college' variation as $5(6^2 + 14^2 + 8^2 + 12^2)$. The expression for the 'total' variation may be expanded as

$$\sum_{i} \sum_{j} (X_{ij} - \overline{X}..)^2 = \sum_{i} \sum_{j} X_{ij}^2 - nk(\overline{X}..^2),$$

which is convenient for calculation. Finally, the 'within college' variation may be obtained as 'total' variation *minus* 'between college' variation.

The hypothesis being tested is $H_0: \mu_1 = \mu_2 = \mu_3 = \mu_4 = \mu$, and let us choose $p = 0.95$ as the probability level at which to make our decision. The appropriate $F$-distribution has 3 and 16 degrees of freedom, and from Table A4 we find $F = 3.24$ cuts off a proportion 0·05 of the area in the right-hand tail of the distribution. Given our chosen probability level, $F = 3.72$ falls in the rejection region and we therefore *reject* $H_0: \mu_1 = \mu_2 = \mu_3 = \mu_4 = \mu$. There is some evidence that mean grades do differ between the colleges.

Within the simple analysis of variance we have considered so far, two rather strict assumptions may be quite easily relaxed. We have taken as our alternative hypothesis $H_1: \mu_1 \neq \ldots \neq \mu_k$, that is, that *all* the population means have different values. This makes for a dramatic contrast between $H_0$ and $H_1$, but is not an essential assumption. The analysis of variance discussed above can be used to test $H_0$ against the alternative hypothesis that *some* of the means of the populations being sampled differ (for example, see problem 6.12 below). The second assumption, which was not essential, but was made for the sake of algebraic simplicity, was that the $k$ samples were all of equal size, $n$. The algebra may be extended to the case in which the samples are of unequal sizes, $n_j$, $j = 1, \ldots, k$. This will be left as an exercise for the reader (see problem 6.13 below).

The partitioning process which was illustrated with this simple example of the analysis of variance could be extended to cover more complex and interesting possibilities, but this subject will not be pursued here. Enough has been said to illustrate the basic ideas and we shall return to this subject in chapters 8 and 9 below.

# problems

**6.1** The standard deviation of a normal population is known to be 25. A random sample of size 100 is to be drawn from the population to test the hypothesis $H_0:\mu = 200$ at the probability level $p = 0.95$. If the alternative hypothesis is $H_1:\mu = 204$, calculate the magnitude of the Type II error in this case. Keeping the probability level $p = 0.95$, how large must the sample size be for the Type II error to be equal to 0.05?

**6.2** A matchmaker advertises that his boxes contain an average of 48 matches per box. A random sample of ten boxes was drawn and it was found that the boxes contained the following number of matches:

42, 41, 52, 47, 43, 50, 45, 46, 51, 47.

Is his claim justified?

**6.3** The variance of a random sample of ten observations is found to be 81.3. Test the hypothesis that the true value of the population variance is 125.

**6.4** The BURP Brewery claims that its new lager can be distinguished by taste from the lager produced by a rival. To test this assertion, a random sample of 25 drinkers tasted the two lagers while blindfolded and 17 correctly chose the new lager. Assess the claim of the BURP Brewery.

**6.5** A drug has been found to provide a cure 60 per cent of the time when applied to a given disease. A new drug is claimed to provide a cure in 75 per cent of cases. In order to test this claim, it is planned to test a random sample of 25 patients who are given the new drug. Discuss the Type I and Type II errors in this case. It is found that 17 of the 25 patients given the new drug recover from the disease. How would you interpret these data?

**6.6** An advertising agency claims a 60 per cent 'cure rate' for a patent medicine that it handles. If 230 people out of a random sample of 400 feel cured after taking the medicine, is the claim justified?

**6.7** An economist has the following data on the values of sales (in £ thousand) for two random samples of firms from two allied industries:

Sample 1: $n_1 = 100, \overline{X}_1 = 350, s_1^2 = 400,$
Sample 2: $n_2 = 400, \overline{X}_2 = 360, s_2^2 = 500.$

Would he be justified in pooling the results obtained from these two samples?
**6.8** A random sample of 25 rural families was found to spend an average of
£18·54 per quarter on fuel consumption, with a standard deviation of £6·42. A
corresponding random sample of 22 urban families spend £19·65 on average
with a standard deviation of £7·10 per quarter, Is there any evidence of a significant
difference between urban and rural spending on fuel?
**6.9** A random sample of 400 voters over the age of 21 showed 200 in favour of
candidate $X$. A random sample of 100 voters under the age of 21 showed 40 in
favour of candidate $X$. Do the proportions supporting candidate $X$ differ significantly with age? Can one predict the outcome of the election on the basis of
these results?
**6.10** A random sample of 250 items produced by one type of wrecking machine
produced 50 defectives, while another sample of 300 items produced by another
type of wrecking machine produced 40 defective units. Test the hypothesis that
the two wrecking machines are equally destructive.
**6.11** Random samples are drawn of 20 shipbuilding firms in Country A and 25
in Country B. It is found that the sample standard deviations of waiting times for
the delivery of the product are respectively 17 months and 9 months. Is the difference between the variances of waiting times
   (a) significant,
   (b) interesting?
**6.12** Use the data on examination grades which are given on p. 109 to test the
hypothesis that the first two colleges have the same mean scores ($\mu_1 = \mu_2$).
Also, test the hypothesis that the third and fourth colleges have the same mean
score ($\mu_3 = \mu_4$). Comment on your results in the light of the analysis of variance
given in Table 6.2.
**6.13** (Analysis of variance with unequal sample size). Check that if the data consist of $k$ samples of unequal sample size, $n_j, j = 1, \ldots, k$, the 'total' sum of squares
can be partitioned into 'within sample' and 'between sample' components

$$\sum_{j}^{k}\sum_{i}^{n_j}(X_{ij} - \overline{X}..)^2 = \sum_{j}^{k}\sum_{i}^{n_j}(X_{ij} - \overline{X}._j)^2 + \sum_{j}^{k}n_j(\overline{X}._j - \overline{X}..)^2.$$

Also check that the degrees of freedom appropriate for the 'total', 'within sample'
and 'between sample' sums of squares are $\sum_{j}^{k}n_j - 1$, $\sum_{j}^{k}(n_j - 1)$ and $(k - 1)$
respectively.
**6.14** First-year pupils in a very large school are divided at random into three
groups, each of which is taught mathematics by a different method. At the end
of the year all the pupils sit the same examination and random samples of the
grades obtained by pupils taught by each of the three methods are drawn:

|     | Method |     |
| --- | --- | --- |
| I | II | III |
| 192 | 263 | 234 |
| 215 | 275 | 232 |
| 258 | 262 | 275 |
| 260 | 220 | 250 |
| 220 | 226 | 210 |
| 230 | 260 | 260 |
| 222 | 280 | 253 |
| 252 | 273 | 268 |
|     |     | 246 |
|     |     | 230 |

Does the method of teaching affect mean performance in the examination?

# part two
# relationships between variables

It does require maturity to realize that models are to be used but not to be believed.

HENRI THEIL                                    *Principles of Econometrics**

\*   Published by the North-Holland Publishing Company, Amsterdam, 1971.

# chapter 7

# correlation analysis and tests of association

## 7.1 introduction

In the first part of this text we developed statistical techniques to solve problems of estimation and hypothesis testing when we were concerned with a single variable. We shall now extend the discussion to consider estimation and hypothesis testing when we are interested in relationships between two or more variables. As we proceed, the reader will find that many of the ideas and concepts he has already met in the earlier discussion of sampling and estimation can be extended very easily to solve the new problems being raised.

Turning now to the relationship between variables, we shall first consider the simplest possible case, which involves only two variables, $X$ and $Y$. In discussing relationships we may distinguish between two different kinds of questions which could be asked:

(1) Is there any evidence of a relationship (or association) between $X$ and $Y$? If so, can we develop a statistical measure of the degree of association?
(2) If we are prepared to specify the form of the relationship between $X$ and $Y$ (for example, a linear relationship $Y = \alpha + \beta X$), how can we estimate the parameters in the relationship from a sample of observations?

The answer to the first question is explored through *correlation analysis*, the technique to be discussed in this chapter. The second type of question is dealt with by *regression analysis*, which is the subject of chapters 8 and 9.

In comparing the two types of questions raised above, it may be noted that the first question does not specify any form for the relationship between $X$ and *Y. In particular, it does not specify the direction of causality between $X$ and $Y.* For example, if we were to draw a large random sample of 18-year-old males, we would probably expect to find some association between height and weight, since on the whole tall men will be heavier than short-statured men. It would be incorrect, however, to conclude that there was a simple causal link between height and weight, so that, for example, being heavy tended to cause a man to be tall! This is a rather obvious nonsense example, but the point being made is an important one. The techniques to be developed in this chapter are designed to measure the degree of association between $X$ and $Y$, *and they cannot be used to prove that a causal link exists between X and Y.*

In discussing association we shall first consider the case in which $X$ and $Y$ are associated in a linear fashion. A more general test for association which does not assume linearity will be discussed later in the chapter.

## 7.2 the correlation coefficient

To illustrate our development of a measure of association, we shall analyse the following (hypothetical) data. A publishing firm employs a large number of door-to-door salesmen selling its encyclopedias. It has a scheme to train potential salesmen which works as follows. Every salesman has to attend the course, for which he pays £10 per week. The course lasts for a maximum of 20 weeks, but he can leave at the end of any week he chooses when he feels ready to start selling. In order to investigate whether there is any association between the time spent on the course and selling success, the sales manager draws a random sample of 20 salesmen who all left the course on a particular date and who have since been selling in comparable districts. For each salesman he records the number of weeks spent on the course $(X)$ and the average level of monthly sales over the first year of selling $(Y)$. The data are presented in Table 7.1.

*Table 7.1.* Hypothetical training and sales data

| Salesman | X (weeks) | Y (£) | Salesman | X (weeks) | Y (£) |
|---|---|---|---|---|---|
| 1 | 4 | 118 | 11 | 10 | 130 |
| 2 | 3 | 108 | 12 | 4 | 113 |
| 3 | 8 | 135 | 13 | 7 | 136 |
| 4 | 9 | 131 | 14 | 20 | 208 |
| 5 | 11 | 146 | 15 | 18 | 201 |
| 6 | 4 | 127 | 16 | 9 | 149 |
| 7 | 16 | 195 | 17 | 10 | 158 |
| 8 | 11 | 156 | 18 | 14 | 172 |
| 9 | 18 | 186 | 19 | 13 | 155 |
| 10 | 20 | 205 | 20 | 19 | 192 |

As a first step in the analysis we may examine the data visually by plotting them in a scatter diagram. This is done in Figure 7.1, where there does appear to be some positive association between the level of sales and the time spent training. To obtain a numerical measure of the degree of linear association between the two variables, we proceed as follows. We calculate the arithmetic means of the two variables, obtain $\overline{X} = 11{\cdot}4$ and $\overline{Y} = 156{\cdot}0$, and transform our observations to deviations from the sample means, defining $x_i = (X_i - \overline{X})$ and $y_i = (Y_i - \overline{Y})$, $i = 1, \dots, 20$. The new variables $x_i$ and $y_i$ may be illustrated geometrically in Figure 7.2, where the $(x_i X y_i)$ combinations are obtained from the

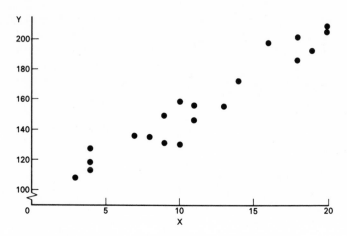

Figure 7.1  Average monthly sales versus length of training period

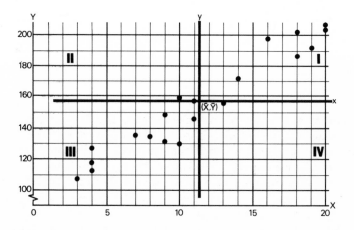

Figure 7.2

original $(X_i, Y_i)$ combinations by shifting the origin of the axes from the point $(0, 0)$ to the point $(\overline{X}, \overline{Y})$. The $(x_i, y_i)$ will be distributed between the four quadrants of the new axes, which quadrants we shall number I through IV. Since $x_i$ and $y_i$ are deviations from the sample means, they will take both positive and negative values depending on the quadrant in which they occur, as will their products, $x_i y_i$. For example, in quadrant I both $x_i$ and $y_i$ will be positive, as will the products $x_i y_i$. In quadrant III both $x_i$ and $y_i$ will be negative, so that the products $x_i y_i$ will be positive in this quadrant. On the other hand, in quadrants II and IV either $x_i$ or $y_i$ will be positive (negative) when the other is negative (positive), so that the products $x_i y_i$ will be negative in these two quadrants.

The sign pattern of $x_i y_i$ suggests a numerical measure of association. If there is a positive association between $X$ and $Y$ we would expect a high proportion of $(x_i, y_i)$ points to fall in quadrants I and III and hence the sum of products $\sum_{i=1}^{20} x_i y_i$ will be positive. If there is a negative association between $X$ and $Y$ we would expect most of the $(x_i, y_i)$ to fall in quadrants II and IV and hence $\sum_{i=1}^{20} x_i y_i$ will be negative. If there is no association between $X$ and $Y$ we would expect to find the $(x_i, y_i)$ scattered over all four quadrants, with positive and negative $x_i y_i$ tending to cancel out, so that $\sum_{i=1}^{20} x_i y_i$ will tend to be close to zero. The latter case may be illustrated by the data given in Table 7.2, where $X$ and $Y$ are in fact two samples of random numbers, which we would not expect to be associated. The scatter diagram corresponding to these data is presented in Figure 7.3. For the training/sales data we calculate $\sum_{i=1}^{20} x_i y_i = 3347 \cdot 6$, while for the random numbers presented in Table 7.2 we obtain $\sum_{i=1}^{20} x_i y_i = -3452$.

Table 7.2.

| X | Y | X | Y | X | Y | X | Y |
|---|---|---|---|---|---|---|---|
| 44 | 55 | 99 | 15 | 94 | 54 | 64 | 45 |
| 1 | 91 | 97 | 48 | 2 | 30 | 26 | 40 |
| 0 | 76 | 7 | 24 | 28 | 92 | 61 | 32 |
| 61 | 74 | 68 | 40 | 73 | 89 | 89 | 62 |
| 7 | 72 | 88 | 51 | 32 | 22 | 91 | 36 |

However, $\sum x_i y_i$ is a rather crude measure of association, since it is difficult to interpret. For example, in both our examples $\sum x_i y_i$ is large and is larger in the case of the two random variables, which we would expect to show less association than the sales and training data. If in one case we obtain $\sum x_i y_i = 800$, while in a second we obtain $\sum x_i y_i = 8$, is the association in the first case 100 times as strong

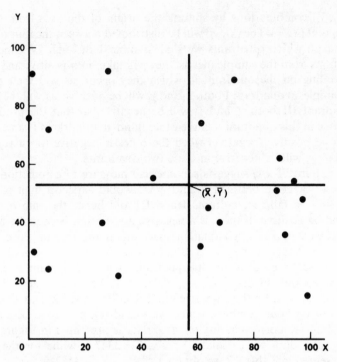

Figure 7.3  Scatter diagram for two samples of random numbers

as in the second? The answer to the first question depends on the size of the sample, since if there is some positive association between $X$ and $Y$ we can make $\sum x_i y_i$ very large by increasing the size of the sample. We may counter this by considering not $\sum x_i y_i$, but its mean value $1/n \sum_{i=1}^{20} x_i y_i$, but this still leaves us with a major problem of interpretation, which is that the magnitude of $\sum x_i y_i$ depends on the units of measurement of the variables. For example, $\sum x_i y_i = 800$ $\sum x_i y_i = 8$ could actually refer to the same variables if in the first case $X$ is measured in new pence while in the second the same series is measured in pounds sterling. The second problem may be solved if the average sum of products is divided by the product of the standard deviations of the variables, $s_X$ and $s_Y$. The ratio which results from these operations is called the *coefficient of correlation, r,*

$$r = \frac{\sum x_i y_i}{n s_X s_Y}$$

where[1]

$$s_X = \sqrt{\sum (X_i - \bar{X})^2/n} \quad \text{and} \quad s_Y = \sqrt{\sum (Y_i - \bar{Y})^2/n},$$

so that

$$r = \frac{\sum x_i y_i}{\sqrt{\sum x_i^2 \cdot \sum y_i^2}}. \qquad (7.1)$$

For computational purposes, the following alternative ways of expressing $r$ are convenient:

$$r = \frac{n \sum X_i Y_i - (\sum X_i)(\sum Y_i)}{\sqrt{n \sum X_i^2 - (\sum X_i)^2} \sqrt{n \sum Y_i^2 - (\sum Y_i)^2}}$$

$$= \frac{\sum X_i Y_i - n\bar{X}\bar{Y}}{\sqrt{\sum X_i^2 - n\bar{X}^2} \sqrt{\sum Y_i^2 - n\bar{Y}^2}} \qquad (7.2)$$

The reader may recall that the expressions in the denominators were derived in chapter 1, and he should check the expressions presented in the numerators as an exercise.

The coefficient of correlation is a pure number lying in the range $-1 \leq r \leq 1$, taking the values $\pm 1$ when all the points lie exactly on a straight line, the sign depending on whether the line has a positive or negative slope. This result will be asserted here and proved in the next chapter.[2]

For our numerical examples we have $r = 0\cdot966$ for the training/sales data and $r = -0\cdot213$ for the two series of random numbers, two results which provide strong numerical confirmation for the visual impressions obtained from Figures 7.2 and 7.3. However, not all cases are as clear-cut as these, and in order to interpret intermediate values of the correlation coefficient it is necessary to develop some statistical theory concerning the sampling distribution of $r$.

### 7.3 testing hypotheses concerning the correlation coefficient

The sampling distribution of $r$ has been investigated extensively for the case in which $X$ and $Y$ are jointly normally distributed, with means $\mu_X$ and $\mu_Y$, and variances $\sigma_X^2$ and $\sigma_Y^2$. If there is an association between them, they are connected by a new parameter, the *population* correlation coefficient, $\rho$, which is defined as

$$\rho = \frac{E[(X - \mu_X)(Y - \mu_Y)]}{\sigma_X \cdot \sigma_Y}, \qquad -1 < \rho < 1. \qquad (7.3)$$

The joint distribution, involving the five parameters $\mu_X, \mu_Y, \sigma_X^2, \sigma_Y^2$ and $\rho$, is usually referred to as a *bivariate* normal distribution. If the variables are independent, then $\rho = 0$.

The sampling distribution of $r$ has been investigated for various values of $\rho$, but in spite of the normality of $X$ and $Y$, the distribution of $r$ is not easy to work with directly. The problem arises because $r$ is constructed to lie in the range $-1 \leq r \leq 1$. Thus when $\rho = 0$, the distribution of $r$ is symmetrical but non-normal, while when $\rho \neq 0$, the distribution of $r$ can become extremely skewed. For example, when $\rho = -0.8$, most of the area under the probability distribution of $r$ is crammed down against its lower bound, $-1$. Some examples are given in Figure 7.4.

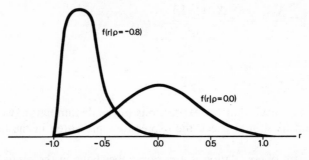

Figure 7.4

Fortunately, it was discovered that a non-linear transformation could be applied to $r$ to produce a new variable which is approximately normally distributed. If we define the new variable

$$z = \tfrac{1}{2}\log_e\left(\frac{1 + r}{1 - r}\right), \tag{7.4}$$

then $z$ is approximately normally distributed, with mean

$$\mu_z = \tfrac{1}{2}\log_e\left(\frac{1 + \rho}{1 - \rho}\right) \tag{7.5}$$

and variance

$$\sigma_z^2 = \frac{1}{(n - 3)}. \tag{7.6}$$

If we now standardize $z$, to obtain

$$Z' = \frac{z - \mu_z}{\sigma_z}, \tag{7.7}$$

then $Z'$ is distributed approximately $N(0, 1)$. Values of the variable $z$ which is defined in 7.4 are tabulated for different values of $r$ in the Appendix, Table A5.

The formulae presented in equations 7.4–7.7 enable us to set up confidence intervals for $\rho$ or to test hypotheses concerning this parameter. To illustrate the first procedure, suppose we have calculated $r = 0.65$ from a sample of size 28 and we wish to set up a confidence interval for $\rho$ at the probability level, $p = 0.95$. We first calculate that $\sigma_z = 1/\sqrt{25} = 0.2$, and from Table A5 we find that when $r = 0.65$, $z = 0.775$. We now use 7.7 to set up the confidence interval for $\mu_z$:

$$(-1.96 < Z' < 1.96) = 0.95$$

$$\left( -1.96 < \frac{z - \mu_z}{\sigma_z} < 1.96 \right) = 0.95$$

$$(z - 1.96\sigma_z < \mu_z < z + 1.96\sigma_z) = 0.95$$

$$(0.775 - 1.96(0.2) < \mu_z < 0.775 + 1.96(0.2)) = 0.95$$

$$(0.383 < \mu_z < 1.167) = 0.95.$$

The final step is to convert our confidence interval for $\mu_z$ into a confidence interval for $\rho$, which we may do by using Table A5 to find the values of $\rho$ which when substituted into 7.5 would yield values of 0.383 and 1.167. Corresponding to $z = 0.383$ we find $r = 0.366$, and for $z = 1.167$, $r = 0.824$, so that our confidence interval for $\rho$ is

$$p(0.366 < \rho < 0.824) = 0.95.$$

The reader may note that the confidence interval is not symmetrical about $r = 0.65$, a fact which reflects the skewness in the distribution of $r$.

Testing hypotheses concerning $\rho$ is also a straightforward matter. If our hypothesis is $\rho = \rho_0$, we have

$$\mu_z = \tfrac{1}{2}\log_e \frac{1 + \rho_0}{1 - \rho_0}$$

and we would substitute this value into 7.7 to obtain our test ratio. One hypothesis which is usually of considerable interest is the hypothesis $H_0 : \rho = 0$, that is, there is no association between the two variables. Under this hypothesis $\mu_z = 0$ and the test ratio becomes

$$Z' = \frac{z}{\sigma_z}.$$

To illustrate this calculation, let us test the hypothesis that there is no associa-tion between the length of the training period and the average level of sales for the data in Table 7.1. Here we have $H_0 : \rho = 0$, and assuming that if training does have an effect it is helpful rather than harmful, our alternative hypothesis is $H_1 : \rho > 0$, which suggests a one-tail test. If we choose 0·95 as our probability level, we shall reject $H_0$ if $Z' > 1·64$. Our data are $n = 20$ and $r = 0·966$, so that $\sigma_z = 1/\sqrt{17} = 0·243$ and from Table A5 corresponding to $r = 0·966$ we find $Z = 3·106$. Our test ratio given $H_0$ is true is $Z' = 3·106/0·243 = 12·8$. This value of $Z'$ falls well inside the rejection region and we reject the hypothesis $H_0$. There is evidence of a strong positive association between the two variables.

One point concerning the use of Table A5 is that it is only tabulated for positive values of $r$. This is simply making use of the symmetry of the normal distribution, and the values of $z$ corresponding to negative values of $r$ are obtained by putting a minus sign before the corresponding positive $z$ entry in the table.

Finally, we may ask what happens to the test procedure outlined above if the variables being analysed are not normally distributed. The answer is that when $\rho = 0$ the approximation is quite good for moderately sized samples, but the test is less robust when $\rho \neq 0$ and may only give a good approximation here for large sample sizes. It would be difficult to specify a number here as a safe large sample size. All one can do is to suggest that not too much weight should be placed on significant values of $r$ if the sample size is small.

### 7.4  some limitations on correlation analysis

The sample correlation coefficient is conceptually simple and is easy to compute, but its interpretation does require some care. In the first place, as has already been mentioned, the correlation coefficient measures the degree of *linear* associa-tion between $X$ and $Y$. Thus if we find that $r$ is non-significant (that is, we accept $H_0 : \rho = 0$), we have not ruled out the possibility of a strong non-linear relationship between $X$ and $Y$. The latter case must be investigated by other methods, the simplest of which is to plot the data in a scatter diagram. As an example, the reader may plot $X$ and $Y$, where $Y = X^2$ and $X = -4, -3, -2, -1, 1, 2, 3, 4$, and calculate the correlation between $X$ and $Y$.

The second problem, which has also been mentioned already, is the dangers involved in interpreting a strong association as a causal relationship. A high correlation between $X$ and $Y$ does not prove that $X$ *causes* $Y$, or vice versa. Here we may contrast two situations:

$$\begin{array}{cc} \text{I} & \text{II} \\ \uparrow X \leftrightarrow Y \uparrow & \uparrow X \underset{Z}{\overset{\nwarrow \nearrow}{}} Y \uparrow \end{array}$$

In situation I, $X$ and $Y$ are causally linked and we observe $Y$ increasing (decreasing) because $X$ is increasing (decreasing). In situation II, $X$ and $Y$ are not causally linked to each other, but both are positively linked to a third variable $Z$. Thus if $Z$ increases, both $X$ and $Y$ increase and we observe a high correlation between them. This is often referred to as a 'spurious' correlation. Often, when the observations on $X$ and $Y$ are collected over time, a good candidate for $Z$ is the growth in population, although it will be left to the reader to decide whether a high positive correlation between the growth of university teachers' salaries and the growth of crimes of violence in London should be assigned to situation I or situation II.

The answer here is that not too much weight should be placed on a significant sample correlation coefficient *in isolation*, since to establish causal linkage may require a considerable amount of further research. For example, if a medical research worker finds a very high correlation between pollutant $X$ and disease $Y$, he has not proved that $X$ *causes* $Y$. However, having found the high correlation, he may be encouraged to direct his research towards investigating the possible link between $X$ and $Y$ and he may establish such a causal link on the basis of theorizing and controlled laboratory experiments. A high correlation indicates a pattern which may be spurious or may be serious; only further research can hope to decide which.

### 7.5 a general test of association
In some situations we may be interested in the association between two variables but may not be able to measure the degree of association by means of the correlation coefficient. These situations can arise in a number of ways; for example, (a) the relationship may be non-linear, (b) the data may be condensed into a form which is not amenable to correlation analysis, or (c) some or all of the data may be qualitative rather than quantitative.

As an example of this kind of situation, consider the following summary of the (hypothetical) results obtained by interviewing a random sample of 500 businessmen concerning their salaries and education in economics.

| Salary \ Education | Studied economics | Studied no economics | Total |
|---|---|---|---|
| Salary < £4000 | 100 | 250 | 350 |
| Salary ≥ £4000 | 100 | 50 | 150 |
| Total | 200 | 300 | 500 |

Here the data are not in a form suitable for correlation analysis since the income data are condensed into only two groups and the education data are non-quantitative. However, a test of association can be developed by using the following argument. If we assume that education and salary are *independent*, we may calculate the number of businessmen we would expect to have a given education and be in a given salary range. We know that 200 members of the sample studied economics, so that if we selected a member of the sample at random, the probability that he would have studied economics is $\frac{200}{500}$, or 0·4. If the two characteristics of education and salary are independent, then we would expect that out of the 350 businessmen earning less than £4000, a total of 350.(0·4) = 140 would have studied economics. Once we have worked out one of the expected values, the remaining three entries may be obtained by subtracting 140 from the row and column totals. Then we obtain the following table of expected values:

| Salary \ Education | Studied economics | Studied no economics | Total |
|---|---|---|---|
| Salary < £4000 | 140 | 210 | 350 |
| Salary ≥ £4000 | 60 | 90 | 150 |
| Total | 200 | 300 | 500 |

The actual frequencies differ from the expected frequencies that we have calculated on the assumption that education and salary are independent, but since our data represents only a sample of information, we would be surprised if the observed frequencies were equal to the expected frequencies, even if the hypothesis were true. What we have to decide is whether the discrepancies between the actual and expected frequencies are small enough to be accounted for in terms of random sampling fluctuations, or whether they are so large that it is more reasonable to reject the hypothesis that the characteristics are independent. For example, if we were able to draw repeated stratified random samples of 500 businessmen, holding the proportions by education and salary constant from sample to sample, we could treat the discrepancies between actual and expected frequencies as random variables and investigate the sampling distribution. At first sight it might appear that each replication would involve four random variables, but since we are imposing the constraints of keeping the sample size and the proportions by education and salary fixed, only one of the cell frequencies can be chosen at random, the other three being determined by the constraints. We thus have one degree of freedom to play with in this conceptual experiment.

The relevant sampling distribution has been investigated for the case in which the characteristics are independent. Let $a_{ij}$ be the actual frequency observed for the $j$th cell entry in the $i$th row, $(i, j = 1, 2)$, and $e_{ij}$ be the corresponding expected frequency, calculated as described above on the assumption that the characteristics salary and education are independent. Suppose that for each of the four cells we calculate $(a_{ij} - e_{ij})^2/e_{ij}$ and then from the sum

$$\sum_{i=1}^{2} \sum_{j=1}^{2} \frac{(a_{ij} - e_{ij})^2}{e_{ij}}. \tag{7.8}$$

It has been shown that this sum has a sampling distribution which is approximately a chi-squared distribution with one degree of freedom.

This result enables us to set up an operational rule for deciding between large and small discrepancies. Suppose we choose a probability level of $p = 0.95$ (that is, a Type I error of $0.05$); then, with one degree of freedom, we find from Table A3 that $p(0 < \chi^2 < 3.84) = 0.95$. That is, if our hypothesis that there is no association between salary and education is correct, then we would expect the statistic defined in 7.8 to lie in the range $0 < \chi^2 < 3.84$ with $p = 0.95$. Hence if the calculated value of 7.8 is greater than 3.84, we would reject the hypothesis that there is no association between education and salary. In this example we calculate

$$\chi^2 = \frac{(100 - 60)^2}{60} + \frac{(50 - 90)^2}{90} + \frac{(100 - 140)^2}{140} + \frac{(250 - 210)^2}{210}$$

$$= 26.7 + 17.8 + 11.4 + 7.6 = 63.5,$$

which is highly significant, so that at the probability level $p = 0.95$ we reject the hypothesis that there is no association between salary and education.

The results presented here for two rows and two columns can be extended to the general case where we have $r$ rows and $c$ columns, as illustrated in Table 7.3. Here one characteristic is represented by the rows $A_1, ..., A_r$, the second characteristic by the columns $B_1, ..., B_c$, and $a_{ij}$ represents the frequency with which the combination $(A_i, B_j)$ occurs.

*Table 7.3.*

| Characteristics | $B_1$ | $B_2$ | ... | $B_c$ | Total |
|---|---|---|---|---|---|
| $A_1$ | $a_{11}$ | $a_{12}$ | ... | $a_{1c}$ | $R_1$ |
| $A_2$ | $a_{21}$ | $a_{22}$ | ... | $a_{2c}$ | $R_2$ |
| . | . | . | ... | . | . |
| . | . | . | ... | . | . |
| . | . | . | ... | . | . |
| $A_r$ | $a_{r1}$ | $a_{r2}$ | ... | $a_{rc}$ | $R_r$ |
| Total | $C_1$ | $C_2$ | ... | $C_c$ | $N$ |

Choosing a member of this sample at random, the probability that it has characteristic $A_i$ is $R_i/N$, $i = 1, ..., r$. Then if there is no association between the $A$ and $B$ characteristics, we would expect that, of those sample members having the characteristic $B_j$, a total of $e_{ij} = (A_i \cdot B_j)/N$ would also have the characteristic $A_i$. Evaluating $(A_i \cdot B_j)/N$ for $i = 1, \ldots, r; j = 1, \ldots, c$, provides the expected frequencies under the null hypothesis of no association. There are $rc$ cells, but only $(r - 1)(c - 1)$ need be computed from the formula, since the remaining $(r + c - 1)$ may then be obtained by subtraction from the row and column totals. The statistic

$$\sum_{i=1}^{r} \sum_{j=1}^{c} \frac{(a_{ij} - e_{ij})^2}{e_{ij}}$$

has an approximate chi-squared distribution, but in this case, since $(r - 1)(c - 1)$ cells may be filled at random given the constraints that the $R_i$ and $C_j$ are fixed, the chi-squared distribution has $(r - 1)(c - 1)$ degrees of freedom.

To illustrate the extension of our results, we shall consider an example concerning the wages of random samples of clerks working in two different insurance companies. The expected frequencies on the assumption that there is no association between wage and company are given in parenthesis.

| Wage | < £20 | £20–£30 | > £30 | Total |
|------|-------|---------|-------|-------|
| Company A | 180 (200) | 155 (150) | 165 (150) | 500 |
| Company B | 220 (200) | 145 (150) | 135 (150) | 500 |
| Total | 400 | 300 | 300 | 1000 |

In this case our calculated chi-squared value is

$$\chi^2 = \frac{(180 - 200)^2}{200} + \frac{(220 - 200)^2}{200} + \frac{(155 - 150)^2}{150} + \frac{(145 - 150)^2}{150}$$

$$+ \frac{(165 - 150)^2}{150} + \frac{(135 - 150)^2}{150}$$

$$= \frac{400}{200} + \frac{400}{200} + \frac{25}{150} + \frac{25}{150} + \frac{225}{150} + \frac{225}{150} = 7 \cdot 33.$$

If we choose $p = 0 \cdot 95$ as our probability, then with $(3 - 1)(2 - 1) = 2$ degrees of freedom we find from Table A3 that the critical value is $\chi^2 = 5 \cdot 99$. Since our calculated $\chi^2$ value $7 \cdot 33$ is greater than $5 \cdot 99$, we reject the hypothesis that there is no association between the clerk's company and the wage he earns.

# problems

**7.1** Given the following sets of data on $X$ and $Y$, plot a scatter diagram and calculate the correlation coefficient. Comment on your results.

| (a) $X$ | $Y$ | | (b) $X$ | $Y$ | | (c) $X$ | $Y$ |
|---|---|---|---|---|---|---|---|
| 1 | 5 | | 1 | 1 | | 1 | 1 |
| 3 | 9 | | 2 | 9 | | 2 | 16 |
| | | | 3 | 2 | | 3 | 81 |
| | | | 4 | 7 | | 4 | 256 |
| | | | 5 | 6 | | 5 | 625 |

**7.2** If the correlation coefficient calculated from a sample of 39 observations is $r = 0.35$, calculate a confidence interval with $p = 0.95$ for the population correlation coefficient $\rho$. Does this confidence interval imply acceptance of the hypothesis that $\rho = 0$?

**7.3** If the correlation coefficient calculated from a sample of 28 observations is equal to 0.8, test the hypothesis that the population correlation coefficient equals 0.6.

**7.4** How large must the sample correlation coefficient calculated from a random sample of 100 observations be to reject the hypothesis $H_0 : \rho = 0$, with the alternative hypothesis $H_1 : \rho \neq 0$, at the probability level $p = 0.95$?

**7.5** Given $H_0 : \rho = 0$, $H_1 : \rho \neq 0$, what are the minimum sample sizes necessary to reject $H_0$ at the probability level $p = 0.95$ if we calculate

(a) $r = 0.99$, (b) $r = 0.9$, (c) $r = 0.5$, (d) $r = 0.1$, (e) $r = 0.01$?

**7.6** An American social scientist wishes to investigate whether there is a close association between impulse buying in supermarkets $(X)$ and either transitory shock $(Y)$ or permanent surprise $(Z)$. From a random sample of 12 observations on each variable he calculates $r_{XY} = 0.5$, $r_{XZ} = 0.4$, $r_{YZ} = 0.9$. How would you advise him to interpret his results?

**7.7** The correlation between the two samples of random numbers presented in Table 7.2 was found to be $-0.213$. Test the hypothesis that $\rho = 0$ in this case.

**7.8** The following data refer to attendance at lectures and examination performance for a random sample of 100 students on a large economics course:

129

|  | Examination Performance | | Total |
|---|---|---|---|
|  | Passed | Failed |  |
| Attended lectures | 40 | 20 | 60 |
| Did not attend lectures | 15 | 25 | 40 |
| Total | 55 | 45 | 100 |

Do the data suggest that attending the lectures is associated with passing the examination?

**7.9**    Given the data presented in the previous problem, test the hypothesis that there is a significant difference between the population proportions passing the examination. Comment on your results.

**7.10**  A random sample of 1000 electors taken just before an election in a two-party country yielded the following data:

|  | Elector supports | | Total |
|---|---|---|---|
|  | Party A | Party B |  |
| Elector probably will vote in the election | 250 | 350 | 600 |
| Elector probably will not vote in the election | 300 | 100 | 400 |
| Total | 550 | 450 | 1000 |

Advise Party A on the likely outcome of the election.

**7.11**  A survey firm wishes to choose between three alternative forms on which to collect sales information so as to maximize the response rate. A random sampling experiment gave the following results:

| Type of form | responded | Did not respond | Total |
|---|---|---|---|
| Typed | 250 | 200 | 450 |
| Mimeographed | 300 | 450 | 750 |
| Computer-printed | 300 | 500 | 800 |
| Total | 850 | 1150 | 2000 |

Does the type of form affect the response rate?

**7.12**  In order to investigate a popular belief that the sun shines more often at weekends than on other days in the land of Ruritania, a meteorologist keeps a record of the number of days on which the sun is visible for at least ten minutes. His record covers a period of 50 complete weeks.

| Number of | Mon. | Tues. | Wed. | Thur. | Fri. | Sat. | Sun. | Total |
|---|---|---|---|---|---|---|---|---|
| sunny days | 34 | 33 | 23 | 20 | 21 | 40 | 39 | 210 |

If we propose the hypothesis that there is no association between sunshine and the day of the week, what values would you calculate as the expected frequencies of sunny days for each day of the week? If you were going to consider a repeated sampling experiment in this case, what would you hold constant and how many degrees of freedom would you have? Calculate an appropriate $\chi^2$ value and test the hypothesis that there is no association between sunshine and day of the week.

# chapter 8
# simple regression analysis

### 8.1 introduction

We turn now to the estimation of the parameters in a relationship between a number of variables, and in this chapter we shall consider the simplest possible case in which some variable, $Y$, is a linear function of a single explanatory variable, $X$. We shall extend our results to include both non-linear relationships and the introduction of more than one explanatory variable in chapter 9. For the economist, the relationships to be investigated are often suggested by economic theory and are usually specified as *exact* relationships between the variables. If the empirical data corresponded completely to these theories, there would be no statistical problem of estimation, but in practice the data on $Y$ and $X$ do not generally lie precisely on a straight line.

How are we to explain the discrepancies between the empirical data and the relationship suggested by the theory? We shall ignore for the moment the possibility that the theory may be incorrect and consider two alternative ways in which the discrepancies might arise.

(1) *Errors of measurement* (errors in variables). One possibility is that the discrepancies represent measurement errors in the recording or processing of the data on $X$ and $Y$. This is a plausible idea, since it is very likely that many economic variables are measured inaccurately. However, it will be shown in chapter 11 that if both variables are subject to measurement errors it is difficult to obtain unbiased and consistent estimates of the parameters in the relationship. To obtain unbiased estimates, we must assume that if the discrepancies are errors of measurement they affect only the $Y$s, and that the $X$s are free from measurement errors. This assumption may be a reasonable one in some situations, for example the accounting data kept by firms on many variables may be very accurate, but in general it is very difficult to accept. For example, a simple economic theory might suggest that the demand for imports was a function of aggregate income, and it would be difficult here to argue that imports contained errors of measurement while aggregate income did not, given aggregate income includes imports.

For this reason we shall ignore errors of measurement until later and consider another explanation of the discrepancies.

(2) *Errors in equations.* Except when we are trying to explain extremely simple phenomena, our theories must always be less complex than reality and concentrate on the most important aspects of a relationship. To the extent that we ignore some factors which influence the variable we are trying to explain, we would not expect our relationship to be exact unless these factors remain constant. For example, a simple economic theory might be that the demand for imports is a function of aggregate income. A more complex theory might well suggest additional variables which influence the demand for imports, such as relative prices, restrictions on international trade, government policy, the state of the balance of payments, the composition of imports, etc. If all these additional factors remained constant while our data were being collected, our simple model ought to give an exact relationship between imports and income. However, if these additional variables which have been excluded from a complete explanation of imports do vary, then we shall observe discrepancies between our relationship and the empirical data. For the present we shall assume that the discrepancies are caused by variables which influence the behaviour of $Y$ but which have not been included in the relationship, that is, they reflect the net effects of excluded variables.

Finally, the reader may recall that in chapter 1 (p. 4) we distinguished between cross-section data and time series data. In what follows we shall represent cross-section data as $X_i$, $Y_i$, $i = 1, \ldots, n$, and time series data as $X_t$, $Y_t$, $t = 1, \ldots, T$. For the moment we shall assume that both kinds of data may be treated as random samples, an assumption we shall scrutinize for time series data in chapters 10 and 11.

## 8.2 estimation and the method of least squares

To illustrate the estimation of a linear relationship, consider the data on average sales ($Y$) and the length of training ($X$) that were reported in Table 7.1. We shall suppose that the sales manager suggests that there may be a linear relationship between them, say $\alpha + \beta X$, but since variables other than length of training affect sales, we expect discrepancies, because of the effects of excluded variables. These errors in the equation we will denote as $\varepsilon$s, and the equation may then be written as

$$Y_i = \alpha + \beta X_i + \varepsilon_i, \qquad i = 1, \ldots, n. \tag{8.1}$$

We have data on the $X$s and $Y$s, but the random errors, $\varepsilon_i$, representing the net effects on the $Y$s of a set of variables we have not examined, are *unobservable*. The parameters $\alpha$ and $\beta$ are unknown and our object is to estimate them from our sample of data on the $X$s and $Y$s.

The scatter diagram for these two variables is reproduced in Figure 8.1 in which the straight line labelled $\alpha + \beta X_i$ represents the (unknown) true relationship.[1] Our object is to draw some straight line which is representative of the data and take this as our estimate of $\alpha + \beta X_i$. The simplest approach is to judge by eye what appears to the best fitting straight line, but this method is very subjective and we shall consider more objective criteria for choosing the estimated line. The reader may recall the earlier discussion of estimation in which we considered criteria for choosing between alternative estimators of a given parameter. In our regression problem we might consider a variety of alternative methods of estimating the line, but if we wish our estimates to be unbiased and have the minimum variance property, we shall choose estimates based on the *method of least squares*.

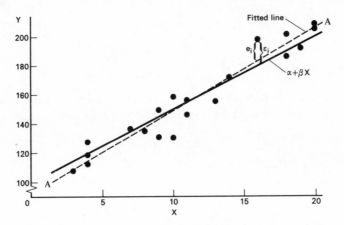

Figure 8.1 Regression of sales on length of training period

To define this criterion, consider Figure 8.1 and any fitted line, such as the dotted line $AA$. Let the equation of this fitted line be

$$\tilde{Y}_i = \tilde{\alpha} + \tilde{\beta} X_i, \tag{8.2}$$

where $\tilde{\alpha}$ and $\tilde{\beta}$ are the estimates of $\alpha$ and $\beta$. Unless this fitted line passes through the point $(X_i, Y_i)$, so that $Y_i = \tilde{Y}_i$, there will be discrepancies between the $Y$s and the $\tilde{Y}$s. We shall call these discrepancies from the fitted line the residuals, denoted by $e_i$, where[2]

$$e_i = Y_i - \tilde{Y}_i. \tag{8.3}$$

The method of least squares involves finding values for the estimates $\tilde{\alpha}$ and $\tilde{\beta}$ which will minimize the sum of the squared residuals, $\sum_{i=1}^{n} e_i$. We shall first derive the appropriate values of $\tilde{\alpha}$ and $\tilde{\beta}$ and then investigate their properties.

Substituting 8.2 into 8.3 gives

$$e_i = Y_i - \tilde{Y} = Y_i - \tilde{\alpha} - \tilde{\beta}X_i,$$

so that

$$\sum_{i=1}^{n} e_i^2 = \sum_{i=1}^{n} (Y_i - \tilde{\alpha} - \tilde{\beta}X_i)^2. \tag{8.4}$$

The least squares estimators of $\alpha$ and $\beta$ may be found by calculus methods, if we differentiate 8.4 with respect to $\tilde{\alpha}$ and $\tilde{\beta}$, set the partial derivatives equal to zero and solve the resulting two simultaneous equations. Thus

$$\frac{\partial \sum}{\partial \tilde{\alpha}} = -2\sum (Y_i - \tilde{\alpha} - \tilde{\beta}X_i) = 0 \tag{8.5}$$

$$\frac{\partial \sum}{\partial \tilde{\beta}} = -2\sum X_i(Y_i - \tilde{\alpha} - \tilde{\beta}X_i) = 0. \tag{8.6}$$

Summing the terms in parentheses in 8.5 and 8.6 and denoting the least squares estimates by $a$ and $b$, we have the so-called *normal* equations

$$\sum Y_i \quad = an \quad + b\sum X_i \tag{8.7}$$

$$\sum Y_iX_i = a\sum X_i + b\sum X_i^2. \tag{8.8}$$

From 8.7 we have

$$na = Y_i - b\sum X_i$$

or

$$a = \bar{Y} - b\bar{X}. \tag{8.9}$$

Substituting for $a$ in 8.8, we have

$$\sum Y_iX_i = (\bar{Y} - b\bar{X})\sum X_i + b\sum X_i^2$$

$$\sum Y_iX_i - \bar{Y}\sum X_i = b\left(\sum X_i^2 - \bar{X}\sum X_i\right),$$

so that

$$b = \frac{\sum Y_iX_i - \bar{Y}\sum X_i}{\sum X_i^2 - \bar{X}\sum X_i}. \tag{8.10}$$

After a little manipulation we may write 8.10 as

$$b = \frac{\sum (Y_i - \bar{Y})(X_i - \bar{X})}{\sum (X_i - \bar{X})^2}.$$  (8.11)

To summarize, the least squares estimators, that is, the values of $\tilde{\alpha}$ and $\tilde{\beta}$ which minimize the sum of the squared residuals, are

$$a = \bar{Y} - b\bar{X}$$  (8.9)

and

$$b = \frac{\sum (Y_i - \bar{Y})(X_i - \bar{X})}{\sum (X_i - \bar{X})^2}.$$  (8.11)

As an example, we may estimate $\alpha$ and $\beta$ for the sales and training data. For the purpose of computing $\beta$, 8.10 is to be preferred to 8.11 and this equation can also be expressed in two other convenient computational forms:

$$b = \frac{\sum Y_i X_i - n\bar{Y}\bar{X}}{\sum X_i^2 - n\bar{X}^2} = \frac{\sum Y_i X_i - (\sum Y_i)(\sum X_i)/n}{\sum X_i^2 - (\sum X_i)^2/n}.$$  (8.12)

From our sales and training data we calculate $\sum Y_i X_i = 38,927, \sum X_i = 228, \sum Y_i = 3121, \sum X_i^2 = 3204, \bar{Y} = 156.05, \bar{X} = 11.4$, and substituting into the second expression in 8.12 we obtain

$$b = \frac{38,927 - (3121)(228)/20}{3204 - (228)^2/20} = \frac{3347.6}{604.8} = 5.535.$$

Substituting for $b$ in 8.9 gives

$$a = \bar{Y} - b\bar{X} = 156.05 - (5.535)(11.4) = 92.95.$$

Hence our estimated line is[3] $Y_i = 92.95 + 5.54X_i$. The relationship suggests that each additional week of training adds about £5.50 to the average level of monthly sales.

In this example we know that the true relationship is $Y_i = 100 + 5X_i$ and we would probably agree that the estimates obtained from our sample are of the right order of magnitude. However, what would happen if we were to replicate our experiment by drawing fresh random samples of $\varepsilon$s from $N(0, 100)$ to generate new sets of $Y$s which with the $X$s were used to estimate $\alpha$ and $\beta$? We shall investigate this problem in the next section and develop the appropriate sampling theory for evaluating estimates obtained by the method of least squares.

### 8.3 the statistical properties of the least squares estimators

In order to simplify the derivation of the results as far as possible, we shall keep to a minimum the number of factors that we vary from sample to sample. The random errors, $\varepsilon_i$, representing as they do the net effects of variables which are excluded from the analysis, we cannot meaningfully hold constant, but in principle we may hold the $X$s constant from sample to sample. Thus if we are investigating the demand for a commodity and the $X$s are the prices and the $Y$s the quantities demanded at those prices, there is no reason in principle why we should not conduct our experiment by observing in repeated samples what quantities would be demanded at the same set of prices given different net effects of the excluded variables.

Having assumed that the $X$s may be held constant, the statistical behaviour of $Y$ depends only on the random variable, $\varepsilon$. Since the least squares estimators $a$ and $b$ are functions of the $Y$s, their distributions in repeated samples will also depend on the $\varepsilon$s. Thus we must now make some assumptions concerning the distribution of the $\varepsilon$s. We shall first list the assumptions and then explain them before discussing the sampling properties of $a$ and $b$.

The assumptions we make concerning the $\varepsilon_i$ are

$$E(\varepsilon_i) = 0, \qquad i = 1, \ldots, n, \tag{8.13}$$

$$E(\varepsilon_i^2) = \sigma^2, \tag{8.14}$$

$$E(\varepsilon_i \varepsilon_j) = 0, \qquad \text{when } i \neq j, \tag{8.15}$$

the $X_i$ may be held constant, so that

$$E(X_i \varepsilon_j) = X_i E(\varepsilon_j) = 0, \qquad \text{for all } i, j = 1, \ldots, n. \tag{8.16}$$

The first assumption, 8.13, is that the $\varepsilon_i$ come from populations having zero means, implying that the $Y$s are randomly distributed around the true relationship so that $E(y_i) = E(\alpha + \beta X_i + \varepsilon_i) = \alpha + \beta X_i$. The simplest interpretation of the second assumption, 8.14, is that the $\varepsilon_i$ are drawn from the same population and that its variance is $\sigma^2$. We are assuming that the random sampling procedure will ensure that our observations are independent of each other, so that we would expect the covariance between them to be zero. This is what is expressed in the third assumption, 8.15. The fourth assumption states that the expected covariance is zero between a set of constants and a random variable having a mean of zero. The roles of these assumptions will emerge in our discussion of the sampling distributions of $a$ and $b$, and we shall consider what happens when these assumptions are relaxed in chapter 11.

1. *Linearity*

The first result we shall establish is that if the $X$s are constant, $a$ and $b$ may be written as *linear* functions of the $Y$s. To consider $b$ first, expanding the numerator of 8.11 gives

$$b = \frac{\sum Y_i(X_i - \bar{X}) - \bar{Y}\sum(X_i - \bar{X})}{\sum(X_i - \bar{X})^2} = \frac{\sum Y_i(X_i - \bar{X})}{\sum(X_i - \bar{X})^2} \tag{8.17}$$

since $\bar{Y}\sum(X_i - \bar{X}) = 0$, because $\sum(X_i - \bar{X}) = 0$. To simplify the algebra we let $x_i = (X_i - \bar{X})$ and $y_i = (Y_i - \bar{Y})$. Now define the set of constants

$$k_i = \frac{(X_i - \bar{X})}{\sum(X_i - \bar{X})^2} = \frac{x_i}{\sum x_i^2}, \qquad i = 1, \ldots, n, \tag{8.18}$$

and from 8.17

$$b = \sum_{i=1}^{n} k_i Y_i. \tag{8.19}$$

Similarly,

$$a = \bar{Y} - b\bar{X} = \frac{1}{n}\sum Y_i - \bar{X}\sum k_i Y_i$$

and collecting up the constants

$$a = \sum\left(\frac{1}{n} - \bar{X}k_i\right) Y_i. \tag{8.20}$$

Thus both $a$ and $b$ have been expressed as *linear* functions of the $Y$s. This result will be of some interest in the next section; at the moment its main interest is that it greatly simplifies the derivation of the results that follow. To this end we shall consider the following properties of the weights defined in 8.18:

$$\sum k_i = \frac{\sum x_i}{\sum x_i^2} = 0, \text{ since } \sum x_i = 0. \tag{8.21}$$

$$\sum k_i^2 = \sum x_i^2 / (\sum x_i^2)^2 = \frac{1}{\sum x_i^2}. \tag{8.22}$$

$$\sum k_i X_i = \sum k_i(X_i - \bar{X}) = \sum x_i^2 / \sum x_i^2 = 1. \tag{8.23}$$

2. *Unbiasedness*

Given these results, it is easy to show that $a$ and $b$ are *unbiased* estimators. Consider first $b = \sum k_i Y_i$ and substitute for $Y_i$ from 8.1

$$b = \sum k_i(\alpha + \beta X_i + \varepsilon_i)$$

$$= \alpha \sum k_i + \beta \sum k_i X_i + \sum k_i \varepsilon_i.$$

Using 8.21 and 8.23, this reduces to

$$b = \beta + \sum k_i \varepsilon_i. \tag{8.24}$$

Taking expected values,

$$E(b) = \beta + E\left(\sum k_i \varepsilon_i\right) = \beta + \sum k_i E(\varepsilon_i) = \beta$$

by virtue of 8.13.

Now substitute for $Y_i$ in 8.20

$$a = \sum \left(\frac{1}{n} - \bar{X} k_i\right)(\alpha + \beta X_i + \varepsilon_i)$$

$$= \alpha + \beta \frac{1}{n}\sum X_i + \frac{1}{n}\sum \varepsilon_i - \alpha \bar{X} \sum k_i - \beta \bar{X} \sum k_i X_i - \bar{X} \sum k_i \varepsilon_i.$$

Using 8.21 and 8.23,

$$a = \alpha + \beta \bar{X} + \frac{1}{n}\sum \varepsilon_i - \beta \bar{X} - \bar{X} \sum k_i \varepsilon_i$$

$$= \alpha + \frac{1}{n}\sum \varepsilon_i - \bar{X} \sum k_i \varepsilon_i.$$

Taking expected values,

$$E(a) = \alpha + \frac{1}{n}\sum E(\varepsilon_i) - \bar{X} \sum k_i E(\varepsilon_i) = \alpha$$

by virtue of 8.13. *Thus we have shown that a and b are unbiased estimators of $\alpha$ and $\beta$.*

### 3. Minimum Variance Properties

The final result we shall establish in this section is that out of the class of linear, unbiased estimators of $\alpha$ and $\beta$, $a$ and $b$ have the smallest sampling variances. We shall first derive a formula for the variance of $b$ and show that it is the minimum variance. The results for $a$ involve more complicated algebra and will therefore be given here without proof.

To derive a formula for the variance of $b$, var$(b)$, we rearrange the terms in (8.24) to give $b - \beta = \sum k_i \varepsilon_i$. Then

$$var (b) = E[(b - \beta)^2] = E[(\sum k_i \varepsilon_i)^2].$$

When we expand $(k_1 \varepsilon_1 + ... + k_n \varepsilon_n)^2$ we may arrange the resulting terms into two groups: those obtained by forming the product of the $i$th term in each bracket, which yield $(k_i \varepsilon_i)^2$, $i = 1, ..., n$, and those obtained as the product of the $i$th term in one bracket and the $j$th term in the other, that is, $k_i k_j \varepsilon_i \varepsilon_j$, $i \neq j$, $i, j = 1, ..., n$. Hence

$$\text{var} (b) = E\left[ \sum (k_i \varepsilon_i)^2 \right] + E\left[ \sum_{i \neq j} k_i k_j \varepsilon_i \varepsilon_j \right]$$

$$= \sum k_i^2 E(\varepsilon_i^2) + \sum k_i k_j E(\varepsilon_i \varepsilon_j) = \sigma^2 \sum k_i^2,$$

using 8.14 and 8.15. Substituting for $\sum k_i^2$ from 8.22 gives

$$\text{var} (b) = \frac{\sigma^2}{\sum x_i^2}. \tag{8.25}$$

The corresponding result for $a$, which we shall state without proof, is

$$\text{var} (a) = \sigma^2 \left( \frac{\sum X_i^2}{n \sum x_i^2} \right). \tag{8.26}$$

To establish that $b$ does possess the minimum variance property, we shall compare its variance with that of some alternative linear, unbiased estimator of $\beta$, say $\tilde{\beta}$, where

$$\tilde{\beta} = \sum c_i Y_i \tag{8.27}$$

and the constants $c_i$ are given by $c_i = k_i + d_i$. Substituting for $Y_i$ in 8.27,

$$\tilde{\beta} = \sum c_i (\alpha + \beta X_i + \varepsilon_i)$$

$$= \alpha \sum c_i + \beta \sum c_i X_i + \sum c_i \varepsilon_i. \tag{8.28}$$

Taking expected values,

$$E(\tilde{\beta}) = \alpha \sum c_i + \beta \sum c_i X_i,$$

since $\sum c_i E(\varepsilon_i) = 0$, by using 8.13. Since $\tilde{\beta}$ is assumed to be unbiased, $E(\tilde{\beta}) = \beta$, which implies that $\sum c_i = 0$ and $\sum c_i X_i = 1$. Further, $\sum c_i = \sum (k_i + d_i) = \sum k_i + \sum d_i$, and since, from 8.21, $\sum k_i = 0$, we have $\sum d_i = 0$. Similarly, $\sum c_i X_i = \sum k_i X_i + \sum d_i X_i$, and since we have just shown $\sum c_i X_i = 1$ and we have, from 8.23, $\sum k_i X_i = 1$, then $\sum d_i X_i = 0$ and hence $\sum d_i x_i = \sum d_i X_i - \bar{X} \sum d_i = 0$.

Since $\tilde{\beta}$ is unbiased, 8.28 may be rewritten as $\tilde{\beta} - \beta = \sum c_i \varepsilon_i$, so that

$$\text{var}\,(\tilde{\beta}) = E[(\tilde{\beta} - \beta)^2] = E\left[\left(\sum c_i \varepsilon_i\right)^2\right] = \sigma^2 \sum c_i^2, \tag{8.29}$$

by following exactly the same arguments that were used to find var $(b)$. Now

$$\sum c_i^2 = \sum (k_i + d_i)^2 = \sum k_i^2 + 2\sum k_i d_i + \sum d_i^2$$

and $\sum k_i d_i = \sum d_i x_i / \sum x_i^2 = 0$, since we have shown $\sum d_i x_i = 0$. Hence $\sum c_i^2 = \sum k_i^2 + \sum d_i^2$. Substituting this result into 8.29,

$$\text{var}\,(\tilde{\beta}) = \sigma^2 \sum k_i^2 + \sigma^2 \sum d_i^2 = \text{var}\,(b) + \sigma^2 \sum d_i^2. \tag{8.30}$$

Now $\sum d_i^2$ must be positive, so that var $(\tilde{\beta}) > $ var $(b)$ unless $d_i = 0$, in which case var $(\tilde{\beta}) = $ var $(b)$, since $\tilde{\beta} = b$. This establishes the minimum variance property for $b$. We shall assert without proof that $a$ also has this property.[4]

Having derived these results, there are a number of features that the reader should notice. Firstly, not all the assumptions concerning the random errors were used to derive each result. For example, linearity required only 8.16. The unbiasedness result used 8.13 and 8.16, but neither 8.14 nor 8.15. The derivation of var $(b)$ and the minimum variance property depended on 8.14 and 8.15. We shall explore the relaxation of these assumptions formally in chapter 11, but it is clear already that not all the properties of the least squares estimators will be affected by relaxing a given one of the assumptions.

The second point to notice is that we have proved that the least squares estimators are best, linear and unbiased without making any assumption about the specific form of the distribution of the random errors. These are very general results, but in order to establish confidence intervals and test hypotheses concerning $\alpha$ and $\beta$, we have to make some specific assumption concerning the distribution of the $\varepsilon$s. This we shall discuss in the next section.

### 8.4 confidence intervals and hypothesis testing for $\alpha$ and $\beta$
Given the important role the normal distribution played in the earlier chapters, the reader will probably not be surprised to discover that the assumption we make is that the random errors are normally distributed. Then, given our assumption that the $X$s are constant, $Y_i = $ a constant $+ \varepsilon_i$, so that $Y_i$ is also normally distributed. We showed in the previous section that $a$ and $b$ were linear functions of the $Y$s and this result now becomes important, since linear functions of normally distributed variables are themselves normally distributed. That is, if the $\varepsilon$s are normally distributed, then $a$ and $b$ are normally distributed in repeated samples.

We have shown that $a$ and $b$ are unbiased estimators, so that the means of these normal distributions are $\alpha$ and $\beta$ respectively. In addition, the variances of these distributions were presented in 8.25 and 8.26, so that we have all the information we need concerning these normal distributions. If we standardize $b$ by subtracting its mean value, $\beta$, and dividing by its standard deviation, $\sigma/\sqrt{\sum x_i^2}$, we obtain

$$Z = \frac{b - \beta}{\sigma \sqrt{\dfrac{1}{\sum x_i^2}}} \tag{8.31}$$

which is a standard normal variable, as is

$$Z = \frac{a - \alpha}{\sigma \sqrt{\dfrac{\sum X_i^2}{n \sum x_i^2}}}. \tag{8.32}$$

Thus the sampling theory of estimation and hypothesis testing developed in the first part of the text may be applied to the regression parameters and their least squares estimators. For example, if we wish to set up a confidence interval for $\beta$ at the probability level $p = 0.95$, then from Table A1 we have

$$p(-1.96 < Z < 1.96) = 0.95.$$

Substituting for $Z$ from 8.31 and rearranging the terms in the parentheses then gives

$$p\left( b - 1.96 \frac{\sigma}{\sqrt{\sum x_i^2}} < \beta < b + 1.96 \frac{\sigma}{\sqrt{\sum x_i^2}} \right) = 0.95. \tag{8.33}$$

However, for 8.33–and the corresponding expression for $\alpha$–to be operational requires that $\sigma^2$, the variance of the unobservable $\varepsilon$s, should be known. In practice we generally do not know the value of this parameter, and the results discussed in the previous paragraph need to be modified to take this into account. The problem is analogous to the one we encountered earlier in which we were concerned with estimating the population mean when the population variance was unknown, and the solution here is very similar to our earlier procedure. If we define an unbiased estimator of $\sigma^2$ and substitute the estimate for the unknown $\sigma$ in 8.31 and 8.32, the resulting variables follow the $t$-distribution with $(n - 2)$ degrees of freedom.

The first problem that arises in estimating $\sigma^2$ is that the $\varepsilon$s are unobservable. However, having estimated 8.1 by the method of least squares, we do have the residuals from our fitted line,

$$e_i = Y_i - a - bX_i, \qquad i = 1, \ldots, n, \tag{8.34}$$

and substituting for $Y_i$ gives $e_i = \alpha + \beta X_i + \varepsilon_i - a - bX_i$, which may be rearranged to give $(e_i - \varepsilon_i) = (\alpha - a) + (\beta - b) X_i$. Taking expected values gives $E(e_i - \varepsilon_i) = 0$, that is, the $e_i$ are unbiased estimators of the $\varepsilon_i$ and so we may base our estimate of $\sigma^2$ on the residuals, $e_i$. If we find the sample mean of both sides of 8.34, we obtain $\bar{e} = \bar{Y} - b\bar{X} - a = 0$, so that $\sum_{i=1}^{n} (e_i - \bar{e})^2 = \sum_{i=1}^{n} e_i^2$.

Now we are in a position to estimate $\sigma^2$, and it can be shown[5] that if we define

$$s^2 = \frac{\sum\limits_{i=1}^{n} e_i^2}{(n - 2)}, \tag{8.35}$$

then $s^2$ is an unbiased estimator of $\sigma^2$. $s$ is often called the *standard error of estimate*. In calculating $\sum e_i^2$, we could obtain $e_i$ from 8.34, but computationally it is more efficient to obtain $\sum e_i^2$ directly by using either of the expressions

$$\sum e_i^2 = \sum y_i^2 - b^2 \sum x_i^2 \tag{8.36}$$

or

$$\sum e_i^2 = \sum y_i^2 - b \sum x_i y_i, \tag{8.37}$$

where $y_i = (Y_i - \bar{Y})$ and $x_i = (X_i - \bar{X})$.

Substituting $s$ for $\sigma$ in 8.31 and 8.32 produces

$$t = \frac{a - \alpha}{s\sqrt{\dfrac{\sum X_i^2}{n \sum x_i^2}}} \tag{8.38}$$

and

$$t = \frac{b - \beta}{s/\sqrt{\sum x_i^2}} \tag{8.39}$$

where these new variables both have a $t$-distribution with $(n - 2)$ degrees of freedom. Now suppose we wish to set up a confidence interval for $\beta$ at the probability level $p = 0.95$ and let $n = 20$. Then with 18 degrees of freedom we find from Table A2 that

$$p(-2.10 < t < 2.10) = 0.95.$$

Substituting for $t$ from 8.39 and rearranging gives

$$p\left( b - 2{\cdot}10\,\frac{s}{\sqrt{\sum x_i^2}} < \beta < b + 2{\cdot}10\,\frac{s}{\sqrt{\sum x_i^2}} \right) = 0{\cdot}95. \tag{8.40}$$

As a numerical example we shall return to the data on the level of sales $(Y)$ and the length of training $(X)$ presented in Table 7.1, from which we estimated (p. 135) that $Y_i = 92{\cdot}95 + 5{\cdot}54X_i$. We have already calculated that $\sum x_i^2 = 604{\cdot}8$ and $\sum x_i y_i = 3347{\cdot}6$. Now we calculate that

$$\sum y_i^2 = \sum Y_i^2 - \left( \sum Y_i \right)^2/n = 506{,}869 - (3121)^2/20 = 19{,}837.$$

Substituting into 8.36, $\sum e_i^2 = 19{,}837 - (5{\cdot}54)^2 (604{\cdot}8) = 1308{\cdot}3$, so that $s^2 = 1308{\cdot}3/18 = 72{\cdot}68$, and $s = 8{\cdot}53$. The standard error of $b$ is s.e.$(b) = s/\sqrt{\sum x_i^2} = 8{\cdot}53/\sqrt{604{\cdot}8} = 0{\cdot}347$, so substituting $b$ and s.e.$(b)$ in 8.40 our confidence interval for $\beta$ is

$$p(5{\cdot}54 - 0{\cdot}73 < \beta < 5{\cdot}54 + 0{\cdot}73) = 0{\cdot}95$$

or

$$p(4{\cdot}81 < \beta < 6{\cdot}27) = 0{\cdot}95.$$

*Hypothesis testing*
The results presented in 8.38 and 8.39 are also of importance if we want to test hypotheses concerning $\alpha$ or $\beta$. One hypothesis of considerable interest is $H_0 : \beta = 0$, since its acceptance would imply that really there is no relationship between $Y$ and $X$. If we assume $H_0 : \beta = 0$ is true, then the test ratio in 8.38 becomes

$$t = \frac{b}{\text{s.e.}(b)}. \tag{8.41}$$

We shall illustrate the testing of the hypothesis $H_0 : \beta = 0$ by means of the following numerical example in which $n = 27$; $\sum x_i y_i = 200$; $\sum y_i^2 = 1025$; $\sum x_i^2 = 100$; $\bar{Y} = 500$ and $\bar{X} = 120$. Given these data, we calculate $b = 200/100 = 2$ and $a = 500 - (2)(120) = 260$, so that the estimated linear relationship between $Y$ and $X$ is

$$Y_i = 260 + 2X_i.$$

To calculate s.e.$(b)$, we have $\sum e_i^2 = 1025 - (2)^2 (100) = 625$, so that $s^2 = \sum e_i^2/(n-2) = 625/25 = 25$, hence s.e.$(b) = s/\sqrt{\sum x_i^2} = 0{\cdot}5$. Substituting for $b$ and s.e.$(b)$ in 8.41, we obtain

$$t = 2/0{\cdot}5 = 4{\cdot}00.$$

If we assume that the probability level $p = 0.95$ and that our alternative hypothesis is $H_1 : \beta \neq 0$, indicating a two-tail test, then we shall reject the hypothesis unless the calculated value of 8.41 falls inside the range $-2.06 < t < 2.06$. Given our calculated value is $t = 4.00$, we reject the hypothesis that $\beta = 0$.

### 8.5 goodness of fit

The theory developed in the previous section enables us to test hypotheses concerning the parameters $\alpha$ and $\beta$. In this section we shall introduce a statistic to measure the goodness of fit of our regression line. A convenient statistic for this purpose may be developed by rearranging the terms in 8.36, so that

$$\sum y_i^2 = b^2 \sum x_i^2 + \sum e_i^2. \tag{8.42}$$

Recalling the principle discussed in the context of the analysis of variance in chapter 6, we may note that the terms in 8.42 can be interpreted as follows. On the left-hand side we have the sum of the squared deviations from $\bar{Y}$, which we may take as a measure of the total variation in $Y$ which is to be explained. On the right-hand side, the term $\sum e_i^2$ measures the variation in $Y$ which remains unexplained by the estimated relationship between $Y$ and $X$. It follows then that the term $b^2 \sum x_i^2$ may be taken as a measure of the variation in the $Y$s which is explained by the fitted line.

Given this partitioning of $\sum y_i^2$ into explained and unexplained components, we may define a measure of goodness of fit as

$$R^2 = \frac{\text{variation explained}}{\text{variation to be explained}}$$

$$= \frac{b^2 \sum x_i^2}{\sum y_i^2} \tag{8.43}$$

$$= \frac{y_i^2 - \sum e_i^2}{\sum y_i^2} = 1 - \frac{\sum e_i^2}{\sum y_i^2}. \tag{8.44}$$

This statistic, $R^2$, is called the *coefficient of determination*, and when it is expressed in the form given in 8.44 it is easy to see that its limits are zero and unity. For example, if the fit is perfect, $\sum e_i^2 = 0$ and $R^2 = 1$. At the other extreme, if our estimated line is horizontal ($b = 0$), then $\sum e_i^2 = \sum y_i^2$ and $R^2 = 0$. Thus

$$0 \leq R^2 \leq 1. \tag{8.45}$$

As a numerical example, we may take the regression of average monthly sales level ($Y$) on length of training ($X$) which we considered above. We have calculated that $\sum y_i^2 = 19{,}837$ and $\sum e_i^2 = 1308{\cdot}3$, from which we obtain $R^2 = 0{\cdot}934$. This suggests that our regression line fits the sales data well, since we can explain about 93 per cent of the variation in the level of sales in terms of the length of the training period.

The advantage of using the coefficient of determination as the measure of goodness of fit is that it is relatively simple to develop a statistical test to see whether the goodness of fit is good enough. We have divided $\sum y_i^2$ into two components which we can interpret as the explained variation and the unexplained variation. Furthermore, we know that the Ys and $b$ are normally distributed, as must be the $es$. Therefore, the terms appearing in 8.42 all involve the squares of normally distributed variables and hence will be chi-squared variables. These results enable us to set up a test for the hypothesis $H_0 : R^2 = 0$ (that is, $\beta = 0$) which is analogous to the analysis of variance developed in chapter 6, section 6.8. The breakdown is presented in Table 8.1.

The components in this breakdown are derived in the following way. Under the null hypothesis $H_0 : \beta = 0$, the ratio in 8.31 becomes

$$Z = \frac{b}{\sigma / \sqrt{\sum x_i^2}} = \frac{b \sqrt{\sum x_i^2}}{\sigma}. \tag{8.46}$$

*Table 8.1.* Partition of variation in regression model

| Source of variation | Sum of squares | Degrees of freedom | Mean sum of squares | F |
|---|---|---|---|---|
| Explained | $b^2 \sum x_i^2$ | 1 | $s_1^2 = b^2 \sum x_i^2$ | |
| Residual | $\sum e_i^2$ | $n - 2$ | $s_2^2 = \bar{F} e_i^2 / (n - 2)$ | $F = \dfrac{s_1^2}{s_2^2}$ |
| Total | $\sum y_i^2$ | $n - 1$ | — | |

Since $Z$ is a standard normal variable, squaring the ratio in 8.46 gives $b^2 \sum x_i^2 / \sigma^2$ as a chi-square variable with one degree of freedom. For $s^2$, the unbiased estimator of $\sigma^2$ defined in 8.36, we have

$$\frac{(n - 2) s^2}{\sigma^2} = \frac{\sum e_i^2}{\sigma^2} \tag{8.47}$$

which has an independent chi-square distribution with $(n - 2)$ degrees of freedom. Combining these results, we may define the ratio

$$F = \frac{b^2 \sum x_i^2}{\sum e_i^2 / (n - 2)} \tag{8.48}$$

which follows the $F$-distribution with 1 and $n - 2$ degrees of freedom. This ratio may be used to test $H_0:\beta = 0$.

The calculation is illustrated in Table 8.2 for the sales/training data. In the last column of the table we have $F = 254.9$, and if we choose a probability level of $p = 0.95$, from Table A4 with 1 and 18 degrees of freedom we obtain $F = 4.41$. Since our calculated $F$-value is greater than $F = 4.41$, we may reject the hypothesis $H_0:\beta = 0$.

*Table 8.2*

| Source of variation | Sum of squares | Degrees of freedom | Mean sum of squares | $F$ |
|---|---|---|---|---|
| Explained | 18528 | 1 | 18528 | |
| Residual | 1308 | 18 | 72·68 | $F = 254.9$ |
| Total | 19837 | 19 | — | |

The $F$-ratio in 8.48 may be related to $R^2$ by the following modifications:

$$F = \frac{b^2 \sum x_i^2}{\sum e_i^2/(n-2)} = \frac{(n-2)\,b^2 \sum x_i^2}{\sum e_i^2} = \frac{(n-2)\,b^2 \sum x_i^2}{(1-R^2)\sum y_i^2},$$

using 8.44, and

$$= \frac{(n-2)\,b^2 \sum x_i^2 / \sum y_i^2}{(1-R^2)} = \frac{(n-2)\,R^2}{(1-R^2)}, \tag{8.49}$$

using 8.43. Hence the $F$-test of $H_0:\beta = 0$ is equivalent to a comparison of $R^2$ with $(1 - R^2)$, that is, $H_0:R^2 = 0$.

One further modification of 8.47 which may be noted is

$$F = \frac{b^2 \sum x_i^2}{\sum e_i^2/(n-2)} = \frac{b^2 \sum x_i^2}{s^2} = \left(\frac{b}{s/\sqrt{\sum x_i^2}}\right)^2.$$

That is, the ratio calculated for the $F$-test of $H_0:\beta = 0$ is the square of the $t$-ratio defined to test $H_0:\beta = 0$ in 8.41. The reader may confirm that the two tests lead to equivalent results by comparing, for a given probability level, the square of the value of $t$ with $(n-2)$ degrees of freedom obtained from Table A2 with the corresponding value of $F$ with 1 and $n-2$ degrees of freedom obtained from Table A4. As an example, we may note that with $p = 0.95$ and 18 degrees of freedom, $t = \pm 2.10$ and $t^2 = (2.1)^2 = 4.41 = F$.

We conclude this section with some comments on the presentation of the results of regression analysis. It is clear from our discussion of hypothesis testing and goodness of fit that while the object of the exercise is to estimate $\alpha$ and $\beta$, it is not sufficient merely to report the estimates $a$ and $b$. In practice nowadays most economists would report their estimates of $\alpha$ and $\beta$ together with the standard errors and the value of $R^2$. The conventional format is to report the estimated equation, together with the value of $R^2$, with the standard errors placed in parentheses below the estimated parameter values. For example, the regression equation for our sales/training data would be reported as

$$Y_i = 92 \cdot 9 + 5 \cdot 54 X_i, \qquad R^2 = 0 \cdot 934.$$
$$\quad (4 \cdot 39) \quad (0 \cdot 35)$$

The reader may then test the significance of $\alpha$ and $\beta$, that is, test the hypotheses $H_0: \alpha = 0$ and $H_0: \beta = 0$. Alternatively, some economists would present the results in a way that makes testing these hypotheses even easier, by reporting not the standard errors but the $t$-ratios corresponding to 8.38 and 8.39 given $H_0: \alpha = 0$ and $H_0: \beta = 0$. In this case the $t$-ratios are presented in parentheses, so that our sales/training results would be reported as

$$\underline{Y_i = 92 \cdot 95 + 5 \cdot 54 X_i} \quad , \quad R^2 = 0 \cdot 934.$$
$$\quad (21 \cdot 17) \quad (15 \cdot 97)$$

The information contained in these two equations is the same and the form is a matter of convenience, but it may be regarded as the minimum information the reader should expect to receive (communicate) when interpreting (presenting) empirical results. In chapters 10 and 11 we shall introduce modifications to the regression model which will require that more information should be reported when regression results are presented.

### 8.6 the relationship between correlation and regression analysis
In presenting the material in chapters 7 and 8, we have stressed the theoretical differences between correlation analysis and regression analysis, but the reader may have noticed some algebraic similarities between the results obtained. In this section we shall explore some of the algebraic connections between the two models.

The most obvious algebraic connection is that

$$b = \frac{\sum x_i y_i}{\sum x_i^2} \quad \text{and} \quad r = \frac{\sum x_i y_i}{\sqrt{\sum x_i^2 \cdot \sum y_i^2}},$$

and in these two ratios the numerators are equal while the denominators are positive, so that $b$ and $r$ must have the same algebraic sign. If we define $r$ as

$$r = \frac{\frac{1}{n}\sum x_i y_i}{s_X . s_Y},$$

where $s_X = \sqrt{\frac{1}{n}\sum x_i^2}$ and $s_Y = \sqrt{\frac{1}{n}\sum y_i^2}$, then

$$b = r\frac{s_Y}{s_X}.$$

A more interesting result emerges if we consider the coefficient of determination, $R^2$. Here, if we substitute for $b$ from 8.11, we have

$$R^2 = \frac{b^2 \sum x_i^2}{\sum y_i^2} = \left(\frac{\sum x_i y_i}{\sum x_i^2}\right)^2 . \frac{\sum x_i^2}{\sum y_i^2} = \left(\frac{\sum x_i y_i}{\sqrt{\sum x_i^2 . \sum y_i^2}}\right)^2 = r^2.$$

Thus the coefficient of determination equals the square of the coefficient of correlation between $X$ and $Y$. At a symbolic level this may explain why $R^2$ was chosen to represent the coefficient of determination, but in addition we may now write 8.49 as

$$F = \frac{(n-2)R^2}{(1-R^2)} = \frac{(n-2)r^2}{(1-r^2)} \tag{8.50}$$

and taking the square root of 8.50 we obtain

$$t = \frac{\sqrt{(n-2)}r}{\sqrt{(1-r^2)}}, \tag{8.51}$$

which follows the $t$-distribution with $(n-2)$ degrees of freedom. Hence the ratio in 8.51 may be used as an alternative to the test proposed in chapter 7 for the hypothesis that the population correlation coefficient, $\rho$, equals zero. However, for hypotheses in which it is assumed that $\rho \neq 0$, the test discussed in chapter 7 is appropriate.

We showed above that $0 \leq R^2 \leq 1$, so that $0 \leq r^2 \leq 1$ and, taking square roots,

$$-1 \leq r \leq 1, \tag{8.52}$$

which provides a simple proof of the assertion made in the last chapter concerning the range of $r$.

Finally, having noted the algebraic linkages between the models, it is worth stressing again the theoretical *differences* between the correlation and regression models:

(i) The correlation model does not specify the direction of a causal link, while the regression model distinguishes between the dependent and independent variables.

(ii) Statistical tests on the correlation model depend upon both $X$ and $Y$ being normally distributed variables, whereas in the regression model the $X$s are assumed to be a set of constants and tests depend upon the assumption that the random errors are normally distributed.

# appendix to chapter 8

*The derivation of an unbiased estimator of $\sigma^2$*
From 8.34

$$e_i = Y_i - a - bX_i$$

and since $\bar{e} = 0$, we have

$$e_i = y_i - bx_i. \tag{8.A1}$$

Averaging $Y_i = \alpha + \beta X_i + \varepsilon_i$ yields $\bar{e} = \alpha + \beta \bar{X} + \bar{\varepsilon}$, so that in terms of deviations from $\bar{Y}$ and $\bar{X}$, $y_i = \beta x_i + (\varepsilon_i - \bar{\varepsilon})$. Substituting this expression for $y_i$ in 8.A1 gives

$$e_i = -(b - \beta)x_i + (\varepsilon_i - \bar{\varepsilon}),$$

so that

$$\sum e_i^2 = \sum [-(b - \beta)x_i + (\varepsilon_i - \bar{\varepsilon})]^2$$
$$= (b - \beta)^2 \sum x_i^2 - 2(b - \beta) \sum x_i(\varepsilon_i - \bar{\varepsilon}) + \sum (\varepsilon_i - \bar{\varepsilon})^2. \tag{8.A2}$$

Now consider the expected values of the terms on the right-hand side of 8.A2:

$E[(b - \beta)^2 \sum x_i^2] = \sigma^2$, by multiplying both sides of 8.25 by $\sum x_i^2$

$E[\sum (\varepsilon_i - \bar{\varepsilon})^2] = E(\sum \varepsilon_i^2 - n\bar{\varepsilon}^2) = (n - 1)\sigma^2$, from 5.2

$E[(b - \beta) \sum x_i(\varepsilon_i - \bar{\varepsilon})] = E[\sum k_i \varepsilon_i \sum x_i(\varepsilon_i - \bar{\varepsilon})]$, from 8.24,

$$= E\left[ \frac{\sum x_i \varepsilon_i}{\sum x_i^2} \left( \sum x_i \varepsilon_i - \bar{\varepsilon} \sum x_i \right) \right]$$

$$= E\left[ \frac{\sum x_i \varepsilon_i}{\sum x_i^2} \sum x_i \varepsilon_i \right], \text{ since } \sum x_i = 0,$$

$$= E\left[ \frac{\sum (x_i \varepsilon_i)^2 + 2 \sum x_i x_j \varepsilon_i \varepsilon_j}{\sum x_i^2} \right]$$

using the argument preceding 8.25. Finally, using 8.14 and 8.15,

$$E\big[(b - \beta)\sum x_i(\varepsilon_i - \bar{\varepsilon})\big] = \sigma^2 \sum x_i^2 / \sum x_i^2 = \sigma^2.$$

Substituting these three results into the expected value of 8.A2, we have

$$E\left(\sum e_i^2\right) = \sigma^2 - 2\sigma^2 + (n - 1)\sigma^2 = (n - 2)\sigma^2. \tag{8.A3}$$

Hence if we define

$$s^2 = \sum e_i^2 / (n - 2), \tag{8.35}$$

then $E(s^2) = \sigma^2$ and $s^2$ is an unbiased estimator of $\sigma^2$.

*Convenient formulae for calculating the sum of squared residuals*
From 8.A1

$$\sum e_i^2 = \sum (y_i - bx_i)^2$$

$$= \sum y_i^2 - 2b \sum x_i y_i + b^2 \sum x_i^2. \tag{8.A4}$$

We may now proceed in one of two ways. If we multiply the middle term on the right-hand side of 8.A4 by $\sum x_i^2 / \sum x_i^2$,

$$\sum e_i^2 = \sum y_i^2 - 2b\left(\frac{\sum x_i y_i}{\sum x_i^2}\right)\sum x_i^2 + b^2 \sum x_i^2$$

$$= \sum y_i^2 - 2b^2 \sum x_i^2 + b^2 \sum x_i^2 = \sum y_i^2 - b^2 \sum x_i^2. \tag{8.36}$$

Alternatively, if we substitute for $b$ in the third term of 8.A4, we obtain

$$\sum e_i^2 = \sum y_i^2 - 2b \sum x_i y_i + b\left(\frac{\sum x_i y_i}{\sum x_i^2}\right)\sum x_i^2$$

$$= \sum y_i^2 - 2b \sum x_i y_i + b \sum x_i y_i = \sum y_i^2 - b \sum x_i y_i. \tag{8.37}$$

# problems

**8.1** The following sums were obtained from 16 pairs of observations on $X$ and $Y$:

$$\sum Y_i^2 = 526, \ \sum X_i^2 = 657, \ \sum X_i Y_i = 492, \ \sum Y_1 = 64, \ \sum X_i = 96.$$

Estimate the parameters in the model $Y_i = \alpha + \beta X_i + \varepsilon_i$ and $R^2$. Test the hypothesis that $\beta = 2 \cdot 0$.

**8.2** If for a sample of 60 observations

$$\sum (X_i - \bar{X})^2 = 130, \ \sum (Y_i - \bar{Y})^2 = 600,$$

$$\sum (X_i - \bar{X})(Y_i - \bar{Y}) = 260, \ \bar{X} = 30, \ \bar{Y} = 80$$

estimate the parameters in the model $Y_i = \alpha + \beta X_i + \varepsilon_i$ and, by means of an analysis of variance, test the significance of the regression equation.

**8.3** Given the following data:

$$\sum (X_i - \bar{X})(Y_i - \bar{Y}) = 200, \ \sum (X_i - \bar{X})^2 = 100, \ \sum (Y_i - \bar{Y})^2 = 500,$$

$$\bar{X} = 100, \ \bar{Y} = 150, \ n = 27,$$

estimate the parameters in the model $Y_i = \alpha + \beta X_i + \varepsilon_i$ and test the hypothesis $H_0: \beta = 1 \cdot 5$ against the alternative hypothesis $H_1: \beta > 1 \cdot 5$.

**8.4** Consider the equations

$$Y_i = \alpha_1 + \beta_1 X_{1i} + \varepsilon_{1i}, \quad i = 1, \ldots, n,$$

$$Y_i = \alpha_2 + \beta_2 X_{2i} + \varepsilon_{2i},$$

where $X_{1i} = 100 - X_{2i}$. Compare the least squares estimates of $\alpha_1$ and $\alpha_2$, $\beta_1$ and $\beta_2$, the standard errors of $b_1$ and $b_2$, and the values of $R^2$ obtained in estimating the two equations.

**8.5\*** The table below contains twenty consecutive observations on the quantity of a good sold ($Q_t$) and its price ($P_t$).

| Period | $Q_t$ | $P_t$ | Period | $Q_t$ | $P_t$ |
|--------|-------|-------|--------|-------|-------|
| 1  | 43·50 | 56·67 | 11 | 43·77 | 59·64 |
| 2  | 55·05 | 44·78 | 12 | 67·57 | 32·14 |
| 3  | 45·49 | 55·37 | 13 | 40·49 | 59·38 |
| 4  | 65·79 | 33·15 | 14 | 65·84 | 37·78 |
| 5  | 41·90 | 55·24 | 15 | 31·13 | 68·62 |
| 6  | 60·52 | 45·43 | 16 | 69·80 | 31·17 |
| 7  | 47·74 | 52·19 | 17 | 23·89 | 79·03 |
| 8  | 60·71 | 39·67 | 18 | 82·94 | 17·10 |
| 9  | 44·61 | 57·27 | 19 | 20·19 | 78·73 |
| 10 | 53·89 | 48·54 | 20 | 81·31 | 16·50 |

Source: Hypothetical data constructed by author.

(i) Estimate the linear regression of $Q_t$ on $P_t$ by the method of least squares.

(ii) If $P_0 = 33.00$, estimate the linear regression of $Q_t$ on $P_{t-1}$.

(iii) Plot scatter diagrams of the data and relationships and discuss the economic interpretation of the results.

**8.6\*** The table below contains annual data on aggregate saving and personal disposable income (both at current prices) for the United Kingdom.

| Year | Saving | Personal disposable income | Year | Saving | Personal disposable income |
|------|--------|---------------------------|------|--------|---------------------------|
| 1950 | 1·6 | 11·0 | 1960 | 4·2 | 21·2 |
| 1951 | 1·7 | 11·9 | 1961 | 5·0 | 22·9 |
| 1952 | 2·0 | 12·7 | 1962 | 5·2 | 24·1 |
| 1953 | 2·1 | 13·5 | 1963 | 5·6 | 25·6 |
| 1954 | 2·2 | 14·4 | 1964 | 6·2 | 27·6 |
| 1955 | 2·5 | 15·6 | 1965 | 7·0 | 29·7 |
| 1956 | 3·0 | 16·8 | 1966 | 7·6 | 31·7 |
| 1957 | 3·1 | 17·7 | | | |
| 1958 | 3·2 | 18·6 | | | |
| 1959 | 3·5 | 19·7 | | | |

Source: Various issues of the *Monthly Digest of Statistics*. Both series are in £ thousand million.

(i) Plot saving against personal disposable income.

(ii) Estimate the marginal propensity to save, by regressing saving on personal disposable income. Does your saving function correspond to economic theory?

(iii) Calculate the residuals from your estimated saving function and plot these against time. Would you describe the plotted series as 'random'?

---

\* This question involves a considerable amount of arithmetic and the reader is advised to make use of a desk calculating machine or an electronic computer.

**8.7** If $e_i$, $i = 1, \ldots, n$, are the residuals from the least squares line

$$\hat{Y}_i = a + bX_i,$$

prove that (i) $\sum e_i = 0$, (ii) $\sum e_i x_i = 0$, and (iii) $\sum e_i \hat{y}_i = 0$.

**8.8** Suppose that our regression model does not contain a constant term, that is, $Y_i = \gamma X_i + \varepsilon_i$, $i = 1, \ldots, n$. Show that the least squares estimator of $\gamma$ is $c = \sum Y_i X_i / \sum X_i^2$. Under what conditions (if any) will $c$ be equal to the least squares estimator of $\beta$ in $Y_i = \alpha + \beta X_i + \varepsilon_i$? How would you define $R^2$ in the estimation of $Y_i = \gamma X_i + \varepsilon_i$?

**8.9** By substituting for $(a - \alpha)$ and $(b - \beta)$ in $E[(a - \alpha)(b - \beta)]$, show that

$$\mathrm{cov}\,(a, b) = -\frac{\sigma^2 \bar{X}}{\sum x_i^2}.$$

How would you interpret this result in terms of (a) geometry and (b) economics?

# chapter 9
# multiple regression analysis

## 9.1 introduction
In the last chapter we explored the simple regression model in which the dependent variable was a linear function of one independent (explanatory) variable. In this chapter we extend our analysis to deal (*a*) with more than one explanatory variable and (*b*) with non-linear relationships. We shall first extend the regression model to consider the case in which the dependent variable is a linear function of two explanatory variables. The algebra involved in this case is not too complicated and this model will be explored in some depth in the next two sections. The results obtained here will provide the basis for the discussion of the estimation of non-linear relationships that follows in section 9.4. Finally, the regression model will be generalized to include any number of explanatory variables in section 9.5. As this general treatment of the regression model involves extensive and tedious algebra, only the outline of important results will be given in the text, a rigorous treatment using matrix algebra being presented in an appendix to this chapter.

## 9.2 the regression model with two independent variables
A simple economic theory of the demand for some commodity which has no very close substitutes or complements might suggest that the quantity demanded, was a function of its price, $X_2$, and the consumers' income, $X_3$. If it is assumed that the relationship is linear, the equation may be written as[1]

$$Y_i = \beta_1 + \beta_2 X_{2i} + \beta_3 X_{3i} + \varepsilon_i, \qquad i = 1, ..., n, \qquad (9.1)$$

where the $\varepsilon$s represent the effects of other variables excluded from the relationship. Geometrically, we may represent the deterministic part of 9.1 as a plane in the three-dimensional space corresponding to $(X_2, X_3, Y)$, the observations deviating from this plane in the $Y$ dimension to the extent that the $\varepsilon$s are not equal to zero. This is illustrated in Figure 9.1.

Given $n$ observations on $Y$, $X_2$ and $X_3$, we wish to estimate the $\beta$s and we shall consider how this may be done by means of the method of least squares. Denoting the estimated parameters as $b_1$, $b_2$ and $b_3$, the fitted plane is

$$\hat{Y}_i = b_1 + b_2 X_{2i} + b_3 X_{3i},$$

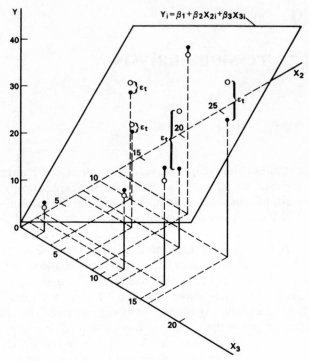

Figure 9.1

the residuals from the fitted plane are

$$e_i = Y_i - \hat{Y}_i = Y_i - b_1 - b_2 X_{2i} - b_3 X_{3i},$$

and the least squares estimators are obtained by minimizing $\sum\limits_{i=1}^{n} e_i^2$, where

$$\sum e_i^2 = \sum (Y_i - b_1 - b_2 X_{2i} - b_3 X_{3i})^2. \tag{9.2}$$

To obtain expressions for the least squares estimators, we partially differentiate 9.2 with respect to $b_1$, $b_2$ and $b_3$, set these partial derivatives equal to zero and solve the resulting three equations for $b_1$, $b_2$ and $b_3$. Specifically,

$$\frac{\partial \sum e_i^2}{\partial b_1} = -2\sum (Y_i - b_1 - b_2 X_{2i} - b_3 X_{3i}) = 0 \tag{9.3}$$

$$\frac{\partial \sum e_i^2}{\partial b_2} = -2\sum X_{2i}(Y_i - b_1 - b_2 X_{2i} - b_3 X_{3i}) = 0 \tag{9.4}$$

$$\frac{\partial \sum e_i^2}{\partial b_3} = -2\sum X_{3i}(Y_i - b_1 - b_2 X_{2i} - b_3 X_{3i}) = 0. \tag{9.5}$$

Summing from 1 to $n$, we obtain the normal equations

$$\sum Y_i = nb_1 + b_2\sum X_{2i} + b_3\sum X_{3i} \tag{9.6}$$

$$\sum X_{2i}Y_i = b_1\sum X_{2i} + b_2\sum X_{2i}^2 + b_3\sum X_{2i}X_{3i} \tag{9.7}$$

$$\sum X_{3i}Y_i = b_1\sum X_{3i} + b_2\sum X_{2i}X_{3i} + b_3\sum X_{3i}^2. \tag{9.8}$$

Solving 9.6 for $b_1$ gives

$$b_1 = \bar{Y} - b_2\bar{X}_2 - b_3\bar{X}_3 \tag{9.9}$$

and substituting this expression for $b_1$ into 9.7 and 9.8 yields

$$\sum X_{2i}Y_i = (\bar{Y} - b_2\bar{X}_2 - b_3\bar{X}_3)\sum X_{2i} + b_2\sum X_{2i}^2 + b_3\sum X_{2i}X_{3i}$$

$$\sum X_{3i}Y_i = (\bar{Y} - b_2\bar{X}_2 - b_3\bar{X}_3)\sum X_{3i} + b_2\sum X_{2i}X_{3i} + b_3\sum X_{3i}^2.$$

Rearranging the terms and using the result that $\sum X_i^2 - \bar{X}\sum X_i = \sum (X_i - \bar{X})^2$ gives

$$\sum (X_{2i} - \bar{X}_2)(Y_i - \bar{Y}) = b_2\sum (X_{2i} - \bar{X}_2)^2 + b_3\sum (X_{2i} - \bar{X}_2)(X_{3i} - \bar{X}_3)$$

$$\sum (X_{3i} - \bar{X}_3)(Y_i - \bar{Y}) = b_2\sum (X_{2i} - \bar{X}_2)(X_{3i} - \bar{X}_3) + b_3\sum (X_{3i} - \bar{X}_3)^2,$$

or, using lower-case letters to denote deviations from sample means,[2]

$$\sum x_2 y = b_2\sum x_2^2 + b_3\sum x_2 x_3 \tag{9.10}$$

$$\sum x_3 y = b_2\sum x_2 x_3 + b_3\sum x_3^2 \tag{9.11}$$

Solving 9.10 and 9.11 for $b_2$ and $b_3$, we obtain

$$b_2 = \frac{\sum x_2 y \cdot \sum x_3^2 - \sum x_2 x_3 \cdot \sum x_3 y}{\sum x_2^2 \cdot \sum x_3^2 - (\sum x_2 x_3)^2} \tag{9.12}$$

$$b_3 = \frac{\sum x_3 y \cdot \sum x_2^2 - \sum x_2 x_3 \cdot \sum x_2 y}{\sum x_2^2 \cdot \sum x_3^2 - (\sum x_2 x_3)^2} \tag{9.13}$$

In order to investigate the statistical properties of the least squares estimates given in 9.9, 9.12 and 9.13, it is necessary to make some assumptions concerning the random errors, $\varepsilon_i$, in equation 9.1. The assumptions we make here will be the same as those made in the simple linear regression model of the last chapter, with one modification. As before, we shall assume that the expected value of the $\varepsilon$s is zero, that they are drawn from a population with variance $\sigma^2$, and that they are uncorrelated among themselves, that is,

$$E(\varepsilon_i) = 0, \qquad i = 1, ..., n, \tag{8.13}$$

$$E(\varepsilon_i^2) = \sigma^2, \tag{8.14}$$

$$E(\varepsilon_i\varepsilon_j) = 0, \quad \text{when} \quad i \neq j \tag{8.15}$$

However, the assumption given in 8.16 has to be modified, since we now have two independent variables, $X_2$ and $X_3$. Our new assumption is that both $X_{2i}$ and $X_{3i}$ may be treated as constants,[3] so that

$$E(X_{2i}\varepsilon_j) = X_{2i}E(\varepsilon_j) = E(X_{3i}\varepsilon_j) = X_{3i}E(\varepsilon_j) = 0 \quad \text{for all} \quad i, j \tag{9.14}$$

by virtue of 8.13.

On the basis of these assumptions, it may be shown that the method of least squares yields estimates which are best, linear and unbiased. Given the complexity of the algebra involved, we shall not prove the minimum variance property, but shall illustrate the results by showing that $b_2$ is an unbiased estimator of $\beta_2$ and deriving its sampling variance.

To show that $b_2$ is an unbiased estimator of $\beta_2$, we first average 9.1 to obtain $\overline{Y} = \beta_1 + \beta_2\overline{X}_2 + \beta_3\overline{X}_3 + \bar{\varepsilon}$. Subtracting this expression from equation 9.1 yields

$$Y_i - \overline{Y} = \beta_2(X_{2i} - \overline{X}_2) + \beta_3(X_{3i} - \overline{X}_3) + \varepsilon_i - \bar{\varepsilon},$$

or

$$y_i = \beta_2 x_{2i} + \beta_3 x_{3i} + (\varepsilon_i - \bar{\varepsilon}).$$

Substituting for $y_i$ in 9.12,

$$b_2 = \frac{\sum x_3^2 \cdot \sum x_2[\beta_2 x_2 + \beta_3 x_3 + (\varepsilon_i - \bar{\varepsilon})]}{\sum x_2^2 \cdot \sum x_3^2 - (\sum x_2 x_3)^2} - \frac{\sum x_2 x_3 \cdot \sum x_3[\beta_2 x_2 + \beta_3 x_3 + (\varepsilon_i - \bar{\varepsilon})]}{\sum x_2^2 \cdot \sum x_3^2 - (\sum x_2 x_3)^2}.$$

Rearranging terms and noting $\sum x_j\bar{\varepsilon} = \bar{\varepsilon}\sum x_j = 0$, since $\sum x_j = 0, j = 2, 3,$

$$b_2 = \frac{\beta_2[\sum x_2^2 \cdot \sum x_3^2 - (\sum x_2 x_3)^2] + \beta_3[\sum x_3^2 \cdot \sum x_2 x_3 - \sum x_3^2 \cdot \sum x_2 x_3]}{\sum x_2^2 \cdot \sum x_3^2 - (\sum x_2 x_3)^2}$$

$$+ \frac{\sum x_3^2 \cdot \sum x_2\varepsilon_i - \sum x_2 x_3 \cdot \sum x_3\varepsilon_i}{\sum x_2^2 \cdot \sum x_3^2 - (\sum x_2 x_3)^2}$$

$$= \beta_2 + \frac{\sum x_3^2 \cdot \sum x_2\varepsilon_i - \sum x_2 x_3 \cdot \sum x_3\varepsilon_i}{\sum x_2^2 \cdot \sum x_3^2 - (\sum x_2 x_3)^2}. \tag{9.15}$$

Taking expected values, remembering that the $X_2$s and $X_3$s are constants,

$$E(b_2) = \beta_2 + \frac{\sum x_3^2 \cdot \sum x_2 E(\varepsilon_i) - \sum x_2 x_3 \cdot \sum x_3 E(\varepsilon_i)}{\sum x_2^2 \cdot \sum x_3^2 - (\sum x_2 x_3)^2}$$

$$= \beta_2,$$

since $E(\varepsilon_i) = 0$. Thus $b_2$ is an unbiased estimator of $\beta_2$, and given the symmetry between 9.12 and 9.13, it obviously follows that $b_3$ is an unbiased estimator of $\beta_3$. It can also be shown that $b_1$ is an unbiased estimator of $\beta_1$.

To derive the sampling variance of $b_2$, we rewrite 9.15 as

$$b_2 - \beta_2 = \frac{\sum x_3^2 \cdot \sum x_2 \varepsilon_i - \sum x_2 x_3 \cdot \sum x_3 \varepsilon_i}{\sum x_2^2 \cdot \sum x_3^2 - (\sum x_2 x_3)^2}.$$

Then the sampling variance of $b_2$ is given by

$$E[(b_2 - \beta_2)^2] = E\left[ \frac{\sum x_3^2 \cdot \sum x_2 \varepsilon_i - \sum x_2 x_3 \cdot \sum x_3 \varepsilon_i}{\sum x_2^2 \cdot \sum x_3^2 - (\sum x_2 x_3)^2} \right]^2. \qquad (9.16)$$

Considering the numerator of the term on the right-hand side of 9.16, multiplying out and collecting like and unlike terms, we have

$$[\sum x_3^2(x_{21}\varepsilon_1 + \ldots + x_{2n}\varepsilon_n) - \sum x_2 x_3(x_{31}\varepsilon_1 + \ldots + x_{3n}\varepsilon_n)]^2$$

$$= (\sum x_3^2)^2 \left[ \sum x_{2i}^2 \varepsilon_i^2 + \sum_{i \neq j} x_{2i} x_{2j} \varepsilon_i \varepsilon_j \right]$$

$$- 2\sum x_3^2 \cdot \sum x_2 x_3 \left[ \sum x_{2i} x_{3i} \varepsilon_i^2 + \sum_{i \neq j} x_{2i} x_{3j} \varepsilon_i \varepsilon_j \right]$$

$$+ (\sum x_2 x_3)^2 \left[ \sum x_{3i}^2 \varepsilon_i^2 + \sum_{i \neq j} x_{3i} x_{3j} \varepsilon_i \varepsilon_j \right]. \qquad (9.17)$$

Taking the expected values of 9.17 and using $E(\varepsilon_i^2) = \sigma^2$ and $E(\varepsilon_i \varepsilon_j) = 0$ gives

$$E(\sum x_3^2 \cdot \sum x_2 \varepsilon_i - \sum x_2 x_3 \cdot \sum x_3 \varepsilon_i)^2 = \sigma^2[(\sum x_3^2)^2 \cdot \sum x_2^2 - 2\sum x_3^2(\sum x_2 x_3)^2$$

$$+ \sum x_3^2(\sum x_2 x_3)^2]$$

$$= \sigma^2 \sum x_3^2[\sum x_2^2 \cdot \sum x_3^2 - (\sum x_2 x_3)^2].$$

Substituting this result into 9.16, we have

$$E[(b_2 - \beta_2)^2] = \frac{\sigma^2 \sum x_3^2 [\sum x_2^2 \sum x_3^2 - (\sum x_2 x_3)^2]}{[\sum x_2^2 \cdot \sum x_3^2 - (\sum x_2 x_3)^2]^2}$$

$$= \frac{\sigma^2 \sum x_3^2}{\sum x_2^2 \cdot \sum x_3^2 - (\sum x_2 x_3)^2} . \tag{9.18}$$

By symmetry,

$$E[(b_3 - \beta_3)^2] = \frac{\sigma^2 \sum x_2^2}{\sum x_2^2 \cdot \sum x_3^2 - (\sum x_2 x_3)^2} \tag{9.19}$$

and it may also be shown that

$$E[(b_1 - \beta_1)^2] = \frac{\sigma^2 [\sum X_2^2 \cdot \sum X_3^2 - (\sum X_2 X_3)^2]}{n[\sum x_2^2 \cdot \sum x_3^2 - (\sum x_2 x_3)^2]} . \tag{9.20}$$

For some of the tests to be discussed below it is necessary to know the covariance between $b_2$ and $b_3$. By substituting for $(b_2 - \beta_2)$ and $(b_3 - \beta_3)$ in $E(b_2 - \beta_2) \cdot (b_3 - \beta_3)]$, it may be shown that

$$\text{cov}(b_2, b_3) = - \frac{\sigma^2 \sum x_2 x_3}{\sum x_2^2 \cdot \sum x_3^2 - (\sum x_2 x_3)^2} . \tag{9.21}$$

We shall refer to this result in section 9.4 below.

Further results extend easily from those obtained in chapter 8. For example, if it is assumed that the εs are normally distributed, then the least squares estimators will have normal sampling distributions, whose means will be the βs being estimated and whose variances are given by the formulae in equations 9.18 to 9.20. This is an important theoretical result, but not one which can be exploited operationally, since the sampling variances are all functions of $\sigma^2$, the variance of the εs and generally $\sigma^2$ is unknown.

However, we may define an unbiased estimator of $\sigma^2$ as

$$s^2 = \frac{\sum e_i^2}{(n - 3)} , \tag{9.22}$$

where $\sum e_i^2$ is the sum of the squared residuals from the fitted relationship. The standard errors of the least squares estimates may then be calculated by substituting $s^2$ for the unknown $\sigma^2$ in 9.18 to 9.20, so that, for example,

$$\text{s.e.}(b_2) = \sqrt{\frac{s^2 \sum x_3^2}{\sum x_2^2 \cdot x_3^2 - (\sum x_2 x_3)^2}} . \tag{9.23}$$

Then it may be shown that the ratio

$$t = \frac{b_j - \beta_j}{\text{s.e.}(b_j)}, \qquad j = 1, 2, 3 \tag{9.24}$$

follows the $t$-distribution with $(n - 3)$ degrees of freedom.[4]

As a numerical example, suppose that our regression of the quantity of a commodity demanded, $Y$, on the price of the commodity, $X_2$, and consumers' income, $X_3$, gives the following estimated relationship,

$$Y_i = 14 \cdot 13 - 0 \cdot 60 X_{2i} + 2 \cdot 14 X_{3i},$$
$$\quad (1.31) \quad (0.15) \quad (0.80)$$

where the figures in parentheses are the standard errors of the parameter estimates and the estimates are based on 21 observations. If we wish to test the hypothesis $\beta_j = 0$, the $t$-ratio defined in 9.24 becomes

$$t = \frac{b_j}{\text{s.e.}(b_j)}. \tag{9.25}$$

A one-tail test is appropriate since we would expect $\beta_2 < 0$ and $\beta_3 > 0$, and if we choose $p = 0 \cdot 95$ as the probability level for our significance test, we shall compare our calculated $t$-ratio with the tabulated $t$-values for 18 degrees of freedom (that is, $t = -1 \cdot 73$ for testing $\beta_2 = 0$ and $t = 1 \cdot 73$ for testing $\beta_3 = 0$). For our estimated demand equation we have $t_1 = 14 \cdot 13/1 \cdot 31 = 10 \cdot 79$, $t_2 = -4 \cdot 00$ and $t_3 = 2 \cdot 67$, and on the basis of these $t$-ratios we may reject the hypothesis that $\beta_j = 0$, $j = 2, 3$. For the constant term in the equation the expected sign of $\beta_1$ is not obvious from economic theory, so that a two-tail test is appropriate if we wish to test the hypothesis $\beta_1 = 0$. With our chosen probability level and 18 degrees of freedom, we find from Table A2 that we would then accept the hypothesis that $\beta_1 = 0$ if $b_1$ fell in the range $t = \pm 2 \cdot 10$. However, since economic theory has little or nothing to say about the constant term in a relationship, this test is usually not very interesting as compared with testing hypotheses concerning the slope coefficients.

*A measure of goodness of fit*
In the last chapter, in equation 8.44, we defined the coefficient of determination as a measure of the proportion of variation in the dependent variable which was explained by the behaviour of the independent variable. This measure extends easily to the new regression model, and we may define the coefficient of determination as

$$R^2 = 1 - \frac{\sum e_i^2}{\sum y_i^2}, \tag{9.26}$$

where $\sum e_i^2$ is the sum of the squared residuals, defined in equation 9.2. Substituting 9.9 for $b_1$ in 9.2 and expressing the variables in terms of deviations from the sample means, we obtain

$$\sum e_i^2 = \sum (y_i - b_2 x_{2i} - b_3 x_{3i})^2$$

$$= \sum e_i(y_i - b_2 x_{2i} - b_3 x_{3i})$$

$$= \sum e_i y_i - b_2 \sum e_i x_{2i} - b_3 \sum e_i x_{3i}.$$

Now $\sum e_i x_{2i}$ and $\sum e_i x_{3i}$ were set equal to zero in 9.4 and 9.5 in deriving the formulae for $b_2$ and $b_3$, so that

$$\sum e_i^2 = \sum e_i y_i.$$

Substituting for $e_i$ in this expression yields

$$\sum e_i^2 = \sum y_i(y_i - b_2 x_{2i} - b_3 x_{3i})$$

$$= \sum y_i^2 - b_2 \sum x_{2i} y_i - b_3 \sum x_{3i} y_i. \tag{9.27}$$

Equation 9.27 represents a convenient formula for calculating $\sum e_i^2$, and in addition the terms may be rearranged to give

$$\sum y_i^2 = \underbrace{b_2 \sum x_{2i} y_i + b_3 \sum x_{3i} y_i} + \sum e_i^2. \tag{9.28}$$

| total = | explained | + residual |
|---------|-----------|------------|
| variation | variation | variation |

Equation 9.28 may be expressed in several other forms. We first note that averaging $Y_i = b_1 + b_2 X_{2i} + b_3 X_{3i} + e_i$ yields

$$\bar{Y} = b_1 + b_2 \bar{X}_2 + b_3 \bar{X}_3 + \bar{e}$$

$$= b_1 + b_2 \bar{X}_2 + b_3 \bar{X}_3$$

$$= \hat{\bar{Y}},$$

since $\bar{e} = 0$.

Thus

$$Y_i - \bar{Y} = \hat{Y}_i - \bar{\hat{Y}} + e_i$$

or

$$y_i = \hat{y}_i + e_i.$$

Substituting this expression in 9.28, we obtain

$$\sum y_i^2 = \sum \hat{y}_i^2 + \sum e_i^2$$

or, since $\hat{y}_i = b_2 x_{2i} + b_3 x_{3i}$,

$$\sum y_i^2 = \sum (b_2 x_{2i} + b_3 x_{3i})^2 + \sum e_i^2. \tag{9.29}$$

The coefficient of determination may also be expressed in alternative ways, so that

$$R^2 = \frac{\sum \hat{y}_i^2}{\sum y_i^2}$$

$$= \frac{b_2 \sum x_{2i} y_i + b_3 \sum x_{3i} y_i}{\sum y_i^2} \tag{9.30}$$

by substituting for $y_i$ from 9.28, or alternatively, using 9.29,

$$= \frac{\sum (b_2 x_{2i} + b_3 x_{3i})^2}{\sum y_i^2}. \tag{9.31}$$

Equation 9.31 enables us to examine one property of the coefficient of determination which arises when we move from simple to multiple regression. Expanding the squared term in the numerator of 9.31, we have

$$R^2 = \frac{b_2^2 \sum x_{2i}^2 + b_3^2 \sum x_{3i}^2 + 2b_2 b_3 \sum x_{2i} x_{3i}}{\sum y_i^2}. \tag{9.32}$$

Looking at the numerator in 9.32, it is clear that in general we cannot think of $R^2$ as being made up of the sum of the separate contributions of $X_2$ and $X_3$ to explaining the variation in $Y$, since the expression $2b_2 b_3 \sum x_{2i} x_{3i}$ cannot be allocated either to variation in $X_2$ alone or in $X_3$ alone; it represents the interaction between $X_2$ and $X_3$. The exception to this result is the case in which $X_2$ and $X_3$ are uncorrelated in the sample, so that $\sum x_{2i} x_{3i} = 0$. In this case $X_2$ and $X_3$ are said to be *orthogonal* and 9.32 reduces to

$$R^2 = \frac{b_2^2 \sum x_2^2 + b_3^2 \sum x_3^2}{\sum y_i^2}. \tag{9.33}$$

However, there are further implications for the assumption that $\sum x_2 x_3 = 0$. For example, under this assumption, equations 9.12 and 9.13 become

$$b_2 = \frac{\sum y_i x_{2i}}{\sum x_{2i}^2} \quad \text{and} \quad b_3 = \frac{\sum y_i x_{3i}}{\sum x_{3i}^2},$$

which the reader will recognize as the least squares estimators in the simple regressions of $Y$ on $X_2$ and $Y$ on $X_3$ respectively. Thus when $X_2$ and $X_3$ are orthogonal, the coefficient of determination defined in 9.32 is the sum of the variation in $Y$ explained by its regression on $X_2$ plus the variation in $Y$ explained by regressing $Y$ on $X_3$.

Referring back to the assumptions that were made concerning the statistical properties of the regression model with two explanatory variables, the reader will recall that we made no assumptions at all concerning the relationship between $X_2$ and $X_3$. Orthogonality is obviously a special case and, while it makes for easy identification of the separate effects of $X_2$ and $X_3$ on $Y$, is not necessary for the model. In experimental situations where one can choose and control the values of the explanatory variables, one would aim at orthogonality by randomizing the $(X_2, X_3)$ combinations. However, in economics, especially when we are dealing with time series data, we shall not be faced with orthogonal variables, but shall tend to meet variables which are not uncorrelated enough.[5]

In the last chapter we developed a statistical test for the goodness of fit which involved the $F$-distribution. This test may also be extended in the context of the multiple regression model. Given the assumption that the $\varepsilon_i$ are normally distributed, the breakdown of the variation in $Y$ into the explained and residual components as represented by 9.28 and 9.29 involves squared normal variables, as before. The only differences are the formulae and the number of the degrees of freedom for the chi-squared variables involved. In this case we may set up our test for the goodness of fit of the regression of $Y$ on $X_2$ and $X_3$ as is shown in Table 9.1.

The ratio computed in the last column of Table 9.1 is the ratio of two chi-squared variables and follows the $F$-distribution with 2 and $n-3$ degrees of freedom. This forms the basis for our test that $R^2 = 0$.

*Table 9.1.* Partition of variation in the multiple regression model

| Source of variation | Sum of squares | Degrees of freedom | Mean sum of squares | $F$ |
|---|---|---|---|---|
| Explained | $\sum (b_2 x_{2i} + b_3 x_{3i})^2$ | 2 | $\sum (b_2 x_{2i} + b_3 x_{3i})^2 = s_1^2$ | $F = \frac{s_1^2}{s_2^2}$ |
| Residual | $\sum e_i^2$ | $n-3$ | $\sum e_i^2/(n-3) = s_2^2$ | |
| Total | $\sum y_i^2$ | $n-1$ | — | |

In the simple regression model which was explored in the last chapter, the $F$-test of the hypothesis that $R^2 = 0$ was equivalent to the $t$-test of the hypothesis that $\beta = 0$. This equivalence does not carry over to the multiple regression model; a statistically significant result to the $F$-test for overall goodness of fit (that is, $R^2 \neq 0$) does not imply obtaining statistically significant results for *both* the two $t$-tests involved in testing $\beta_2 = 0$ and $\beta_3 = 0$. If we reject the hypothesis that $R^2 = 0$, then there are three possible outcomes to our $t$-tests, namely (i) $\beta_2 \neq 0$, $\beta_3 \neq 0$, (ii) $\beta_2 \neq 0$, $\beta_3 = 0$, or (iii) $\beta_2 = 0$, $\beta_3 \neq 0$. The other possibility that $\beta_2 = 0$ and $\beta_3 = 0$ can be ruled out on logical grounds: if the $F$-test shows that $R^2 \neq 0$, then our regression model must be explaining some of the variation in $Y$, which is not possible if both $\beta_2 = 0$ and $\beta_3 = 0$.

While we may test $\beta_2 = 0$ or $\beta_3 = 0$ by means of a $t$-test, it is possible to develop an equivalent $F$-test by modifying the partitioning given in Table 9.1. There we have the variation explained by $X_2$ and $X_3$ is given by $\sum (b_2 x_{2i} + b_3 x_{3i})^2$. We may isolate the variation in $Y$ explained by $X_2$ by estimating the regression of $Y$ on $X_2$, that is, if we estimate $\hat{Y}_i = a + bX_{2i}$, the variation explained by $X_2$ is $b^2 \sum x_{2i}^2$. The variation explained by the addition of $X_3$ to the regression analysis may then be obtained by subtraction. The new partitioning is presented in Table 9.2. The ratio calculated in the last column of Table 9.2 follows the $F$-distribution with 1 and $n - 3$ degrees of freedom and enables us to test whether the addition of $X_3$ contributes significantly to explaining the variation in $Y$.

| Source of variation | Sum of squares | Degrees of freedom | Mean sum of squares | $F$ |
|---|---|---|---|---|
| Explained by $X_2$ | $b^2 \sum x_{2i}^2$ | 1 | | |
| Addition of $X_3$ | $[\sum(b_2 x_{2i} + b_3 x_{3i})^2 - b^2 \sum x_{2i}^2]$ | 1 | $[\sum(b_2 x_{2i} + b_3 x_{3i})^2 - b^2 \sum x_{2i}^2] = s_3^2$ | |
| Explained by $X_2$ and $X_3$ | $\sum(b_2 x_{2i} + b_3 x_{3i})^2$ | 2 | | $F = \dfrac{s_3^2}{s_2^2}$ |
| Residual | $\sum e_i^2$ | $n - 3$ | $\sum e_i^2 (n - 3) = s_2^2$ | |
| Total | $\sum y_i^2$ | $n - 1$ | — | |

As an example to illustrate the interpretation of these test procedures, we consider the following data for a selection of twenty-eight United Kingdom industries.[6] Here $Y_i$ = total employment, $X_{2i}$ = gross output and $X_{3i}$ = output per head. The variables measure the change between 1924 and 1950 expressed on the base 1924 = 100. To examine the effects of changes in output and productivity on total employment, the following regressions were estimated:

(1)    $Y_i = 61 \cdot 2 + 0 \cdot 277 X_{2i}$              $R^2 = 0 \cdot 861$
               $(7 \cdot 0)$   $(0 \cdot 022)$

(2)    $Y_i = 17 \cdot 6 + 0 \cdot 651 X_{3i}$              $R^2 = 0 \cdot 388$
               $(29 \cdot 0)$   $(0 \cdot 161)$

(3)    $Y_i = 111 \cdot 6 + 0 \cdot 382 X_{2i} - 0 \cdot 445 X_{3i}$    $R^2 = 0 \cdot 920$
               $(12 \cdot 9)$   $(0 \cdot 030)$    $(0 \cdot 103)$

the figures in parentheses being the standard errors of the estimated coefficients.

Before we look at the numerical results, we should first ask ourselves what algebraic signs we might expect on the basis of economic theory. The answer here is that these equations do not correspond to any clear-cut economic theory and so it is difficult to say what signs we would expect. For example, we might well expect changes in total employment to be positively related to changes in output *ceteris paribus*, but changes in technology and any development towards more capital-intensive methods of production might reverse the sign of this relationship. The sign in the second case is even more difficult to decide upon. For example, higher labour productivity might lead to an increased demand for labour and hence a positive sign. Alternatively, if output were unchanging or not changing very much and there was not much scope for substituting labour for other inputs of production, increasing labour productivity might lead to a decrease in the demand for labour and thus a negative sign. No doubt the reader can think of other possibilities. Ideally, the regression equation should be based on some economic theory or be concerned with exploring some economic hypothesis, but on occasion, as illustrated by our current example, the equations being reported are somewhat arbitrary and should therefore be scrutinized with suspicion.

Looking at the two simple regression equations, we find a significant positive relationship between total employment and gross output ($t = 0 \cdot 277/0 \cdot 022 = 12 \cdot 6$) and between total employment and productivity ($t = 0 \cdot 651/0 \cdot 161 = 4 \cdot 04$), although the 'significance' is less in the second case than in the first—a result which is also reflected in the relative magnitudes of the two coefficients of determination. Thus if we had to choose between these two equations, we would choose the regression of total employment on output on the basis of goodness of fit.

Turning now to the multiple regression equation, we note that combining both variables increases the coefficient of determination and that both of the estimated parameters are significant ($t_2 = 0 \cdot 382/0 \cdot 030 = 12 \cdot 7$ and $t_3 = -0 \cdot 445/0 \cdot 103 = -4 \cdot 3$). The appropriate partition of the variation to be explained in terms of tests involving the $F$-distribution is given in Table 9.3. If we choose a probability level of $p = 0 \cdot 95$ for our signifance tests, then from Table A4 we find that when we have 1 and 25 degrees of freedom the critical value of $F$ is $4 \cdot 24$, while

*Table 9.3.*

| Source of variation | Sum of squares | Degrees of freedom | Mean sum of squares | $F$ |
|---|---|---|---|---|
| Explained by $X_2$ | 96,084 | 1 | 96,084 | |
| Addition of $X_3$ | 6,588 | 1 | 6,588 $(= s_3^2)$ | |
| Explained by $X_2$ and $X_3$ | 102,672 | 2 | 51,336 | $F = \frac{s_3^2}{s_1^2} = 18\cdot5$ |
| Residual | 8,904 | 25 | 356.1 $(= s_1^2)$ | |
| Total | 111,576 | 27 | — | |

if we have 2 and 25 degrees of freedom the critical value of $F$ is $3\cdot39$. The overall significance of the regression of $Y$ on $X_2$ and $X_3$ may be tested by comparing $F = 51{,}336/356\cdot1 = 144$ with $F = 3\cdot39$. The regression is clearly significant. To test whether the addition of $X_3$ significantly increases the explanatory power of the relationship, we compute $F = 18\cdot5$. Since this value is greater than $4\cdot24$, we conclude that $X_3$ does add significantly to the explanation of $Y$. We may also note that $[b_3/\text{s.e.}(b_3)]^2 = (-4\cdot3)^2 = 18\cdot5 = F$, showing the equivalence of the two tests.

Returning to the multiple regression, we may note that the coefficient on $X_3$ is negative, whereas the simple regression of total employment on productivity was positive. We have already noted that there is no economic theory supporting this equation, but one interpretation which could explain the multiple regression result is that the estimated coefficients can be treated in a *ceteris paribus* way, so that the coefficient on $X_3$ measures the relationship between total employment and productivity, having allowed for changes in output. Given this interpretation, we would expect a negative sign for the relationship between employment and productivity. The possibility of and justification for a *ceteris paribus* interpretation of multiple regression coefficients will be explored in the next section.

### 9.3 partial regression and correlation

The material to be discussed in this section will involve relating the estimates of parameters in multiple regression equations to those in a number of simple regression equations. To simplify this process, it will be convenient to change our parameter notation in a way which makes clear not only which variable is being regressed on which, but also what other variables are included in the equation. We shall write the simple regressions of $Y$ on $X_2$, $Y$ on $X_3$, $X_2$ on $X_3$ and the multiple regression of $Y$ on $X_2$ and $X_3$ as

$$Y_i = \beta_{1.2} + \beta_{12}X_{2i} + \varepsilon_{(1.2)i} \tag{9.34}$$

$$Y_i = \beta_{1.3} + \beta_{13}X_{3i} + \varepsilon_{(1.3)i} \tag{9.35}$$

$$X_{2i} = \beta_{2.3} + \beta_{23}X_{3i} + \varepsilon_{(2.3)i} \tag{9.36}$$

$$Y_i = \beta_{1.23} + \beta_{12.3}X_{2i} + \beta_{13.2}X_{3i} + \varepsilon_{(1.23)}, \tag{9.37}$$

where the first subscript number refers to the variable to be explained, the remaining subscripts refer to the explanatory variables and the full stop is used to separate the explanatory variable being considered from other explanatory variables in the model. For example, $\beta_{12.3}$ is the coefficient on $X_2$ when $X_3$ also appears in the regression, while $\beta_{1.23}$ is the constant term when $X_2$ and $X_3$ also appear in the regression. We shall denote the least squares estimates of these parameters as $b_{12.3}, b_{1.23}$, etc.

The first result, which we merely note, is that the least squares estimates of the parameters in a multiple regression equation can be expressed as functions of the least squares estimates of parameters in simple regression equations. From 9.12 we have

$$b_{12.3} = \frac{\sum x_2 y \sum x_3^2 - \sum x_2 x_3 \cdot \sum x_3 y}{\sum x_2^2 \cdot \sum x_3^2 - (\sum x_2 x_3)^2},$$

and dividing both numerator and denominator by $\sum x_2^2 \cdot \sum x_3^2$ gives

$$b_{12.3} = \frac{\dfrac{\sum x_2 y}{\sum x_2^2} - \left(\dfrac{\sum x_2 x_3}{\sum x_2^2}\right)\left(\dfrac{\sum x_3 y}{\sum x_3^2}\right)}{1 - \left(\dfrac{\sum x_2 x_3}{\sum x_2^2}\right)\left(\dfrac{\sum x_2 x_3}{\sum x_3^2}\right)},$$

that is,

$$b_{12.3} = \frac{b_{12} - b_{13}b_{32}}{1 - b_{23}b_{32}}. \tag{9.38}$$

By symmetry, it follows that

$$b_{13.2} = \frac{b_{13} - b_{12}b_{23}}{1 - b_{23}b_{32}}. \tag{9.39}$$

A second and more important relationship between the estimates in the simple regressions and those in the multiple regression may be derived as follows. Suppose we regress $Y$ on $X_2$ and, by the method of least squares, estimate $\hat{Y}_i = b_{1.3} + b_{13}X_{2i}$, with residuals $e_{(1.3)i}$. Suppose also we regress $X_2$ on $X_3$, estimating $X_{2i} = b_{2.3} + b_{23}X_{3i}$, with residuals $e_{(2.3)i}$. Now $e_{(1.3)i}$ represents the variation in $Y$ which is not explained by the regression of $Y$ on $X_3$ and $e_{(2.3)i}$ represents the variation in $X_2$ not explained by the regression of $X_2$ on $X_3$. If we now regress

$e_{(1.3)i}$ on $e_{(2.3)i}$ and estimate the coefficient[7] in

$$e_{(1.3)i} = \beta_{(1.3)(2.3)} e_{(2.3)i} + \varepsilon_{(1.3)(2.3)i},$$

then we can prove that $b_{(1.3)(2.3)} = b_{12.3}$.

To establish this result, we may express the residuals in terms of deviations from the sample means as $e_{(1.3)i} = y_i - b_{13}x_{3i}$ and $e_{(2.3)i} = x_{2i} - b_{23}x_{3i}$. The least squares estimator of $\beta_{(1.2)(2.3)}$ is

$$b_{(1.3)(2.3)} = \frac{\sum e_{(1.3)i} e_{(2.3)i}}{\sum e^2_{(2.3)i}},$$

and substituting for $e_{(1.3)i}$ and $e_{(2.3)i}$ we obtain

$$b_{(1.3)(2.3)} = \frac{\sum (y_i - b_{13}x_{3i})(x_{2i} - b_{23}x_{3i})}{\sum (x_{2i} - b_{23}x_{3i})^2}$$

$$= \frac{\sum (y_i x_{2i} - b_{23} y_i x_{3i} - b_{13}x_{2i}x_{3i} + b_{13}b_{23}x^2_{3i})}{\sum (x^2_{2i} - 2b_{23}x_{2i}x_{3i} + b^2_{23}x^2_{3i})}.$$

Now $b_{13} = \sum y_i x_{3i} / \sum x^2_{3i}$ and $b_{23} = \sum x_{2i}x_{3i} / \sum x^2_{3i}$, so that substituting for $b_{13}$ and $b_{23}$ we obtain

$$b_{(1.3)(2.3)} = \frac{\sum x_2 y \sum x_3^2 - \sum x_2 x_3 \sum x_3 y}{\sum x_2^2 \sum x_3^2 - (\sum x_2 x_3)^2}$$

$$= b_{12.3},$$

as the reader will see by referring back to equation 9.12. By symmetry, it should be obvious that $b_{13.2} = b_{(1.2)(3.2)}$. That is, the estimate of the multiple regression coefficient on $X_3$ when $Y$ is regressed on $X_2$ and $X_3$ is equal to the estimated coefficient obtained by regressing the residuals from the regression of $Y$ on $X_2$ on the residuals obtained from the regression of $X_3$ on $X_2$.

The least squares estimate $b_{13.2}$ thus allows for the *linear* effects of $X_2$ on both $Y$ and $X_3$, so that we may give a *ceteris paribus* interpretation to estimated coefficients. To return to our economic example, $b_{13.2}$ measures the effects of changes in productivity on employment, having allowed for the linear effects of changes in output on both employment and productivity. Given the obvious analogy with multivariate calculus, the coefficients in the multiple regression model are often referred to as *partial* regression coefficients.

*Partial correlation*

Having obtained two sets of least squares residuals by regressing $Y$ on $X_3$ and $X_2$ on $X_3$, we then considered regressing one set of residuals on the other. A further possibility would be to calculate the correlation between the two sets of residuals. Before developing this operation algebraically, we shall consider the interpretation of this new correlation coefficient. The residuals $e_{(1.3)i}$ represent the variation in $Y$ which is not explained by the linear effects of $X_3$, while $e_{(2.3)i}$ represents the variation in $X_2$ not explained by the linear effects of $X_3$. Hence the correlation between the two sets of residuals measures the linear association between $Y$ and $X_2$ having allowed for the linear effects of $X_3$ on both of them. This correlation coefficient is called the *partial* correlation coefficient.

In practice we do not need to run the simple regressions and correlate the residuals, as the partial correlation coefficient may be expressed in terms of the coefficients of correlation between $Y$ and $X_2$, $Y$ and $X_3$ and $X_2$ and $X_3$. We shall introduce the following subscript notation: let the simple correlation coefficients between $Y$ and $X_2$ be $r_{12}$, between $X_2$ and $X_3$ be $r_{23}$, etc. The coefficient of partial correlation between $Y$ and $X_2$ given the linear effects of $X_3$ we shall denote as $r_{12.3}$, with the full stop being used to separate the variables being correlated from the variable whose linear effects are being allowed for. The partial correlation coefficient may be expressed in terms of the residuals as

$$r_{12.3} = \frac{\sum e_{(1.3)i}e_{(2.3)i}}{\sqrt{\sum e_{(1.3)i}^2 \cdot \sum e_{(2.3)i}^2}}$$

$$= \frac{\sum (y - b_{13}x_3)(x_2 - b_{23}x_3)}{\sqrt{\sum (y - b_{13}x_3)^2 \sum (x_2 - b_{23}x_3)^2}},$$

and substituting for $b_{13}$ and $b_{23}$ we have

$$= \frac{\sum x_2 y - (\sum x_2 x_3 \sum x_3 y)/\sum x_3^2}{\sqrt{[\sum y^2 - (\sum x_3 y)^2/\sum x_3^2][\sum x_2^2 - (\sum x_2 x_3)^2/\sum x_3^2]}}$$

$$= \frac{\left(\dfrac{\sum x_2 y}{\sqrt{\sum x_2^2 \sum y^2}}\right) - \left(\dfrac{\sum x_2 x_3}{\sqrt{\sum x_2^2 \sum x_3^2}}\right)\left(\dfrac{\sum x_3 y}{\sqrt{\sum x_3^2 \sum y^2}}\right)}{\sqrt{\left[1 - \left(\dfrac{\sum x_3 y}{\sqrt{\sum x_3^2 \sum y^2}}\right)^2\right]\left[1 - \left(\dfrac{\sum x_2 x_3}{\sqrt{\sum x_2^2 \sum x_3^2}}\right)^2\right]}}.$$

Now each of the expressions in parentheses defines a simple correlation coefficient, so that we may write

$$r_{12.3} = \frac{r_{12} - r_{13}r_{23}}{\sqrt{(1 - r_{13}^2)(1 - r_{23}^2)}}. \tag{9.40}$$

As an example, we may return to the relationship between total employment, $Y$, output, $X_2$, and productivity, $X_3$. Here the simple correlation coefficients are $r_{12} = 0.93$, $r_{13} = 0.61$ and $r_{23} = 0.81$. Using these values, we may calculate the partial correlation between employment and productivity, given the linear effects of changes in output. Here, by symmetry with 9.40, we have

$$r_{13.2} = \frac{r_{13} - r_{12}r_{32}}{\sqrt{(1 - r_{12}^2)(1 - r_{32}^2)}}$$

$$= \frac{0.61 - (0.93)(0.81)}{\sqrt{[1 - (0.93)^2][1 - (0.81)^2]}}$$

$$= -\frac{0.143}{0.216} = -0.66. \tag{9.41}$$

The simple coefficient of correlation between employment and productivity is 0.61. Having allowed for the linear effects of changes in output on both variables, the partial correlation between them is $-0.66$, the negative sign agreeing with the *ceteris paribus* argument suggested above.[8] We may also note the perfectly general result that the algebraic sign of $b_{ij.k}$ and that of $r_{ij.k}$ are the same.

In our discussion of the simple correlation coefficient we noted the difficulty of deciding whether a high correlation indicated a genuine linear association between two variables or whether it reflected the common influence of some third variable. Clearly, calculating the partial correlation coefficient may help us towards discriminating between these two cases, since we are able to remove the linear effects of the third variable and then see what correlation exists between the other two variables. However, it must be stressed again that association does not prove causation, so that a high partial correlation still has to be interpreted with great care.

One further connection between $r_{ij.k}$ and $b_{ij.k}$ may be noted. In the last chapter we showed that algebraically the square of the simple correlation coefficient was equal to the coefficient of determination and thus measured the proportion of the variation in the dependent variable explained by the regression line. In the same way it may be shown that $r_{12.3}^2$ measures the proportion of the variation in $Y$ not explained by $X_3$ alone, which is explained by adding $X_2$ to the relationship.[9]

### 9.4 the estimation of non-linear relationships
Non-linear relationships arise frequently in economics. As examples we may take:
(a)   A demand curve with unit elasticity may be written as

$$Q_i = \beta/P_i,$$

where $Q$ represents the quantity demanded and $P$ the price.

(b) The conventional textbook representation of the average cost curve is a $U$-shaped function of the quantity produced,

$$C_i = \alpha + \beta Q_i + \gamma Q_i^2.$$

(c) The Cobb–Douglas production function relates output, $Q_i$, to the inputs of capital, $K_i$, and labour, $L_i$, according to

$$Q_i = \gamma K_i^\alpha L_i^\beta.$$

To see what difference non-linearity makes to the process of estimation, we may take these three cases in order. For the demand function we assume that the other variables which influence the quantity demanded but are excluded from the equation are represented by the error term, so that

$$Q_i = \beta \left( \frac{1}{P_i} \right) + \varepsilon_i.$$

The least squares estimator of $\beta$ is $b$ in

$$Q_i = b \left( \frac{1}{P_i} \right) + e_i,$$

where $b$ is determined by minimizing $\sum e_i^2$. Then

$$\sum e_i^2 = \sum \left[ Q_i - b \left( \frac{1}{P_i} \right) \right]^2$$

and differentiating $\sum e_i^2$ with respect to $b$ gives

$$\frac{d \sum e_i^2}{db} = -2 \sum \frac{1}{P_i} \left[ Q_i - b \left( \frac{1}{P_i} \right) \right] = 0,$$

so that

$$b \sum \left( \frac{1}{P_i} \right)^2 = \sum \frac{Q_i}{P_i}$$

and

$$b = \frac{\sum (Q_i/P_i)}{\sum (1/P_i)^2}.$$

In this case the least squares estimate of $\beta$ may be defined and so the non-linearity of the function does not cause any estimation problem.

For the average cost function we have

$$C_i = \alpha + \beta Q_i + \gamma Q_i^2 + \varepsilon_i$$

and the least squares estimates are obtained from

$$\frac{\partial \sum e_i^2}{\partial a} = -2 \sum (C_i - a - bQ_i - cQ_i^2) = 0$$

$$\frac{\partial \sum e_i^2}{\partial b} = -2 \sum Q_i (C_i - a - bQ_i - cQ_i^2) = 0$$

$$\frac{\partial \sum e_i^2}{\partial c} = -2 \sum Q_i^2 (C_i - a - bQ_i - cQ_i^2) = 0.$$

These partial derivatives yield the normal equations

$$\sum C_i = na + b \sum Q_i + c \sum Q_i^2 \qquad (9.42)$$

$$\sum C_i Q_i = a \sum Q_i + b \sum Q_i^2 + c \sum Q_i^3 \qquad (9.43)$$

$$\sum C_i Q_i^2 = a \sum Q_i^2 + b \sum Q_i^3 + c \sum Q_i^4, \qquad (9.44)$$

so that we have three linear equations to solve for three parameter estimates. Hence there is no problem in this case either, since if we write $Y_i = C_i$, $X_{2i} = Q_i$, and $X_{3i} = Q_i^2$, it is clear that these equations are identical with 9.6, 9.7 and 9.8.

In the case of the production function we could write

$$Q_i = \gamma K_i^\alpha L_i^\beta + \varepsilon_i \qquad (9.45)$$

and find the least squares estimators by minimizing

$$\sum e_i^2 = \sum (Q_i - cK_i^a L_i^b)^2. \qquad (9.46)$$

Unfortunately in this case partial differentiation leads to non-linear normal equations which cannot be solved for the least squares estimators.[10] Thus unlike the first two examples we considered, the non-linearity of the production function does lead to problems of estimation. To see why this is so, we may note one important difference between the first two examples and the third: while the first two examples are non-linear in the variables they are *linear* in the parameters, whereas the third example is non-linear in the parameters. It is the non-linearity in the parameters which is causing the difficulty and which has to be overcome.

One solution to the problem, which is possible in some cases, is to transform the equation so that it is linear in the parameters to be estimated. In the case of the production function, such a transformation is to take the logarithm of the function, since

$$\log_e(\gamma K_i^\alpha L_i^\beta) = \log_e \gamma + \alpha \log_e K_i + \beta \log_e L_i,$$

which is now linear in $\alpha$, $\beta$ and $\log_e v$. However, we cannot solve the problem by a logarithmic transformation if we have written the production function to be estimated as 9.45, since that yields

$$\log_e Q_i = \log_e (\gamma K_i^\alpha L_i^\beta + \varepsilon_i),$$

which does not help. Clearly, if the logarithmic transformation is to work, the random error must appear in the equation in a multiplicative form. Let us therefore redefine the production function to be estimated as

$$Q_i = v K_i^\alpha L_i^\beta + \varepsilon_i), \tag{9.47}$$

where $e$ is the base of natural logarithms. Now applying the logarithmic transformation, we have

$$\log_e Q_i = \log_e \gamma + \alpha \log_e K_i + \beta \log_e L_i + \varepsilon_i, \tag{9.48}$$

which is linear in $\alpha$, $\beta$ and $\log_e \gamma$. If we write $Y_i = \log_e Q_i$, $X_{2i} = \log_e K_i$ and $X_{3i} = \log_e L_i$ we may estimate $\alpha$, $\beta$ and $\log_e \gamma$ by substituting our numerical data into equations 9.9, 9.12 and 9.13.

This transformation solves the problem of estimating $\alpha$ and $\beta$, but raises two questions, namely how do we estimate $\gamma$, and secondly, what are the properties of our estimators given the multiplicative error term in 9.47? The answer to the first question is that we estimate $\gamma$ by taking the antilogarithm of the constant term in 9.48. The answer to the second question depends upon what we assume about the random errors in the regression model. Ideally, we would like to assume that the $\varepsilon$s in 9.48 have the same statistical properties as the random errors in 9.1, which would make the least squares estimators best, linear and unbiased. Under these assumptions our estimators of $\alpha$ and $\beta$ are unbiased, but the estimator of $\gamma$ is biased.[11]

### A production function example
To illustrate a number of the points raised in this section, we shall now consider a hypothetical example of the estimation of a Cobb–Douglas production function.[12] Let us assume that we set out to estimate the production function for an industry on the basis of the data concerning a random sample of 28 firms, and that when the data are converted into logarithms we calculate

$$\sum Q_i^2 = 25213\cdot6; \qquad \sum K_i L_i = 5610\cdot0; \qquad \bar{Q} = 30\cdot0;$$

$$\sum K_i^2 = 2830\cdot0; \qquad \sum K_i Q_i = 8415\cdot0; \qquad \bar{K} = 10\cdot0;$$

$$\sum L_i^2 = 11220\cdot0; \qquad \sum L_i Q_i = 16815\cdot0; \qquad \bar{L} = 20\cdot0.$$

These data may be converted into deviations from the sample means to yield

$$\sum q_i^2 \;=\; \sum Q_i^2 - n\bar{Q}^2 \quad = 25213\cdot6 - 25200\cdot0 = 13\cdot6$$

$$\sum k_i^2 \;=\; \sum K_i^2 - n\bar{K}^2 \quad = 2830\cdot0 - 2800\cdot0 \;\;= 30\cdot0$$

$$\sum l_i^2 \;=\; \sum L_i^2 - n\bar{L}^2 \quad = 11220\cdot0 - 11200\cdot0 = 20\cdot0$$

$$\sum k_i l_i \;=\; \sum K_i L_i - n\overline{KL} = 5610\cdot0 - 5600\cdot0 \;\;= 10\cdot0$$

$$\sum k_i q_i \;=\; \sum K_i Q_i - n\overline{KQ} = 8415\cdot0 - 8400\cdot0 \;\;= 15\cdot0$$

$$\sum l_i q_i \;=\; \sum L_i Q_i - n\overline{LQ} = 16815\cdot0 - 16800\cdot0 = 15\cdot0.$$

Substituting these data into equations 9.12, 9.13 and 9.9, we obtain the least squares estimates of the parameters in 9.48:

$$a = \frac{\sum kq \sum l^2 - \sum kl \sum lq}{\sum k^2 \sum l^2 - (\sum kl)^2} = \frac{(15)(20) - (10)(15)}{(30)(20) - (10)^2}$$

$$= 0\cdot3.$$

$$b = \frac{\sum lq \sum k^2 - \sum kl \sum kq}{\sum k^2 \sum l^2 - (\sum kl)^2} = \frac{(15)(30) - (10)(15)}{(30)(20) - (10)^2}$$

$$= 0\cdot6.$$

$$\log_e \gamma = \bar{Q} - a\bar{K} - b\bar{L} = 30 - (0\cdot3)(10) - (0\cdot6)(20)$$

$$= 15\cdot0.$$

Our estimated production function is thus

$$Q_i = e^{15.0} K_i^{0.3} L_i^{0.6}.$$

To measure the success of our estimation, we may calculate the coefficient of determination, $R^2$, which by adapting the formula given in 9.32 is

$$R^2 = \frac{a^2 \sum k_i^2 + b^2 \sum l_i^2 + 2ab \sum k_i l_i}{\sum q_i^2}$$

$$= \frac{(0\cdot3)^2 (30) + (0\cdot6)^2 (20) + 2(0\cdot3)(0\cdot6)(10)}{13\cdot6} = \frac{13\cdot5}{13\cdot6} = 0\cdot993.$$

Clearly, the fit of our production function is extremely good, but we should still go on to see how the two input variables have contributed statistically to the explanation of output. To do this we may calculate the standard errors of $a$ and $b$, for which purpose we need to calculate the sum of the squared residuals, $\sum e_i^2$. Since the total variation $\sum q_i^2 = 13\cdot6$ and the explained variation is $13\cdot5$, we have $\sum e_i^2 = 0\cdot1$. Substituting this value into 9.22, we have

$$s^2 = \frac{0\cdot1}{25} = 0\cdot004,$$

so that substituting $s^2$ into 9.19 and 9.20 gives

$$\mathrm{var}(a) = \frac{s^2 \sum l_i^2}{\sum k_i^2 \sum l_i^2 - (\sum k_i l_i)^2} = \frac{(0\cdot004)\,(20)}{(30)\,(20) - (10)^2} = 0\cdot00016$$

and

$$\mathrm{var}(b) = \frac{s^2 \sum k_i^2}{\sum k_i^2 \sum l_i^2 - (\sum k_i l_i)^2} = \frac{(0\cdot004)\,(30)}{(30)\,(20) - (10)^2} = 0\cdot00024$$

Calculating the square root of these expressions, we obtain s.e. $(a) = 0\cdot0126$ and s.e. $(b) = 0\cdot0155$.

The first tests we consider are for the hypotheses that $\alpha = 0$ or $\beta = 0$. Economic theory suggests that our alternative hypotheses should be that $\alpha > 0$ and $\beta > 0$, so that a one-tail test is appropriate here. If we choose the probability level $p = 0\cdot95$, then with 25 degrees of freedom $t = 1\cdot71$ will be the rejection level for our hypotheses. Substituting our data into 9.25, we obtain

$$t_a = \frac{0\cdot3}{0\cdot013} = 23\cdot1 \quad \text{and} \quad t_b = \frac{0\cdot6}{0\cdot016} = 37\cdot5.$$

Since these two values are greater than $1\cdot71$, we reject the hypotheses that $\alpha = 0$ and $\beta = 0$.

### Estimating a linear combination of parameters

The results obtained are satisfactory from the point of view of economic theory, but of more interest is the question of whether the production process shows increasing, constant or decreasing returns to scale, that is, $\alpha + \beta \gtreqless 1$. To explore this possibility we need to estimate $\alpha + \beta$ and test hypotheses concerning this linear combination of $\alpha$ and $\beta$. The intuitive estimator to choose would be $a + b$, and this choice is in fact correct.[13] It is easy to show that this estimator is unbiased, since

$$E(a + b) = E(a) + E(b) = \alpha + \beta. \tag{9.49}$$

To be able to test hypotheses concerning $\alpha + \beta$, we need to know the variance of $a + b$. By definition,

$$\text{var}(a + b) = E[(a + b) - E(a + b)]^2$$

and substituting from 9.49,

$$\begin{aligned}\text{var}(a + b) &= E[(a + b) - (\alpha + \beta)]^2 = E[(a - \alpha) + (b - \beta)]^2 \\ &= E(a - \alpha)^2 + E(b - \beta)^2 + 2E[(a - \alpha)(b - \beta)] \\ &= \text{var}(a) + \text{var}(b) + 2\text{cov}(a, b). \end{aligned} \tag{9.50}$$

We have already calculated $\text{var}(a) = 0.00016$ and $\text{var}(b) = 0.00024$, and by modifying 9.21 we may write

$$\text{cov}(a, b) = - \frac{s^2 \sum k_i l_i}{\sum k_i^2 \sum l_i^2 - (\sum k_i l_i)^2} = - \frac{(0.004)(10)}{(30)(20) - (10)^2} = -0.00008.$$

Hence $\text{var}(a + b) = 0.00016 + 0.00024 - 2(0.00008) = 0.00024$ and s.e.$(a + b) = \sqrt{0.00024} = 0.0155$.

Finally, we need to say something about the sampling distribution of $(a + b)$. It was stated earlier that sampling distribution for both $a$ and $b$ was the $t$-distribution with $(n - 3)$ degrees of freedom. As $(a + b)$ is a linear combination of these two variables, it is possible[14] to prove that

$$t = \frac{(a + b) - (\alpha + \beta)}{\text{s.e.}(a + b)} \tag{9.51}$$

also follows the $t$-distribution with $(n - 3)$ degrees of freedom. We now have all the results we need to investigate the returns to scale of our production function. We may take as the hypothesis to be tested $H_o : \alpha + \beta = 1$, with $H_1 : \alpha + \beta \neq 1$ as the alternative hypothesis when we have no reason to expect decreasing rather than increasing returns. This alternative hypothesis would suggest a two-tail test. We have $a + b = 0.9$, and substituting in 9.51 we obtain

$$t = \frac{0.9 - 1.0}{0.016} = -6.25.$$

If we have chosen the $p = 0.95$ probability level, with 25 degrees of freedom we have $t = \pm 2.06$ as our decision values, and since $-6.25 < -2.06$ we reject the hypothesis $H_0 : \alpha + \beta = 1$. There would appear to be decreasing returns to scale.

*Constrained estimation*

Before leaving our production function, we shall consider the problem of estimation when there are linear restrictions imposed on the parameters. For example, suppose we had believed *a priori* that there were constant returns to scale and had wished to impose the constraint that $\alpha + \beta = 1$ when estimating the parameters of the production function. In the case of this simple linear restriction, the modifications to our estimating procedure is easily made, for now $\beta = 1 - \alpha$ and substituting for $\beta$ in 9.48 gives

$$\log Q_i = \log \gamma + \alpha \log K_i + (1 - \alpha) \log L_i + \varepsilon_i,$$

that is,

$$\log Q_i - \log L_i = \log \gamma + \alpha (\log K_i - \log L_i) + \varepsilon_i \qquad (9.52)$$

or

$$\log \left( \frac{Q}{L} \right)_i = \log \gamma + \alpha \log \left( \frac{K}{L} \right)_i + \varepsilon_i. \qquad (9.53)$$

Thus having imposed the restriction that $\alpha + \beta = 1$, we obtain our least squares estimates of $\log v$ and $\alpha$ by running the simple regression of output per head on capital per head. Given our previous results, we have unbiased estimators since $E(a) = \alpha$ and hence $E(1 - a) = 1 - E(a) = 1 - \alpha = \beta$.

To illustrate the calculation of the constrained parameters of the production function, we return to our hypothetical data and the notation in which we may express the production function 9.52 as

$$Q_i - L_i = \log \gamma + \alpha (K_i - L_i) + \varepsilon_i$$

or

$$Q_i' = \log \gamma + \alpha K_i' + \varepsilon_i, \qquad (9.54)$$

where $Q_i' = Q_i - L_i$ and $K_i' = K_i - L_i$. It is obvious that $\bar{Q}' = \bar{Q} - \bar{L}$ and $\bar{K}' = \bar{K} - \bar{L}$, so that denoting deviations from sample means with lower-case letters, we have $q_i' = q_i - l_i$ and $k_i' = k_i - l_i$. The least squares estimators of the parameters in 9.54 are

$$\text{est}(\log \gamma) = \bar{Q}' - a\bar{K}'$$

and

$$a = \frac{\sum k_i' q_i'}{\sum (k_i')^2} = \frac{\sum (k_i - l_i)(q_i - l_i)}{\sum (k_i - l_i)^2} = \frac{\sum k_i q_i - \sum l_i q_i - \sum k_i l_i + \sum l_i^2}{\sum k_i^2 - 2 \sum k_i l_i + \sum l_i^2}.$$

Substituting our numerical data, we obtain

$$a = \frac{15 - 15 - 10 + 20}{30 - 2(10) + 20} = \frac{10}{30} = 0.33$$

$$\text{est}(\log \gamma) = 10 - (0.33)(-10) = 13.3$$

and given our constraint, $b = (1 - a) = 0.67$. Thus our estimated constant returns to scale production function is $Q_i = e^{13.3} K_i^{0.33} L_i^{0.67}$.

To complete the regression exercise, we may calculate the standard error of $a$ and the coefficient of determination for the constrained production function. First we calculate $\sum e_i^2 = \sum (q_i')^2 - a^2 \sum (k_i)^2$, noting that $\sum (q_i')^2 = \sum q_i^2 - 2\sum l_i q_i + \sum l_i^2$, to obtain $\sum e_i^2 = (13.6 - 2(15) + 20) - (0.33)^2 (30) = 3.6 - 3.27 = 0.33$. Then s.e. $(a) = \sqrt{\sum e_i^2 (n - 2) \sum (k_i)^2} = \sqrt{0.33/(26)(30)} = 0.0206$. We also have $r^2 = 3.27/3.6 = 0.908$, indicating that a high proportion of the variation in output per head is explained by our regression relation. Our estimate of $\alpha$ is clearly significant, since $a/\text{s.e.}(a) = 0.33/0.021 = 15.7$.

This production function example has enabled us to widen our concepts of estimation for the model with two independent variables. We shall now return to our main theme and extend the discussion of the method of least squares to cover the general linear regression model.

## 9.5 the general linear model

Having discussed regression models containing one or two independent variables, we now consider the general linear model, which may be written as

$$Y_i = \beta_1 + \beta_2 X_{2i} + \dots + \beta_k X_{ki} + \varepsilon_i, \qquad i = 1, \dots, n. \tag{9.55}$$

Compared with the results we have obtained for the cases where $k = 2$ and $k = 3$, the least squares estimation of the parameters in 9.55 produces no novelties and presents no problems beyond a tedious increase in algebraic complexity. The most efficient way to discuss the general model is to use linear algebra, and the reader with some knowledge of matrices and vectors will find such a discussion in the appendix to this chapter. For the reader without a knowledge of matrix algebra, we shall sketch out the main results for the general linear model without getting involved in algebraic detail.[15]

1. The least squares estimators of the parameters in 9.55 are obtained by minimizing the sum of the squared residuals

$$\sum e_i^2 = \sum (Y_i - b_1 - b_2 X_{2i} - \dots - b_k X_{ki})^2 \tag{9.56}$$

with respect to $b_j, j = 1, \dots, k$. Partially differentiating $\sum e_i^2$ and setting the $k$ partial derivatives equal to zero, we have

$$\frac{\partial \sum}{\partial b_1} = -2 \sum (Y_i - b_1 - b_2 X_{2i} - \ldots - b_k X_{ki}) = 0 \tag{9.57}$$

$$\frac{\partial \sum}{\partial b_j} = -2 \sum X_{ji}(Y_i - b_1 - b_2 X_{2i} - \ldots - b_k X_{ki}) = 0, \qquad j = 2, \ldots, k. \tag{9.58}$$

Rearranging 9.57, we get

$$\sum Y_i = nb_1 + b_2 \sum X_{2i} + \ldots + b_k \sum X_{ki}$$

or

$$b_1 = \bar{Y} - b_2 \bar{X}_2 - \ldots - b_k \bar{X}_k, \tag{9.59}$$

a result which is an obvious extension of 9.9. Rearranging 9.58 gives

$$\sum X_{ji} Y_i = b_1 \sum X_{ji} + b_2 \sum X_{2i} X_{ji} + \ldots + b_k \sum X_{ji} X_{ki}, \qquad j = 2, \ldots, k,$$

then substituting for $b_1$ from 9.59 and using lower-case letters for deviations from sample means, we have

$$\sum x_{ji} y_i = b_2 \sum x_{2i} x_{ji} + \ldots + b_k \sum x_{ji} x_{ki}, \qquad j = 2, \ldots, k, \tag{9.60}$$

a set of $(k - 1)$ linear equations to be solved for the $(k - 1)$ unknowns $b_2, \ldots, b_k$.

2. The assumptions we have previously made concerning the statistical properties of the model are easily extended. As before, we assume that

$$E(\varepsilon_i) = 0, \qquad i = 1, \ldots, n \tag{8.13}$$

$$E(\varepsilon_i^2) = \sigma^2, \tag{8.14}$$

$$E(\varepsilon_i \varepsilon_m) = 0, \qquad i \neq m, i, m = 1, \ldots, n. \tag{8.15}$$

The extension is that we now assume that all the independent variables may be treated as constants, so that

$$E(X_{ji} \varepsilon_m) = X_{ji} E(\varepsilon_m) = 0 \quad \text{for all} \quad i, m = 1, \ldots, n \quad \text{and} \quad j = 2, \ldots, k. \tag{9.61}$$

We also need to add one restriction to the model, namely that there are more observations than independent variables,[16] so that

$$k < n. \tag{9.62}$$

Given these assumptions, it may be shown that the least squares estimators are best, linear and unbiased. The sampling variances of the $b_j$ are functions of $\sigma^2$ and the $X_{ji}$ which appear in the regression model.

3. If we make the further assumption that the $\varepsilon_i$ are normally distributed, then the sampling distributions of the $b_j$ are normal. If $\sigma^2$ were known, the areas under

the normal curve could be used to set up confidence intervals for or test hypotheses concerning $\beta_j$.

4. In general, $\sigma^2$ is not known and has to be estimated. If we calculate

$$s^2 = \frac{\sum e_i^2}{n - k},$$
(9.63)

where $\sum e_i^2$ is the sum of the squared residuals, then it may be shown that $s^2$ is an unbiased estimator of $\sigma^2$. If $s^2$ is substituted for $\sigma^2$ in the formulae for the standard errors of the estimators s.e.$(b_j)$, then

$$t = \frac{b_j - \beta_j}{\text{s.e.}(b_j)}, \qquad j = 1, \dots k,$$
(9.64)

follows the $t$-distribution with $(n - k)$ degrees of freedom. This result, which is an obvious extension of 8.38, 8.39 and 9.24, makes our sampling theory operational in the context of the general linear model.

5. As a measure of goodness of fit, the coefficient of determination may be defined as

$$R^2 = 1 - \frac{\sum e_i^2}{\sum y_i^2}.$$
(9.65)

This is a simple statistic to interpret, but it has one property that disturbs some statisticians and economists. Suppose that, having computed the value of $R^2$ for a particular multiple regression in which $k_1$ independent variables are included, we now introduce further $k_2$ independent variables into the equation and re-compute $R^2$. Given the new variables cannot reduce the value of $R^2$ (i.e. they cannot remove what has already been explained), and it is likely that some of them may be related to the dependent variable, we would expect the goodness of fit to improve. However, the value of $R^2$ does not reflect the fact that we have lost a further $k_2$ degrees of freedom by introducing these variables. For this reason some economists and statisticians would compute the *adjusted* coefficient of determination, $\bar{R}^2$, which is defined to take degrees of freedom into account as

$$\bar{R}^2 = 1 - \frac{\sum e_i^2/(n - k)}{\sum y_i^2/(n - 1)},$$
(9.66)

where $n$ equals the number of observations and $k$ the number of parameters being estimated. Rearranging the right-hand side of 9.66, we obtain

$$\bar{R}^2 = 1 - \frac{(n - 1)\sum e_i^2}{(n - k)\sum y_i^2} = 1 - \frac{(n - 1)}{(n - k)}(1 - R^2).$$
(9.67)

Clearly, $(n-1)/(n-k)$ increases as $k$ increases, and unless this increase is offset by a decrease in $(1-R^2)$, the value of $R^2$ decreases as we add more explanatory variables to the equation.[17]

To conclude our discussion of the general linear model, we shall look at some estimated economic relationships. The following examples are taken from Evans (1969), a macro-economic model of the French economy which was based on annual data for the period 1952–65. The components of income and also capital are measured in constant terms, in billions of 1958 frances, employment is in millions and prices are based on $1958 = 1·00$.

### 1. *The demand for durable goods and clothing*
This equation may be expressed as

$$\dot{C}_t^d = \beta_{11} + \beta_{12}\dot{Y}_t + \beta_{13}C_{t-1}^* + \beta_{14}\dot{P}_t^d + \beta_{15}\Delta(RU - UV)_{t-\frac{1}{2}} + \varepsilon_{1t},$$

where $\dot{C}_t^d$ = the percentage change in the consumption of durable goods and clothing,

$\dot{Y}_t$ = the percentage change in personal disposable income,

$C_{t-1}^*$ = a measure of the beginning period stock of durables and clothing,

$\dot{P}_t^d$ = the percentage change in the price of durables and clothing,

$\Delta(RU - UV)_{t-\frac{1}{2}}$ = the change in registered unemployed minus unfilled vacancies lagged six months, a measure of cyclical variation.

On *a priori* grounds we would expect a positive relationship between $\dot{C}_t^d$ and $\dot{Y}_t$, and since both variables are measured in terms of percentage changes, the coefficient gives an estimate of the income elasticity of the demand for durable goods. The higher the existing stock of durable goods, the lower we might expect the demand to be, so we would expect $\beta_{13}$ to be negative, as we would $\beta_{14}$, the own price elasticity. The cyclical variable is defined in such a way that positive values indicate downswings in the cycle and hence we would expect a negative value for $\beta_{15}$.

The estimated consumption function is

$$\dot{C}_t^d = 0·057 + 0·894\dot{Y}_t - 0·262\dot{C}_{t-1}^* - 0·324\dot{P}_t^d - 0·535\Delta(RU - UV)_{t-\frac{1}{2}},$$
$$\quad\quad\quad (0·289)\quad\;\;(0·518)\quad\quad\;\;(0·269)\quad\;\;(0·235)$$

$$\bar{R}^2 = 0·778$$

where the figures in parentheses are the standard errors of the parameter estimates. In examining these results we note that a large proportion of variation is explained and all the estimated coefficients have the correct signs in terms of the argument presented above. The estimated income elasticity is 0·894 and the estimated price elasticity is $-0·324$. The estimated income elasticity is less than unity, which may seem somewhat surprising for durable goods, but this is a point estimate and to evaluate this result we must consider the statistical properties of our estimates.

For this equation we have 14 observations and are estimating five parameters, so that $(n - k) = 9$. We have indicated the expected signs of the parameters in this equation, so that our alternative to the hypothesis $H_0 : \beta_{12} = 0$ is $H_1 : \beta_{12} > 0$, indicating that a one-tail test is appropriate. The same argument implies a one-tail test for the other parameters in this equation. If we choose the probability level $p = 0.95$, we find from Table A2 that $t = 1.83$ for testing $\beta_{12} = 0$ and $t = -1.83$ for testing $\beta_{13} = 0$, $\beta_{14} = 0$ and $\beta_{15} = 0$. To test the hypotheses that $\beta_{1j} = 0$, $j = 2, ..., 5$, we calculate the $t$-ratios $t_{12} = 0.894/0.289 = 3.1$, $t_{13} = -0.262/0.518 = -0.51$, $t_{14} = -0.324/0.269 = -1.2$, $t_{15} = -0.535/0.235 = -2.3$. Given our chosen probability level of $p = 0.95$, we would reject the hypotheses that $\beta_{12} = 0$ and $\beta_{15} = 0$, but we cannot reject the hypotheses that $\beta_{13} = 0$ and $\beta_{14} = 0$. This analysis would suggest that the major factors explaining the percentage change in the demand for durable goods are the percentage change in current income and the cyclical variable. Since the $t$-ratio for $\dot{P}_t^d$ is greater than unity, we may conclude that this variable is contributing to an increase in $\bar{R}^2$.

We may use the standard error of $b_{12}$ to set up a confidence interval for $\beta_{12}$, the income elasticity. Choosing $p = 0.95$ as our probability level with 9 degrees of freedom, $t = \pm 2.26$ and our confidence interval is $\beta_{12} = 0.894 \pm 2.26(0.289)$, that is,

$$0.241 < \beta_{12} < 1.547.$$

## 2. The inventory investment function
In Evans's model this equation is defined as

$$\Delta I_t^i = \beta_{21} + \beta_{22}\Delta S_t + \beta_{23}\Delta S_{t-1} + \beta_{24}\Delta UO_t + \beta_{25}I_{t-1}^i + \varepsilon_{2t},$$

where $\Delta I_t^i$ = the change in inventories in period $t$,

$\Delta S_t$ = the change in sales in the private sector, excluding agriculture, in period $t$,

$\Delta UO_t$ = the change in unfilled orders,

$I_{t-1}^i$ = the level of inventories in the previous period.

On *a priori* grounds we might argue that we would expect $\beta_{22}$ to be negative, since unexpectedly high sales would deplete inventories. Given inventories take time to build up, we would expect a positive relationship between sales during last period and the growth of inventories this period. To the extent that unfilled orders are growing, we would expect a positive correlation here. Finally, high inventories in the previous period would tend to act against inventory accumulation in this period and thus lead to $\beta_{15}$ being negative.

The estimated inventory investment equation is

$$\Delta I_t^i = 3\cdot883 - 0\cdot142\Delta S_t + 0\cdot210 S_{t-1} + 0\cdot0639\Delta UO_{t-1} - 1\cdot209 I_{t-1}^i, \quad \bar{R}^2 = 0\cdot747.$$
$$\qquad\quad (0\cdot127) \qquad (0\cdot100) \qquad\quad (0\cdot0301) \qquad\qquad (0\cdot291)$$

Again, $n = 14$ and $k = 5$, so that $(n - k) = 9$ and the same values of the $t$-distribution are appropriate here as were in the previous example. We note that the estimated parameters all have the expected signs. Calculating the $t$-ratios, we obtain $t_{22} = -0\cdot142/0\cdot127 = -1\cdot12$, $t_{23} = 0\cdot210/0\cdot100 = 2\cdot10$, $t_{24} = 0\cdot0639/0\cdot0301 = 2\cdot12$, $t_{25} = -1\cdot209/0\cdot291 = -4\cdot15$. We see that although $\beta_{22}$ has the correct sign it is not significantly different from zero, while $\beta_{23}$ is significant. This may reflect a lag in the process of building up inventories. The most significant term is the beginning period stock of inventories, a result which may reflect a strong economic relationship, or may partially reflect the fact that $\Delta I_t^i = I_t^i - I_{t-1}^i$ and hence we are regressing $I_{t-1}^i$ on itself.

### 3. The production function

In Evans's model this relationship is basically a Cobb–Douglas production function, modified to include a variable reflecting cyclical variation to allow for capacity utilization, and a time trend which may be interpreted as a measure of the rate of technical progress.[18] The equation is

$$\log X_t = \beta_{31} + \beta_{32} \log N_t + \beta_{33} \log K_t + \beta_{34}(TU - 2UV)_t + \beta_{35}t + \varepsilon_{3t},$$

where $X_t =$ private sector output excluding agriculture,

$\qquad N_t =$ total employment, excluding agriculture and government,

$\qquad K_t =$ the capital stock, excluding government and housing,

$\qquad (TU - 2UV) =$ the difference between total unemployment and unfilled
$\qquad\qquad$ vacancies, a measure of cyclical variation, and

$\qquad\quad t =$ a linear time trend with 1952 $= 1$.

On *a priori* grounds we would obviously expect $\beta_{32}$ and $\beta_{33}$ to be positive. The parameter $\beta_{34}$ reflects cyclical functions in output and we would expect it to be negative. To the extent that the time trend is measuring technical progress, we would expect $\beta_{35}$ to be positive.

The estimated production function is

$$\log X_t = 1\cdot229 + 0\cdot488 \log N_t + 0\cdot193 \log K_t - 0\cdot0791 (TU - 2UV)_t + 0\cdot0170t,$$
$$\qquad\quad (0\cdot297) \qquad\quad (0\cdot152) \qquad\qquad (0\cdot0217) \qquad\qquad\qquad (0\cdot0027)$$
$$\bar{R}^2 = 0\cdot999.$$

Again we find that all the estimates have the correct signs, but when we calculate the $t$-ratios we obtain $t_{32} = 0\cdot488/0\cdot297 = 1\cdot64$, $t_{33} = 0\cdot193/0\cdot152 = 1\cdot27$, $t_{34} = -0\cdot0791/0\cdot0217 = -3\cdot65$, $t_{35} = 0\cdot0170/0\cdot0027 = 6\cdot30$. Although the fit is good

($\bar{R}^2 = 0.999$), we cannot reject the hypotheses that $\beta_{32} = 0$ and $\beta_{33} = 0$, which is not a very satisfactory state of affairs for a production function![19] The sum of the estimated coefficients on labour and capital is $0.488 + 0.193 = 0.681$, which would imply decreasing returns to scale, but Evans suggests that when the cyclical term is taken into account, the fitted production function indicates constant returns to scale.[20]

In assessing the equations presented here we must remember that the number of observations is small and so the number of degrees of freedom is low. This leads to relatively large standard errors and wide confidence intervals for the estimated parameters. We shall need to develop more statistical theory to be in a position to assess such empirical results, and this development will occupy the next two chapters. In chapter 10 we shall consider the implications of the fact that many economic data are collected over time, while in chapter 11 we shall consider what happens to the properties of the least squares estimators when we relax the various assumptions in the regression model.

# appendix to chapter 9

In order to derive the least squares estimators of the parameters in the general linear model, we shall now introduce matrix algebra.[21] The model is

$$Y_i = \beta_1 + \beta_2 X_{2i} + ... + \beta_k X_{ki} + \varepsilon_i, \qquad i = 1, ..., n \tag{9.55}$$

and we shall define the following vectors and matrix. Let

$$\mathbf{y} = \begin{bmatrix} Y_1 \\ Y_2 \\ \vdots \\ Y_n \end{bmatrix}, \qquad \mathbf{X} = \begin{bmatrix} 1 & X_{21} & ... & X_{k1} \\ 1 & X_{22} & ... & X_{k2} \\ \vdots & \vdots & ... & \vdots \\ 1 & X_{2n} & ... & X_{kn} \end{bmatrix}, \qquad \boldsymbol{\beta} = \begin{bmatrix} \beta_1 \\ \beta_2 \\ \vdots \\ \beta_k \end{bmatrix}, \qquad \boldsymbol{\varepsilon} = \begin{bmatrix} \varepsilon_1 \\ \varepsilon_2 \\ \vdots \\ \varepsilon_n \end{bmatrix}.$$

Then we may rewrite 9.55 as

$$\mathbf{y} = \mathbf{X}\boldsymbol{\beta} + \boldsymbol{\varepsilon}. \tag{9.A1}$$

To derive the least squares estimator of $\boldsymbol{\beta}$, we define the vectors of estimated coefficients, $\mathbf{b}$, and residuals, $\mathbf{e}$, as

$$\mathbf{b} = \begin{bmatrix} b_1 \\ \vdots \\ b_k \end{bmatrix} \quad \text{and} \quad \mathbf{e} = \begin{bmatrix} e_1 \\ \vdots \\ e_n \end{bmatrix}.$$

Then

$$\mathbf{y} = \mathbf{X}\mathbf{b} + \mathbf{e}$$

or

$$\mathbf{e} = \mathbf{y} - \mathbf{X}\mathbf{b}. \tag{9.A2}$$

In this notation the sum of the squared residuals may be rewritten as

$$\sum e_i^2 = \mathbf{e'e} = (\mathbf{y} - \mathbf{Xb})'(\mathbf{y} - \mathbf{Xb})$$

$$= \mathbf{y'y} - \mathbf{b'X'y} - \mathbf{y'Xb} + \mathbf{b'X'Xb}$$

$$= \mathbf{y'y} - 2\mathbf{b'X'y} + \mathbf{b'X'Xb}, \tag{9.A3}$$

since $\mathbf{b'X'y} = \mathbf{y'Xb}$ as $\mathbf{y'Xb}$ is scalar and hence equal to its transpose.

To find the least squares estimator of $\boldsymbol{\beta}$ we minimize $\mathbf{e'e}$ with respect to the the elements in $\mathbf{b}$. This involves partially differentiating 9.A3 with respect to the elements in $\mathbf{b}$ and setting the vector of partial derivatives equal to the null vector, $\mathbf{0}$, to give

$$\frac{\partial \mathbf{e'e}}{\partial \mathbf{b}} = -2\mathbf{X'y} + 2\mathbf{X'Xb} = \mathbf{0}. \tag{9.A4}$$

Rearranging this equation, we have

$$\mathbf{X'Xb} = \mathbf{X'y}$$

and provided $\mathbf{X'X}$ is non-singular, premultiplication by $(\mathbf{X'X})^{-1}$ yields

$$\mathbf{b} = (\mathbf{X'X})^{-1}\mathbf{X'y} \tag{9.A5}$$

as the vector of least squares estimators.

*The statistical properties of the model*
The assumptions listed in 8.13, 8.14, 8.15, 9.61 and 9.62 on p. 180 may be stated in matrix terms as

$$E(\boldsymbol{\varepsilon}) = \mathbf{0} \quad , \qquad \text{corresponding to 8.13.} \tag{9.A6}$$

$$E(\boldsymbol{\varepsilon\varepsilon'}) = \sigma^2\mathbf{I}, \qquad \text{corresponding to 8.14 and 8.15.} \tag{9.A7}$$

The elements of $\mathbf{X}$ are constants, corresponding to 9.61. $\tag{9.A8}$

The rank of $\mathbf{X}$ is $k < n$, corresponding to 9.62. $\tag{9.A9}$

Equation 9.A6 is clearly equivalent to $E(\varepsilon_i) = 0$, $i = 1, ..., n$. The second assumption may be expanded to give

$$E(\varepsilon\varepsilon') = \begin{bmatrix} E(\varepsilon_1^2) & E(\varepsilon_1\varepsilon_2) & \ldots & E(\varepsilon_1\varepsilon_n) \\ E(\varepsilon_1\varepsilon_2) & E(\varepsilon_2^2) & \ldots & E(\varepsilon_2\varepsilon_n) \\ \vdots & \vdots & \cdots & \vdots \\ E(\varepsilon_1\varepsilon_n) & E(\varepsilon_2\varepsilon_n) & \ldots & E(\varepsilon_n^2) \end{bmatrix}$$

and given $E(\varepsilon_i^2) = \sigma^2$ and $E(\varepsilon_i\varepsilon_j) = 0$ for $i \neq j$,

$$E(\varepsilon\varepsilon') = \begin{bmatrix} \sigma^2 & 0 & \ldots & 0 \\ 0 & \sigma^2 & \ldots & 0 \\ \vdots & \vdots & \cdots & \vdots \\ 0 & 0 & \ldots & \sigma^2 \end{bmatrix} = \sigma^2\mathbf{I}.$$

Given 9.A8, we have $E(\mathbf{X}'\varepsilon) = \mathbf{X}'E(\varepsilon) = \mathbf{0}$, by virtue of 9.A6, while 9.A9 is a necessary condition for $\mathbf{X}'\mathbf{X}$ to be non-singular.

Given these assumptions, we may now explore the statistical properties of $\mathbf{b}$. Substituting from 9.A1 for $\mathbf{y}$ in 9.A5 gives

$$\mathbf{b} = (\mathbf{X}'\mathbf{X})^{-1}\mathbf{X}'(\mathbf{X}\boldsymbol{\beta} + \varepsilon)$$

$$= (\mathbf{X}'\mathbf{X})^{-1}\mathbf{X}'\mathbf{X}\boldsymbol{\beta} + (\mathbf{X}'\mathbf{X})^{-1}\mathbf{X}'\varepsilon$$

$$= \boldsymbol{\beta} + (\mathbf{X}'\mathbf{X})^{-1}\mathbf{X}'\varepsilon, \tag{9.A10}$$

since $(\mathbf{X}'\mathbf{X})^{-1}\mathbf{X}'\mathbf{X} = \mathbf{I}$. Now taking expected values,

$$E(\mathbf{b}) = \boldsymbol{\beta} + E(\mathbf{X}'\mathbf{X})^{-1}\mathbf{X}'\varepsilon$$

$$= \boldsymbol{\beta} + (\mathbf{X}'\mathbf{X})^{-1}\mathbf{X}'E(\varepsilon) = \boldsymbol{\beta}. \tag{9.A11}$$

This establishes that $\mathbf{b}$ is an unbiased estimator of $\boldsymbol{\beta}$.

To derive the sampling variances of the $b$s, we define the variance–covariance matrix for $\mathbf{b}$ as

$$\text{var}(\mathbf{b}) = E[(\mathbf{b} - \boldsymbol{\beta})(\mathbf{b} - \boldsymbol{\beta})']. \tag{9.A12}$$

Equation 9.A10 may be rewritten as $(\mathbf{b} - \boldsymbol{\beta}) = (\mathbf{X}'\mathbf{X})^{-1}\mathbf{X}'\varepsilon$, and substituting this result into (9.A12) we have

$$\text{var}(\mathbf{b}) = E\left[(\mathbf{X'X})^{-1}\mathbf{X'\varepsilon}\right]\left[(\mathbf{X'X})^{-1}\mathbf{X'\varepsilon}\right]'$$

$$= E\left[(\mathbf{X'X})^{-1}\mathbf{X'\varepsilon\varepsilon'X}(\mathbf{X'X})^{-1}\right]$$

$$= (\mathbf{X'X})^{-1}\mathbf{X'}E(\mathbf{\varepsilon\varepsilon'})\mathbf{X}(\mathbf{X'X})^{-1} \tag{9.A13}$$

$$= (\mathbf{X'X})^{-1}\mathbf{X'}\sigma^2\mathbf{IX}(\mathbf{X'X})^{-1}, \text{ given 9.A7,}$$

$$= \sigma^2(\mathbf{X'X})^{-1}\mathbf{X'X}(\mathbf{X'X})^{-1}, \text{ since } \sigma^2 \text{ is scalar,}$$

$$= \sigma^2(\mathbf{X'X})^{-1}. \tag{9.A14}$$

The sampling variance for $b_j$ is obtained by multiplying $\sigma^2$ by the $j$th element on the main diagonal of $(\mathbf{X'X})^{-1}$, while the covariance between $b_i$ and $b_j$ is given by multiplying $\sigma^2$ by the $ij$th element of $(\mathbf{X'X})^{-1}$. When $\sigma^2$ is unknown, the unbiased estimator $s^2$ may be substituted for $\sigma^2$ in 9.A14 to provide estimates of the standard errors and covariances.[22]

Finally, we may note that in matrix algebra the coefficient of determination may be written as

$$R^2 = 1 - \frac{\sum e_i^2}{\sum y_i^2} = 1 - \frac{\mathbf{e'e}}{\mathbf{y'y} - n\bar{Y}^2}. \tag{9.A15}$$

# problems

**9.1** The following data are based on a random sample of 28 observations:

$$\sum (Y_i - \bar{Y})^2 = 1000; \sum (X_{2i} - \bar{X}_2)^2 = 200; \sum (X_{3i} - \bar{X}_3)^2 = 100;$$
$$\sum (Y_i - \bar{Y})(X_{2i} - \bar{X}_2) = 400; \sum (Y_i - \bar{Y})(X_{3i} - \bar{X}_3) = -100;$$
$$\sum (X_{2i} - \bar{X}_2)(X_{3i} - \bar{X}_3) = 0; \bar{Y} = 50; \bar{X}_2 = 15; \bar{X}_3 = 10.$$

Use these data to estimate the parameters in the equation

$$Y_i = \beta_1 + \beta_2 X_{2i} + \beta_3 X_{3i} + \varepsilon_i$$

and test the hypothesis that $\beta_2 = 1\cdot0$.

**9.2** Given the following data on Ruritania, where

$X_1 =$ stocks of television sets held by manufacturers,
$X_2 =$ aggregate disposable income,
$X_3 =$ the rate of interest on bank advances, and
$X_4 =$ the rate of interest on bank deposits:

$$\sum (X_{1t} - \bar{X}_1)^2 = 625; \qquad \sum (X_{3t} - \bar{X}_3)^2 = 25$$

$$\sum (X_{2t} - \bar{X}_2)^2 = 200; \qquad \sum (X_{4t} - \bar{X}_4)^2 = 16$$

$$\sum (X_{1t} - \bar{X}_1)(X_{2t} - \bar{X}_2) = 340; \qquad \sum (X_{2t} - \bar{X}_2)(X_{3t} - \bar{X}_3) = 0$$

$$\sum (X_{1t} - \bar{X}_1)(X_{3t} - \bar{X}_3) = 30; \qquad \sum (X_{2t} - \bar{X}_2)(X_{4t} - \bar{X}_4) = 0$$

$$\sum (X_{1t} - \bar{X}_1)(X_{4t} - \bar{X}_4) = 4; \qquad \sum (X_{3t} - \bar{X}_3)(X_{4t} - \bar{X}_4) = 0$$

$$\bar{X}_1 = 1000; \qquad \bar{X}_2 = 500; \qquad \bar{X}_3 = 8; \qquad \bar{X}_4 = 6,$$

where $t = 1, 2, ..., 29$, estimate an equation to explain the holding of stocks of television sets and test the significance of your results.

**9.3** Given the equation

$$Y_i = \beta_1 + \beta_2 X_{2i} + \beta_3 X_{3i} + \varepsilon_i,$$

compare the least squares estimates when

    (a) the parameters are unconstrained,
    (b) $\beta_2 = \beta_3$, and
    (c) $\beta_2 + \beta_3 = 1$.

**9.4**  Using annual data for the U.K., 1949–63, the following correlations were calculated:

$$r_{12} = 0\cdot975, \qquad r_{13} = 0\cdot956, \qquad r_{23} = 0\cdot994,$$

where $X_1$ = aggregate personal saving,
$\quad\ X_2$ = aggregate personal income,
$\quad\ X_3$ = a linear time trend.

Calculate the partial correlation coefficient $r_{12.3}$ and comment on these correlations.

**9.5**  If $X_1$ = the level of deposit accounts (£ million),
$\quad\ X_2$ = a time trend measured in months, and
$\quad\ X_3$ = the rate of interest paid on deposit account,

(monthly data for the U.K., 1951–63), given that $r_{12} = 0\cdot954$, $r_{23} = 0\cdot471$ and $r_{13} = 0\cdot380$, calculate $r_{13.2}$ and comment.

**9.6**  For the following non-linear models, find suitable transformations to enable the parameters to be estimated by the method of least squares:

(a) $Y_i = \gamma X^\beta \varepsilon_i$;  $\qquad\qquad$  (b) $Y_i = \gamma e^{(\alpha + \beta X)} \varepsilon_i$;

(c) $\log Y_i = \alpha - \beta/X_i + \varepsilon_i$;  $\qquad$  (d) $Y_i = (X_i^\alpha + Z_i^\beta)\varepsilon_i$.

**9.7**  Suppose the 'true' equation is

$$Y_i = \beta_1 + \beta_2 X_{2i} + \beta_3 X_{3i} + \varepsilon_i$$

and all the least squares assumptions are satisfied. If we estimate a misspecified equation by least squares to obtain

$$\hat{Y}_i = b_1 + b_2 X_{2i} \qquad (i = 1, \dots, n),$$

is $b_2$ (a) an unbiased, (b) a consistent estimator of $\beta_2$?

**9.8**  Suppose the 'true' equation is

$$Y_i = \beta_1 + \beta_2 X_{2i} + \varepsilon_i$$

but we estimate

$$\hat{Y}_i = b_1 + b_2 X_{2i} + b_3 X_{3i}.$$

Is $b_2$ an unbiased estimator of $\beta_2$?

**9.9**  Let $R_1^2$ be the measure of explained variation for $Y_i = \alpha_1 + \alpha_2 X_{2i} + \varepsilon_{1i}$ $(i = 1, \dots, n)$, $R_2^2$ for $Y_i = \beta_1 + \beta_2 X_{2i} + \beta_3 X_{3i} + \varepsilon_{2i}$ and $R_3^2$ for $Y_i = \gamma_1 + \gamma_3 X_{3i} + \varepsilon_{3i}$. Prove that $R_2^2 \geq R_1^2$. Under what conditions will $R_2^2 = R_1^2 + R_3^2$?

**9.10\*** Show that the partial derivatives of $\sum (Q_i - cK_i^a L_i^b)^2$ with respect to $a$, $b$, and $c$ are

$$\frac{\partial \sum e_i^2}{\partial a} = -2 \sum cK_i^a L_i^b \log_e K_i (Q_i - cK_i^a L_i^b)$$

$$\frac{\partial \sum e_i^2}{\partial b} = -2 \sum cK_i^a L_i^b \log_e L_i (Q_i - cK_i^a L_i^b)$$

$$\frac{\partial \sum e_i^2}{\partial c} = -2 \sum K_i^a L_i^b (Q_i - cK_i^a L_i^b)$$

**9.11\***   If $E(\boldsymbol{\varepsilon\varepsilon'}) = \mathbf{V} \neq \sigma^2 \mathbf{I}$, derive the variance–covariance matrix for $\mathbf{b}$.

---

\*   Mathematically the starred questions are rather more difficult than the others.

# chapter 10

# time series problems

## 10.1 introduction

At the beginning of this text, in section 1.1, we drew a distinction between cross-section data and time series data, using the subscript $i$ to denote the former and the subscript $t$ to denote the latter. We have not yet considered why this distinction is important, but time series data may present problems when regression analysis is applied to them and we shall discuss some of these problems in this chapter and the next.

In considering types of data we may distinguish between two kinds of variables: stocks and flows. Stock variables can be measured at a point of time, but flow variables can only be measured over some specified period of time. For example, we can draw a random sample of wine stores and measure their stock of whisky at midnight on 31 December 1973, but to measure their sales of whisky we must decide upon whether we want sales per day, per week or per month. Thus unless we are dealing with stock variables, cross-section data must be collected over time.

However, the important difference between the two types of data arises when we ask whether repeated sampling is possible–a situation which is assumed as the basis for our sampling theory of estimation and hypothesis testing. With cross-sectional data we can often make this assumption, since clearly we can draw random samples of individual consumers, shops, firms, industries, or even countries. In other cases this is not so apparent. For example, in problem 8.6 the reader was asked to estimate the marginal propensity to save from annual data on aggregate saving and personal disposable income for the U.K. for the period 1950–66. At first sight it is difficult to see these data in the context of our sampling model, since here we are dealing with a slice of the post-war history of the U.K., which is a sequence of unique events.[1]

However, at the conceptual level we may resolve this problem by considering the possibility of 'hypothetical' history; actual historical events only happen once, but we can imagine that they might have happened differently. Thus if the saving relationship is

$$S_t = \beta_1 + \beta_2 Y_t + \varepsilon_t,$$

the errors, which represent the net effects of the excluded variables which affect saving, could have been different if, for example, strikes, stock market changes, or variations in the demand for U.K. exports had or had not occurred.[2] If we make this assumption, then we may regard our historical data as one sample from a population of 'hypothetical' histories of the U.K. This assumption enables us to work within the assumptions of a sampling model, but does not imply that the historical data constitute a *random* sample. Whether or not this assumption is justified depends on the properties of the $\varepsilon_t$, since we have not assumed that the values of the independent variable have to be chosen at random. We shall develop a test for the assumption that the $\varepsilon_t$ are random and explore the effects of the violation of this assumption in the next chapter and defer of further discussion until section 11.3.

The reader who has observed the behaviour of economic variables over time will have noticed that they exhibit considerable variety. However, it is possible to suggest a number of characteristics, some but not necessarily all of which most economic time series will possess. For example, Figure 10.1 is the scatter diagram of the U.K. saving and income data of problem 8.6 against time, and the dominant feature of both series is the steady, though not completely regular, growth over the period. This growth we may refer to as a strong *trend* in both series, and the dominant feature of many other economic time series is the trend.

It is not easy to give a formal definition of a trend, since whether or not a variable exhibits a trend depends on the period being considered. For example, Figure 10.2 shows the behaviour of the annual volume of U.K. imports and exports from 1900 to 1969. Certainly, the variables grew during this period, but not in a steady fashion along a trend. Both show the effects of the fluctuations of trade in the inter-war period and the effects of the Second World War and the Korean War. We shall refer to fluctuations of this nature as *cyclical variation* and note that, over a period of several decades, many economic series could be described in terms of trend and cyclical variation. We may also note that had we only considered exports over the period 1958–65, we would have described the variable as being predominantly trend; that is, what we define as a trend in a short period may be part of a cycle when the longer run is considered.

All the variables plotted in Figures 10.1 and 10.2 are flow variables measured on an annual basis, but many economic series are available on a quarterly or monthly basis and these series may exhibit a third kind of variation. For example, in Figure 10.3 we have plotted a quarterly index of the industrial production of food, drink and tobacco for the U.K. from the first quarter of 1948 to the fourth quarter of 1964. The series shows a strong trend, since it grows fairly steadily from year to year and there is no visual evidence of cyclical variation. What is most striking about the series is the 'spiky' within-year pattern which is super-

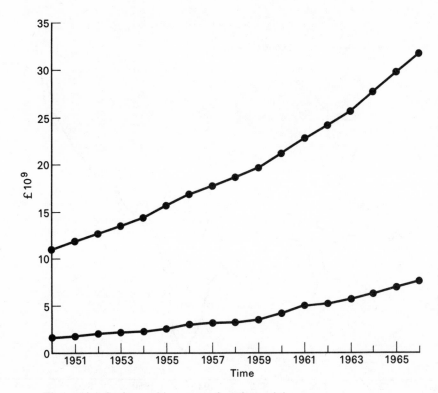

Figure 10.1  Saving and income as functions of time

imposed on the trend. This type of variation we shall call *seasonal variation*, and it is a phenomenon we might well expect to observe in the production of food. The weather can affect many other variables, producing seasonal variation in, for example, house building and the sale of barbecue kits. Some seasonal variation may be man-made, so that, for example, the regular timing of holidays and religious festivals produces seasonal fluctuations in output and expenditure.

Since many economic time series show some combination of these factors, the time series model proposed by early economic statisticians took the form

$$X_t = T_t + C_t + S_t + R_t, \tag{10.1}$$

where $X_t$ represents either the value of some economic variable at time $t$ or the flow over some interval $t$,

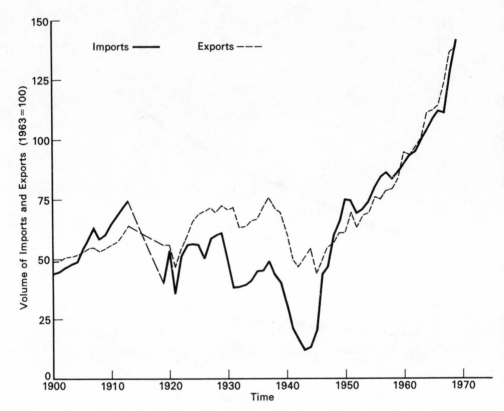

Figure 10.2 Indices of the volume of U.K. imports and exports (1963 = 100)
Source: *The British Economy, Key Statistics 1900–1970,* p. 14, Table K

$T_t$ represents the trend component,

$C_t$ represents the cyclical component,

$S_t$ represents seasonal variation, and

$R_t$ represents 'random' variation which is not explained by the other factors.

This model assumes that the influence of the components is additive and gives constant absolute changes in the variables. An alternative multiplicative form, which gives constant percentage changes, may be written as

$$X_t = (T_t)(C_t)(S_t)(R_t). \tag{10.2}$$

However, by taking logarithms of both sides of 10.2 we obtain a version of 10.1 which is additive in the logarithms, so that 10.2 need not be considered separately.

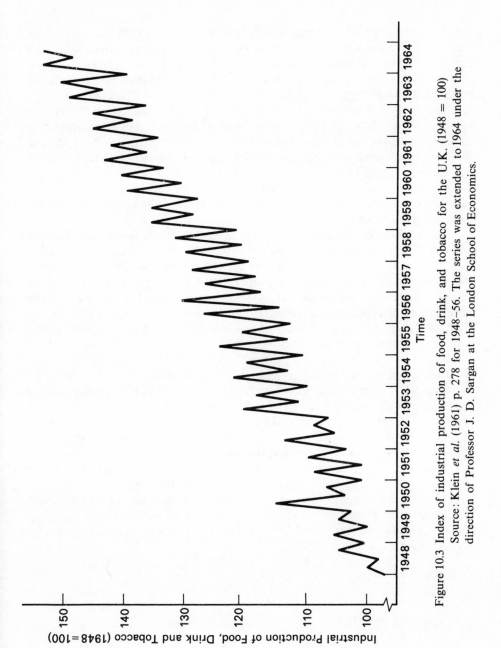

Figure 10.3 Index of industrial production of food, drink, and tobacco for the U.K. (1948 = 100)
Source: Klein *et al.* (1961) p. 278 for 1948–56. The series was extended to 1964 under the direction of Professor J. D. Sargan at the London School of Economics.

The early work with this model was largely concerned with attempts to decompose $X_t$ into its components, attempts which were mainly unsuccessful for a number of reasons. For example, seasonal variation, even when it is not as regular and obvious as it is in Figure 10.3, is reasonably easy to eliminate, but it is much more difficult to distinguish the trend from cyclical variation. Most of the methods that could be used to remove the trend tended to affect the cyclical properties of the series. Secondly, most of the mathematical models of the cyclical component assumed complete, or at least considerable, regularity in the cyclical pattern. While some of the early work suggested considerable regularity, this assumption had to be given up in the face of cumulative cyclical experience.[3] Thirdly, the model does not relate the components to economic variables and is therefore basically descriptive rather than explanatory, which reduces its usefulness for forecasting purposes.[4]

With the development and estimation of macro-economic models after the Second World War, the emphasis has been on explanation and cyclical analysis tended to shift from the attempts to isolate the cyclical component described above to the examination of the dynamic properties of interdependent simultaneous equation macro-economic models.[5] However, while attempts to isolate the cycle have largely disappeared, the other two components often feature in empirical analysis. For this reason we shall discuss trends in the next two sections and consider seasonal variation in section 10.4.

**10.2  trends and the use of 'time' as an explanatory variable**
We described the time series model in 10.1 as a descriptive relationship, and we shall consider the descriptive use of trends before discussing the interpretation of time as an explanatory variable. If we take as examples the saving and personal disposable income data portrayed in Figure 10.1, the data are annual (so that there is no seasonal variation present) and there is no evidence of cyclical fluctuations in either series during the period. Thus we might describe the series as $S_t = f(t) + \varepsilon_{1t}$ and $Y_t = g(t) + \varepsilon_{2t}$, where $f(t)$ and $g(t)$ represent the trends in the two series. From Figure 10.1 it would appear that the trend in both series is probably linear, so that we might write the functions $S_t = \alpha_1 + \beta_1 t + \varepsilon_{1t}$ and $Y_t = \alpha_2 + \beta_2 t + \varepsilon_{2t}$. The time series covers the period 1950–66 and we could substitute these values for $t$ in computing

$$a_1 = \bar{S} - b_1 \bar{t} \tag{10.3}$$

and

$$b_1 = \frac{\sum (S_t - \bar{S})(t - \bar{t})}{\sum (t - \bar{t})^2}, \tag{10.4}$$

with $\bar{t} = 1958$. However, these numbers are larger than we need deal with, since if we subtract a constant, $\lambda$, from all values of $t$, to obtain $(t - \lambda)$, the mean of the new series will be $(\bar{t} - \lambda)$ and the deviations from the mean of the new series $(t - \lambda) - (\bar{t} - \lambda) = (t - \bar{t})$. Hence the value of $b_1$ is independent of which set of consecutive values we choose for $t$, although after transforming $t$ to $(t - \lambda)$ the estimate of the intercept becomes

$$a_1' = \bar{S} - b_1(\bar{t} - \lambda) = a_1 + b_1\lambda.$$

In particular, if we centre $t$ by putting $\lambda = \bar{t}$, we obtain $a_1 = \bar{S}$.

The estimated trends in $S_t$ and $Y_t$, with $t$ set equal to unity in 1950, are

$$S_t = 0.554 + 0.368t, \qquad R^2 = 0.946$$
$$\phantom{S_t = }(0.232) \quad (0.023)$$

$$Y_t = 8.403 + 1.254t, \qquad R^2 = 0.979.$$
$$\phantom{Y_t = }(0.491) \quad (0.048)$$

Judged by the values of $R^2$, the two functions provide good fits and the statistically significant slope coefficients imply that saving increases by £368 million each year, while personal disposable income increases by £1254 million per annum. Thus in 1959, when $t = 10$, the trend value of income is £2094 million as compared with the actual value that year of £1970 million

However, the residuals from the fitted equation can be computed and should always be examined, since they provide additional information about the estimated relationship. In this case the residuals are plotted in Figure 10.4 and we observe that they hardly constitute any random scatter.[6] For both equations the first four observations are under-estimated, the next nine or ten observations are over-estimated, while the last three observations are under-estimated. This pattern suggests that our linear trend is only an approximation and that a quadratic trend would provide a better fit. To test this possibility we have computed the following regressions:

$$S_t = 1.637 + 0.026t + 0.019t^2, \qquad R^2 = 0.994$$
$$\phantom{S_t = }(0.132) \quad (0.034) \quad (0.002)$$

$$Y_t = 10.746 + 0.514t + 0.041t^2, \qquad R^2 = 0.998.$$
$$\phantom{Y_t = }(0.215) \quad (0.055) \quad (0.003)$$

We see that the new values of $R^2$ indicate an improvement in the fit and the quadratic term is highly significant in each case. The new sets of residuals are also plotted in Figure 10.4 and they appear to be somewhat more 'random' than did the residuals from the previous regressions. This is particularly so for the residuals

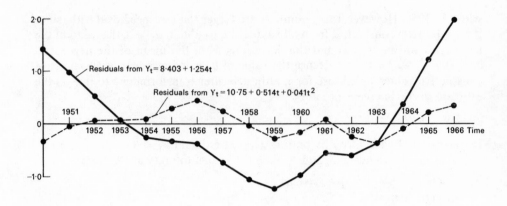

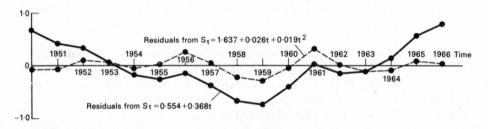

Figure 10.4  Residuals from regression of saving and income on time

from the saving function, a result we shall establish more formally in section 11.3 below.

The two trends fitted above were both quadratic and fitted well. Other time series might require higher-order polynomials which we may represent in general as

$$X_t = \beta_0 + \beta_1 t + \beta_2 t^2 + \dots + \beta_k t^k + \varepsilon_t, \qquad t = 1, \dots, T.$$

The problem that may arise here is that for some time series $(k + 1)$ may be close to $T$, the number of observations, so that the estimated coefficients become imprecise because of the small number of degrees of freedom. One alternative to fitting a single high-order polynomial to the whole time series is to fit lower-order polynomials to sections of the series. This might be done by dividing the time series up into a number of sub-periods to which lower-order polynomials were fitted by least squares. This has the disadvantage that there may be no obvious criterion for splitting the series into sub-periods and that there would be problems with the discontinuities in moving from one sub-period to another. One alternative, which avoids both problems by not specifying separate sub-periods, is the method of *moving averages*.

The simplest application of this technique may be described as follows. Suppose we have a total of $T$ equally spaced observations on some variable, $X_t$, and choose a sub-period of length $r$ (where $r < T$). We then take the first $r$ observations and calculate the arithmetic mean $1/r \sum_{t=1}^{r} X_t$. Having computed this value, we move on by dropping the first observation and introducing the $(r + 1)$th. We then compute $1/r \sum_{t=2}^{r+1} X_t$. Now we move on, dropping the second observation and bringing in the $(r + 2)$th observation, computing $1/r \sum_{t=3}^{r+2} X_t$. If the length of of our sub-period is an even number, the first observation in our new series is located half-way between the $r/2$th and $(r/2 + 1)$th observations in the original series, the second is half-way between the $(r/2 + 1)$th and $(r/2 + 2)$th observations in the original series, and so on. This is only a minor problem, since by taking the average of successive pairs of values of our new series, we may centre them on the original observations. If $r$ is an odd number, say $r = 2m + 1$, then no problem arises, since the first observation in the new series will be located at the same point in time as the $(m + 1)$th observation in the original series. Let us denote the new series by $\bar{X}_t$, $t = (r/2 + 1), (r/2 + 2), \cdots, (T - r/2)$.

The averaging process tends to eliminate period to period fluctuations and produces a much smoother series than the original observations. If the reader recalls that the straight line estimated by least squares passes through the means of the variables, he will see that the process is equivalent to fitting a linear trend by least squares to successive blocks of $r$ observations and only using the point on the trend corresponding to the mean of the block of observations. By choosing other sets of weights which are not all equal to $1/r$, it is possible to fit quadratic or higher-order polynomials to successive blocks of observations.[7] One question which remains to be answered is how the value of $r$, the length of the moving average, is to be chosen. There is no simple rule here, but the longer is the moving average, the smoother the resulting series will be, so that the moving average should be 'long' relative to the frequency of the fluctuations in the series which are to be eliminated. However, $r$ observations are lost, since there are no values of the smoothed series corresponding to the first $r/2$ or last $r/2$ observations in the original series. For this reason the moving average should be 'short' relative to the total length of the time series, $T$. The choice of $r$ involves a trade-off between these two criteria.

To illustrate the estimation of a trend by means of the method of moving averages, we shall use the data on food production which are shown in Figure 10.3. Here the fluctuations to be eliminated in estimating the trend are the within-year seasonal fluctuations, and in this case, therefore, it is reasonable to choose a

Table 10.1.

| Period | Food production $P_t$ | Moving average $\bar{P}$ | Centred moving average $\bar{P}_t$ | $P_t - \bar{P}_t$ |
|---|---|---|---|---|
| 1948 I | 97·1 | | | |
| II | 99·9 | | | |
| | | 100·0 | | |
| III | 98·2 | | 100·5 | −2·3 |
| | | 101·0 | | |
| IV | 104·8 | | 101·7 | 3·1 |
| | | 102·4 | | |
| 1949 I | 100·9 | | 102·6 | −1·7 |
| | | 102·7 | | |
| II | 105·5 | | 102·7 | 2·8 |
| | | 102·7 | | |
| III | 99·7 | | 103·0 | −3·3 |
| | | 103·2 | | |
| IV | 104·8 | | 104·4 | 0·4 |
| | | 105·6 | | |
| 1950 I | 102·7 | | 106·1 | −3·4 |
| | | 106·6 | | |
| II | 115·0 | | 106·8 | 8·2 |
| | | 107·0 | | |
| III | 103·7 | | 106·8 | −3·1 |
| | | 106·6 | | |
| IV | 106·7 | | 105·8 | 0·9 |
| | | 105·0 | | |
| 1951 I | 101·0 | | 104·6 | −3·6 |
| | | 104·3 | | |
| II | 108·7 | | 104·7 | 4·0 |
| | | 105·1 | | |
| III | 101·0 | | 105·4 | −4·4 |
| | | 105·8 | | |
| IV | 109·7 | | 106·4 | 3·3 |
| | | 107·0 | | |
| 1952 I | 103·7 | | 107·6 | −3·9 |
| | | 108·1 | | |
| II | 113·7 | | 108·0 | 5·7 |
| | | 107·8 | | |
| III | 105·3 | | 108·2 | −2·9 |
| | | 108·5 | | |
| IV | 108·7 | | 109·3 | −0·6 |
| | | 110·2 | | |
| 1953 I | 106·3 | | 111·0 | −4·7 |
| | | 111·6 | | |
| II | 120·3 | | 112·8 | 7·5 |
| | | 114·0 | | |
| III | 111·3 | | 114·4 | −1·1 |
| | | 114·8 | | |
| IV | 118·0 | | 115·0 | 3·0 |
| | | 115·2 | | |
| 1954 I | 109·7 | | 115·4 | −5·7 |
| | | 115·7 | | |
| II | 122·0 | | 115·9 | 6·1 |
| | | 116·2 | | |
| III | 113·0 | | 116·3 | −3·3 |
| | | 116·4 | | |
| IV | 120·0 | | 116·7 | 3·3 |
| | | 117·0 | | |
| 1955 I | 110·7 | | 117·0 | −6·3 |
| | | 117·0 | | |
| II | 124·3 | | 117·1 | 7·2 |
| | | 117·2 | | |
| III | 113·0 | | 117·4 | −4·4 |
| | | 117·7 | | |
| IV | 120·7 | | 118·0 | 2·7 |
| | | 118·3 | | |
| 1956 I | 112·7 | | 118·5 | −5·8 |
| | | 118·7 | | |
| II | 126·9 | | 119·9 | 7·0 |

| Period | Food production $P_t$ | Moving average $\bar{P}$ | Centred moving average $\bar{P}_t$ | $P_t - \bar{P}_t$ |
|---|---|---|---|---|
| III | 114·5 | | 121·8 | −7·3 |
| | | 121·2 | | |
| IV | 130·6 | | 122·4 | 8·2 |
| | | 122·4 | | |
| 1957 I | 117·6 | | 122·7 | −5·1 |
| | | 122·3 | | |
| II | 126·4 | | 123·0 | 3·4 |
| | | 123·2 | | |
| III | 118·3 | | 123·0 | −4·7 |
| | | 122·8 | | |
| IV | 128·9 | | 123·7 | 5·2 |
| | | 123·3 | | |
| 1958 I | 119·6 | | 124·5 | −4·9 |
| | | 124·2 | | |
| II | 129·9 | | 125·2 | 4·7 |
| | | 124·8 | | |
| III | 120·7 | | 125·8 | −5·1 |
| | | 125·5 | | |
| IV | 131·9 | | 126·8 | 5·1 |
| | | 126·1 | | |
| 1959 I | 121·9 | | 128·5 | −6·6 |
| | | 127·5 | | |
| II | 135·5 | | 129·9 | 5·6 |
| | | 129·5 | | |
| III | 128·6 | | 131·1 | −2·5 |
| | | 130·4 | | |
| IV | 135·5 | | 132·2 | 3·3 |
| | | 131·8 | | |
| 1960 I | 127·6 | | 132·9 | −5·3 |
| | | 132·7 | | |
| II | 139·2 | | 134·2 | 5·0 |
| | | 133·2 | | |
| III | 130·7 | | 135·3 | −4·6 |
| | | 134·5 | | |
| IV | 140·5 | | 136·5 | 4·0 |
| | | 136·0 | | |
| 1961 I | 133·5 | | 137·6 | −4·1 |
| | | 137·0 | | |
| II | 143·2 | | 138·5 | 4·7 |
| | | 138·3 | | |
| III | 136·1 | | 138·9 | −2·8 |
| | | 138·7 | | |
| IV | 142·0 | | 139·6 | 2·4 |
| | | 139·2 | | |
| 1962 I | 135·7 | | 140·4 | −4·7 |
| | | 140·1 | | |
| II | 146·5 | | 141·1 | 5·4 |
| | | 140·7 | | |
| III | 138·6 | | 141·6 | −3·0 |
| | | 141·5 | | |
| IV | 145·2 | | 142·0 | 3·2 |
| | | 141·7 | | |
| 1963 I | 136·4 | | 142·9 | −6·5 |
| | | 142·3 | | |
| II | 149·0 | | 144·3 | 4·7 |
| | | 143·6 | | |
| III | 143·6 | | 145·3 | −1·7 |
| | | 145·0 | | |
| IV | 150·8 | | 146·2 | 4·6 |
| | | 145·7 | | |
| 1964 I | 139·5 | | 147·4 | −7·9 |
| | | 146·8 | | |
| II | 153·3 | | 148·5 | 4·8 |
| | | 148·0 | | |
| III | 148·3 | | | |
| | | 149·1 | | |
| IV | 155·3 | | | |

moving average of length four quarters to average out the seasonal variation and any random variation in the series. The calculations are laid out in Table 10.1 where the second column contains the data which were plotted in Figure 10.3. The third column contains the values of the moving average, and since here $r = 4$, they do not correspond to the timing of the original series. The moving average series are then centred by taking the average of successive pairs of values and the resulting values are presented in the fourth column of the table, now corresponding to the timing of the original variables. The trend estimated by the method of moving averages is shown in Figure 10.5 and presents a reasonably smooth version of the original series. The final column in the table contains the deviations of the original series from the moving average trend, numbers which we shall analyse in section 10.4 below.

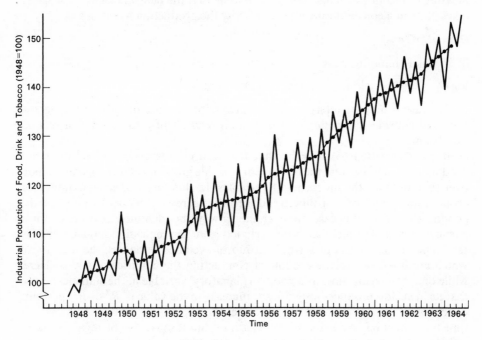

Figure 10.5 Fitting a trend to industrial food production by the method of moving averages

So far we have considered using a time trend descriptively, but in some situations 'time' is used as an explanatory variable in the multiple regression model. Thus we may write

$$Y_t = \beta_1 + \beta_2 X_{2t} + \ldots + \beta_{k-1} X_{(k-1)t} + \beta_k t + \varepsilon_t, \qquad t = 1, \ldots, T,$$

including a linear trend as an independent variable. The interpretation of the coefficient on time in this model is not obvious and can vary considerably. On the one hand it may be that the economic theory underlying our model requires the inclusion of a variable which has grown in a steady manner over time but on which we have no data and for which we use the time trend as a proxy. An example of this kind arises in the estimation of the Cobb–Douglas production function with neutral technical progress, which may be written as

$$Q_t = \gamma L_t^\alpha K_t^\beta A(t)\, \varepsilon_t,$$

where $A(t)$ represents technical change. In general we do not have an independent measure of technical change, but if we assume that the neutral technical progress takes place at a constant rate, we may rewrite the production function as

$$Q_t = \gamma L_t^\alpha K_t^\beta e^{\delta t} \varepsilon_t$$

or in logarithmic form as

$$\log Q_t = \log \gamma + \alpha \log L_t + \beta \log K_t + \hat{\delta} t + \log \varepsilon_t.$$

In this case $\hat{\delta}$ may be interpreted as a proxy for the rate of technical progress. We shall consider an example of this interpretation of a time trend in section 11.3 below.

In other situations, it may be that the multiple regression model does not produce a good fit because the dependent variable grows in a more regular fashion over time than do the independent variables. In this case it is not unknown for economists to introduce a time trend into the regression model to improve the goodness of fit. In this case 'time' cannot always be identified as a proxy for a particular variable which has been excluded from the regression model, but may represent the net effects of a whole group of excluded variables, the identity of which are unknown to the economist conducting the empirical analysis. Here, while one may regard time as a good 'explanatory' variable in the statistical sense that its inclusion produces a significant increase in the value of $R^2$ or $\bar{R}^2$, in terms of economic understanding it cannot be said to explain very much. This use of a time trend may be justified in some situations, but it must also be regarded as an admission that economic theory has failed to provide an adequate hypothesis to explain a particular phenomenon.

One final point on the interpretation of the coefficient on time in the multiple regression model. Suppose we estimate

$$Y_t = \beta_1 + \beta_2 X_{2t} + \beta_3 t + \varepsilon_t, \qquad t = 1, \ldots, T,$$

then from what was said about partial regression coefficients in section 9.3 above, it is clear that $b_2$, the estimate of $\beta_2$ in this equation, is equivalent to removing a linear trend from both $Y_t$ and $X_{2t}$ and then regressing the deviations of $Y_t$ from its trend on the deviations of $X_{2t}$ from its trend.

As an example, we may consider the data on saving and personal disposable income introduced in problem 8.6 above, where the reader estimated the marginal propensity to save. The linear regressions of each of these variables on time were reported on p. 199. If we now regress saving on personal disposable income and time, we obtain

$$S_t = -3.291 + 0.458 Y_t - 0.206t, \qquad R^2 = 0.996,$$
$$(0.286) \quad (0.033) \quad (0.042)$$

and note that the coefficient on income is significant even when linear trends have been removed from both variables. In other words, the deviations from the linear trend in income contribute significantly to the explanation of the variation in saving around its linear trend. The reader may verify the relationship directly by regressing one set of residuals on the other, using the data given in problem 10.3 below.

In this example the time trend has no obvious economic interpretation, unless one wishes to rationalize it *ex post* by noting that, since the estimated coefficient on time is negative, it may be acting as a proxy for wealth.

## 10.3 the problem of multicollinearity
So far we have considered time trends in individual economic series and introducing time as an explanatory variable in the multiple regression model, but now we must consider a problem which may arise if some of the explanatory variables in the multiple regression model are highly correlated with each other.

We shall illustrate how the problem can arise by examining the model

$$Y_i = \beta_1 + \beta_2 X_{2i} + \beta_3 X_{3i} + \varepsilon_i, \qquad i = 1, \cdots, n.$$

We shall first consider the extreme case in which there is an exact linear relationship between $X_2$ and $X_3$, say $X_{2i} = \gamma + \delta X_{3i}$. If these observations are measured as deviations from their sample means, $x_{2i} = \delta x_{3i}$, and substituting for $x_{2i}$ in 9.12 and 9.13 gives

$$b_2 = \frac{\delta \sum yx_3 \sum x_3^2 - \delta \sum yx_3 \sum x_3^2}{\delta^2 (\sum x_3^2)^2 - \delta^2 (\sum x_3^2)^2} = \frac{0}{0} \tag{10.5}$$

$$b_3 = \frac{\delta^2 \sum yx_3 \sum x_3^2 - \delta^2 \sum yx_3 \sum x_3^2}{\delta^2 (\sum x_3^2)^2 - \delta^2 (\sum x_3^2)^2} = \frac{0}{0}. \tag{10.6}$$

Since these ratios are indeterminate, the method of least squares breaks down in this case. We may also note that the formulae for the variances of $b_2$ and $b_3$, given in 9.18 and 9.19, may be written as

$$\text{var}(b_2) = \frac{\sigma^2 \sum x_3^2}{\sum x_2^2 \sum x_3^2 - (\sum x_2 x_3)^2} = \frac{\sigma^2}{\sum x_2^2 (1 - r_{23}^2)} \tag{10.7}$$

$$\text{var}(b_3) = \frac{\sigma^2 \sum x_2^2}{\sum x_2^2 \sum x_3^2 - (\sum x_2 x_3)^2} = \frac{\sigma^2}{\sum x_3^2 (1 - r_{23}^2)}, \qquad (10.8)$$

where $r_{23}$ is the coefficient of correlation between $X_2$ and $X_3$. Clearly, as $r_{23} \to \pm 1$, $\text{var}(b_2)$, $\text{var}(b_3) \to \infty$.

The reason why the method of least squares breaks down in this case is probably easiest to understand if we look at the problem geometrically. Consider the case in which the true relationship is $Y_t = 1 + 2X_{2t} + 3X_{3t} + \varepsilon_t$ and there is an exact linear relationship between $X_2$ and $X_3$, namely $X_{2t} = 1 + 2X_{3t}$. Suppose we have five observations $(Y_t, X_{2t}, X_{3t})$, say $(10 \cdot 32, 3, 1)$, $(13 \cdot 68, 5, 2)$, $(32 \cdot 27, 7, 3)$, $(21 \cdot 38, 9, 4)$ and $(27 \cdot 66, 11, 5)$. Figure 10.6(a) shows these plotted in a scatter diagram and we see that, given the exact relationship between $X_2$ and $X_3$, the five observations all lie in the plane $ABCD$ which cuts the horizontal $X_2$–$X_3$ plane along the line $X_{2t} = 1 + 2X_{3t}$. In other words, our least squares estimation

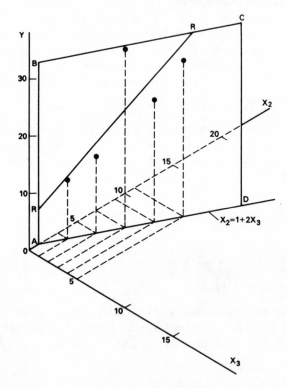

Figure 10.6 (a) Scatter diagram for $Y_t = 1 + 2X_2 + X_3 + \varepsilon_t$ when $X_2 = 1 + 2X_3$

involves trying to determine the position of a plane in three-dimensional space using a two-dimensional scatter of points. Clearly, the missing dimension makes the position of the best fitting estimated plane indeterminate; the line $RR$ minimizes the sum of the squared residuals in the $Y$ dimension, and *any* plane drawn through $RR$ produces the same sum of squared residuals.

In contrast to this extreme case of multicollinearity, the data presented in Table 10.2(b) were constructed in the following way: $X_{3t}$ is the same linear trend variable that was used in Table 10.2(a), but now the values of $X_{2t}$ were chosen from a table of random numbers, giving a correlation between $X_{2t}$ and $X_{3t}$ of 0·127. These data were used to generate the $Y_t$ from $Y_t = 1 + 2X_{2t} + X_{3t} + \varepsilon_t$, using the same random errors that were used to generate the $Y_t$ in Table 10.2(a). The scatter diagram for these data is presented in Figure 10.6(b), and in contrast with Figure 10.6(a) we note here that the points are widely scattered through the three-dimensional space, making it reasonably easy to determine the position of the best fitting plane.

In practice, the linear interdependence between economic variables is usually not complete, and we must therefore see what happens when the correlation between the explanatory variables is 'high' but not unity. To illustrate the high correlation case, we have modified and extended the data which are plotted in

*Table 10.2.*

| $Y$ | $X_2$ | $X_3$ | $Y$ | $X_2$ | $X_3$ |
|---|---|---|---|---|---|
| 6·32 | 1 | 1 | 36·32 | 16 | 1 |
| 9·68 | 3 | 2 | 5·68 | 1 | 2 |
| 30·27 | 7 | 3 | 72·27 | 29 | 3 |
| 21·38 | 9 | 4 | 23·38 | 10 | 4 |
| 23·66 | 9 | 5 | 33·66 | 14 | 5 |
| 30·48 | 11 | 6 | 78·48 | 35 | 6 |
| 36·56 | 15 | 7 | 20·56 | 7 | 7 |
| 49·49 | 17 | 8 | 69·49 | 27 | 8 |
| 45·20 | 17 | 9 | 35·20 | 12 | 9 |
| 44·22 | 19 | 10 | 76·22 | 35 | 10 |
| 54·30 | 21 | 11 | 16·30 | 2 | 11 |
| 50·37 | 25 | 12 | 16·37 | 8 | 12 |
| 65·35 | 27 | 13 | 57·35 | 23 | 13 |
| 68·03 | 27 | 14 | 30·03 | 13 | 14 |
| 76·71 | 29 | 15 | 24·71 | 3 | 15 |
| 71·21 | 31 | 16 | 29·21 | 10 | 16 |
| 88·98 | 35 | 17 | 64·98 | 23 | 17 |
| 87·05 | 37 | 18 | 83·05 | 35 | 18 |
| 98·11 | 39 | 19 | 98·11 | 37 | 19 |
| 105·62 | 41 | 20 | 37·62 | 7 | 20 |

(a) High correlation between $X_2$ and $X_3$.

(b) Low correlation between $X_2$ and $X_3$.

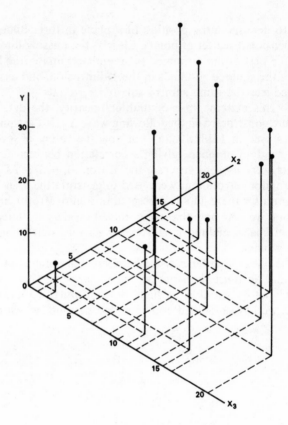

Figure 10.6 (b)  Scatter diagram for $Y_t = 1 + 2X_{2t} + X_{3t} + \varepsilon_t$

Figure 10.6(a). To reduce the correlation between $X_2$ and $X_3$ we have redefined $X_2$ as $X_{2t} = 1 + 2X_{3t} + \omega_t$, where $\omega_t$ was chosen at random to be $\pm 1$. The number of observations was also extended to twenty and the correlation between $X_2$ and $X_3$ was 0·99. The resulting data were used to generate the values of $Y_t$ using $Y_t = 1 + 2X_{2t} + X_{3t} + \varepsilon_t$ and the values were presented in Table 10.2(a). These data were then used in running the following regressions:

$$Y_t = 3·88 + 2·346X_{2t} \qquad , \qquad R^2 = 0·972, \qquad (1a)$$
$$\phantom{Y_t = 3·88} (2·25) \quad (0·093)$$

$$Y_t = 2·74 \qquad\qquad + 4·801X_{3t}, \qquad R^2 = 0·964, \qquad (2a)$$
$$\phantom{Y_t = 2·74} (2·62) \qquad\qquad (0·219)$$

$$Y_t = 4 \cdot 10 + 2 \cdot 695 X_{2t} - 0 \cdot 719 X_{3t}, \qquad R^2 = 0 \cdot 972. \tag{3a}$$
$$(2 \cdot 44) \quad (1 \cdot 183) \qquad (2 \cdot 431)$$

To interpret these results, we note that for the simple regressions of $Y_t$ on $X_{2t}$ and $Y_t$ on $X_{3t}$, using $X_{2t} = 1 + 2X_{3t} + \omega_t$ to substitute for $X_2$ in $Y_t = 1 + 2X_{2t} + X_{3t} + \varepsilon_t$ produces $Y_t = 3 + 4X_{3t} + (\varepsilon_t + 2\omega_t)$, while substituting for $X_3$ gives $Y_t = 1 \cdot 5 + 2 \cdot 5 X_{2t} + (\varepsilon_t - \frac{1}{2}\omega_t)$. Looking at the regressions reported in 1a and 2a, we see that the coefficients on $X_{2t}$ and $X_{3t}$ are statistically significant and are reasonably close to the theoretical values. In both cases the value of $R^2$ is very high. The regression reported in 3a illustrated the effects of multicollinearity:

(i) While the estimated coefficient on $X_{2t}$ is still of the right order of magnitude, the estimated coefficient on $X_{3t}$ is now negative. We have a case in which the indeterminancy suggested by 10.5 and 10.6 is reflected in the instability of the estimated coefficient of $X_3$.

(ii) Comparing 1a or 2a with 3a, we note that the standard errors of the estimated coefficients on $X_2$ and $X_3$ increase dramatically in 3a. This reflects the results presented in 10.7 and 10.8 and is a very good indicator of the presence of multicollinearity: if the standard errors explode when new variables are included in the regression model, we should suspect multicollinearity and check for it by looking at the correlation coefficients between the explanatory variables. In some regressions, multicollinearity can lead to the apparently paradoxical situation in which we obtain a value of $R^2$ which is close to unity, indicating a very high degree of explanatory power, while the inflation of the standard errors produces low $t$-ratios which may indicate that none of the explanatory variables are statistically significant!

By way of contrast, we may consider the regressions of $Y_t$ on $X_{2t}$ and $X_{3t}$ for the data in Table 10.2(b) and Figure 10.6(b). We have

$$Y_t = 8 \cdot 41 + 2 \cdot 123 X_{2t} \qquad , \qquad R^2 = 0 \cdot 944 \tag{1b}$$
$$(2 \cdot 57) \quad (0 \cdot 123)$$

$$Y_t = 32 \cdot 72 \qquad\qquad + 1 \cdot 193 X_{3t}, \qquad R^2 = 0 \cdot 071 \tag{2b}$$
$$(12 \cdot 19) \qquad\qquad (1 \cdot 018)$$

$$Y_t = 2 \cdot 26 + 2 \cdot 083 X_{2t} + 0 \cdot 653 X_{3t}, \qquad R^2 = 0 \cdot 964. \tag{3b}$$
$$(2 \cdot 87) \quad (0 \cdot 101) \qquad (0 \cdot 207)$$

In equation 1b, $X_2$ explains a high proportion of the variation in $Y$, unlike 2b where $X_3$ alone is insignificant and explains a negligible proportion of the variation in $Y$. In 3b, when $X_2$ and $X_3$ are included in the regression, both make contributions to the explanation of the variation in $Y$, and given there is no problem of multicollinearity between $X_2$ and $X_3$ ($r_{23} = 0 \cdot 127$), the standard errors of the estimated coefficients actually *decrease*.

The problem of multicollinearity is reasonably easy to recognize, but much more difficult to solve. Given a high correlation between $X_2$ and $X_3$, the data will tend to fall fairly closely around a straight line such as $RR$ in Figure 10.6(a), and our estimated values of $Y$ will correspond to points on such a line. If the relationship between $Y$ and $X_2$ and $X_3$, and also that between $X_2$ and $X_3$,' both remain stable, then the estimated relationship may provide reasonable forecasts of the values of $Y$ in spite of the multicollinearity. On the other hand, if we are interested in the estimates of the individual parameters, there may be no satisfactory solution. The high inter-correlation among the independent variables can be reduced by excluding some of them from the regression and concentrating on those which we think are the most important, but in that case the resulting estimates may not correspond very closely to the parameters being estimated.[8]

### 10.4  seasonal variation and seasonal adjustment

An example of an economic time series which showed a strong pattern of seasonal variation was given in Figure 10.3. In some situations the seasonal variation may be a serious nuisance requiring special statistical attention. For example, an economic policy-maker who is trying to determine the underlying movement in a time series may find that on a quarter to quarter basis the longer-term trend or cyclical movements are completely dominated by the seasonal variation, as is the case in Figure 10.3. In such a case it may be desirable to seasonally adjust the time series, that is, to eliminate the seasonal variation, in order to emphasize the longer-run components in the series. Seasonal adjustments may be carried out in a number of different ways, and we shall begin by considering seasonal adjustment by means of moving averages.[9]

In section 10.2 we showed how the method of moving averages could be used to estimate the trend and cycle in a time series and illustrated the calculations with the data presented in Table 10.1, the results being presented graphically in Figure 10.5. In general, if we wish to average out the seasonal variation, we choose the length of the moving average to be a year, that is, averaging four quarters if we are dealing with quarterly data, averaging twelve months if we are dealing with monthly data, and so on. In terms of the model presented in equation 10.1, the moving average, $\overline{X}_t$, represents an estimate of $(T_t + C_t)$, and if we subtract $\overline{X}_t$ from $X_t$ we obtain an estimate of $(S_t + R_t)$. If we assume that the seasonal variation remains constant for a given quarter or month over time and that on average the random variation is zero, so that $E(R_t) = 0$, then if we average all the first quarter deviations from the moving average, all the second quarter deviations, etc., we should obtain an estimate of the seasonal variation in that quarter. For monthly data we average all the January deviations from the monthly average, all the February deviations from the monthly average, and so on, to obtain twelve seasonal factors, one for each month. The final step in the seasonal adjustment is to add or subtract (depending on its sign) the seasonal factor to $X_t$ to produce the seasonally adjusted series, $X_t^s$.

To illustrate this method, we shall seasonally adjust the data on the industrial production of food, $P_t$, presented in Table 10.1. Here the trend was estimated by means of a four-quarter moving average, $\bar{P}_t$, and the deviations from the trend, $(P_t - \bar{P}_t)$, were presented in the last column of the table. In Table 10.3 the deviations from the trend, which are estimates of $(S_t + R_t)$, have been grouped by quarter for the seventeen years. The total of the sixteen first quarter deviations from the moving average is $-80{\cdot}2$ and the average first quarter deviation is $-80{\cdot}2/16 = -5{\cdot}01$. By averaging the second, third and fourth quarter deviations, we obtain seasonal factors of $5{\cdot}42$ for the second quarter, $-3{\cdot}53$ for the third quarter and $3{\cdot}26$ for the fourth quarter.

*Table 10.3.* Calculation of seasonal factors for the industrial production of food

| Year | Deviation from moving average $(P_t - \bar{P}_t)$ by quarter | | | |
| | First | Second | Third | Fourth |
|---|---|---|---|---|
| 1948 | | | −2·3 | 3·1 |
| 1949 | −1·7 | 2·8 | −3·3 | 0·4 |
| 1950 | −3·4 | 8·2 | −3·1 | 0·9 |
| 1951 | −3·6 | 4·0 | −4·4 | 3·3 |
| 1952 | −3·9 | 5·7 | −2·9 | −0·6 |
| 1953 | −4·7 | 7·5 | −1·1 | 3·0 |
| 1954 | −5·7 | 6·1 | −3·3 | 3·3 |
| 1955 | −6·3 | 7·2 | −4·4 | 2·7 |
| 1956 | −5·8 | 7·0 | −7·3 | 8·2 |
| 1957 | −5·1 | 3·4 | −4·7 | 5·2 |
| 1958 | −4·9 | 4·7 | −5·1 | 5·1 |
| 1959 | −6·6 | 5·6 | −2·5 | 3·3 |
| 1960 | −5·3 | 5·0 | −4·6 | 4·0 |
| 1961 | −4·1 | 4·7 | −2·8 | 2·4 |
| 1962 | −4·7 | 5·4 | −3·0 | 3·2 |
| 1963 | −6·5 | 4·7 | −1·7 | 4·6 |
| 1964 | −7·9 | 4·8 | | |
| | | | | |
| Total | −80·2 | 86·8 | −56·5 | 52·1 |
| Average | −5·01 | 5·42 | −3·53 | 3·26 |

We could now use these seasonal factors to complete the seasonal adjustment, but the sum of the seasonal factors is $(-5{\cdot}01 + 5{\cdot}42 - 3{\cdot}53 + 3{\cdot}26) = 0{\cdot}14$, so that the effect of the seasonal adjustment would be to increase the sum of every four consecutive adjusted values by $0{\cdot}14$. Since it seems reasonable that the process of seasonal adjustment should leave the average level of the series unchanged, we make a further adjustment to the seasonal factors to ensure that they sum to zero. This may be achieved by subtracting $0{\cdot}14/4 = 0{\cdot}035$ from each of the four seasonal factors to give $-5{\cdot}04, 5{\cdot}38, -3{\cdot}56$ and $3{\cdot}22$ as the final seasonal factors to be used in the process of seasonal adjustment.[10] The seasonally adjusted series is then obtained by adding 5·04 to every first quarter value of $P_t$, subtracting 5·38 from every second quarter observation on $P_t$, and so on. The resulting seasonally adjusted series is presented in Table 10.4. It will be left as an exercise for the reader to plot the seasonally adjusted values against the original data which are presented in the first column of Table 10.1.[11]

*Table 10.4.* Seasonally adjusted industrial production of food

| Year | First | Quarter Second | Third | Fourth |
|------|-------|-------|-------|--------|
| 1948 | 102·1 | 94·5  | 101·8 | 101·6 |
| 1949 | 105·9 | 100·1 | 103·3 | 101·6 |
| 1950 | 107·7 | 109·6 | 107·3 | 103·5 |
| 1951 | 106·0 | 103·3 | 104·6 | 106·5 |
| 1952 | 108·7 | 108·3 | 108·9 | 105·5 |
| 1953 | 111·3 | 114·9 | 114·9 | 114·8 |
| 1954 | 114·7 | 116·6 | 116·6 | 116·8 |
| 1955 | 115·7 | 118·9 | 116·6 | 117·5 |
| 1956 | 117·7 | 121·5 | 118·1 | 127·4 |
| 1957 | 122·6 | 121·0 | 121·9 | 125·7 |
| 1958 | 124·6 | 124·5 | 124·3 | 128·7 |
| 1959 | 126·9 | 130·1 | 132·2 | 132·3 |
| 1960 | 132·6 | 133·8 | 134·3 | 137·3 |
| 1961 | 138·5 | 137·8 | 139·7 | 138·8 |
| 1962 | 140·7 | 141·1 | 142·2 | 142·0 |
| 1963 | 141·4 | 143·6 | 147·2 | 147·6 |
| 1964 | 144·5 | 147·9 | 151·9 | 152·1 |

We have illustrated the moving average method of seasonal adjustment on a single time series, but we must also consider seasonal variation in the context of regression analysis. For example, if the time series of the variables to be used in regression analysis exhibit strong seasonal patterns, the seasonal variation may distort or even mask any relationship between them.

We shall develop this point by means of an economic example. We have already carried out some time series analysis of the data on the industrial production of food, $P_t$, which was presented in Table 10.1, and we shall now try to explain the variation in the industrial production of food by means of regression analysis, using data on the consumption of food in the previous quarter, $C_{t-1}$, as the independent variable.[12] The data on this variable are given in problem 10.3 at the end of this chapter, and the reader will note that the series shows a marked seasonal pattern. If we ignore the seasonal variation and regress $P_t$ on $C_{t-1}$, we obtain

$$P_t = -2·47 + 1·090C_{t-1}, \qquad R^2 = 0·601.$$
$$\quad\;\;(12·58) \;\; (0·109)$$

Looking at these results, the estimated coefficient on $C_{t-1}$ is highly significant ($t = 1{\cdot}090/0{\cdot}109 = 10{\cdot}0$), but as $R^2$ is only equal to $0{\cdot}601$, a high proportion of the variation in $P_t$ is not explained by this regression. This latter result is illustrated graphically in Figure 10.7(a), where $P_t$ is plotted against $C_{t-1}$ and the fitted regression line is shown.

In terms of our earlier discussion of seasonality, one way to allow for seasonal variation would be to work with seasonally adjusted data. We have presented a seasonally adjusted series for the industrial production of food in Table 10.4, and the data on the consumption of food may be seasonally adjusted in the same way (see problem 10.3 below). If we now regress the seasonally adjusted series for

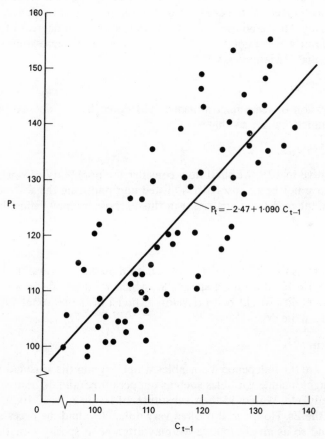

Figure 10.7 (a)

the industrial production of food, $P_t^s$, on the seasonally adjusted data of the consumption of food lagged one quarter, $C_{t-1}^s$, we obtain

$$P_t^s = -42\cdot03 + 1\cdot435 C_{t-1}^s, \qquad R^2 = 0\cdot941.$$
$$\phantom{P_t^s =} (5\cdot10) \quad (0\cdot044)$$

The result of eliminating the seasonal variation is to increase the coefficient on food consumption from $1\cdot090$ to $1\cdot435$. It would be tempting to compare the values of $R^2$, but since $P_t$ and $P_t^s$ are different variables, the temptation should be avoided. The new regression is based on 68 observations and we have estimated 2 parameters, so that there would appear to be 66 degrees of freedom. However, the prior seasonal adjustment involves the implicit fitting of a seasonal model and hence some degrees of freedom are lost in the process, though it is not easy to say how many. Hence 66 *overstates* the actual number of degrees of freedom.

In terms of our general notation we have considered the regression of $Y$ on $X$ ignoring seasonal variation, say

$$Y_t = \alpha + \beta X_t + \varepsilon_{1t},$$

and the regression of seasonally adjusted $Y$, denoted by $Y_t^s$, on seasonally adjusted, $X$, denoted by $X_t^s$, so that

$$Y_t^s = \gamma + \delta X_t^s + \varepsilon_{2t}. \tag{10.9}$$

As an alternative to 10.9, we shall now consider the possibility of working with data which have not been seasonally adjusted and removing the seasonal variation as part of the regression analysis. For this purpose we may define the regression model

$$Y_t = \alpha + \beta X_t + S_t + \varepsilon_{3t}, \tag{10.10}$$

where $S_t$ represents the effects of seasonal variation on the two variables.

In order to use 10.10 we must specify $S_t$, which we might do in a number of ways. One possibility would be to develop a model of the seasonal variation in the model and write

$$S_t = f(Z_{1t}, Z_{2t}, \cdots), \tag{10.11}$$

where the $Z$s are the independent variables which explain the seasonal variation and which might include variables such as temperature, rainfall, number of days of snow in July, etc. We could then substitute 10.11 for $S_t$ in 10.10 and regress $Y_t$ on $X_t$ and the $Z$s. However, it is often very difficult to find the necessary $Z$s to explain $S_t$, and so as an alternative we may attempt to specify a mathematical model to proxy for the variables which cause the seasonal variation, in the same way that a time trend was used to represent unknown or unmeasurable variables in section 10.2 above.

To show how such a model may be developed, let us return to the data on the production and consumption of food which are plotted in Figure 10.7(a). This diagram does not indicate which point corresponds to which quarter to each year, but once we replot the data in Figure 10.7(b) using different symbols to indicate each of the four quarters, some sort of order begins to emerge. Looking at the diagram, we note that the points corresponding to the first quarter of each year lie in a band below the other observations, while those corresponding to the second quarter of each year lie in a band above the other observations. Taking the group of points corresponding to a given quarter separately, the relationship between $P_t$ and $C_{t-1}$ appears to be linear, and visually it would seem a reasonable approximation to assume that the four straight lines are parallel.

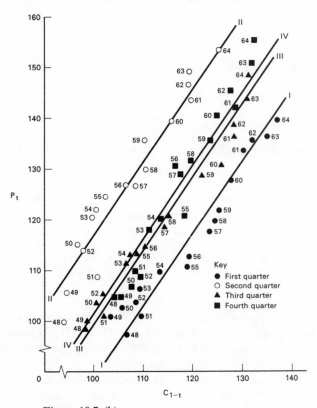

Figure 10.7 (b)

This visual inspection suggests a model in which the slope of the relationship between $P_t$ and $C_{t-1}$ is the same in all four quarters while the effect of the seasonal variation is to produce a parallel shift and hence a different intercept in each quarter. One way of representing this model would be as

$$P_t = \begin{cases} \alpha_1 \text{ in first quarter} \\ \alpha_2 \text{ in second quarter} \\ \alpha_3 \text{ in third quarter} \\ \alpha_4 \text{ in fourth quarter} \end{cases} + \beta C_{t-1} + \varepsilon_t. \qquad (10.12)$$

The estimation of this model may be achieved by introducing some artifical variables which are analogous to the trend. We define four variables $Q_{1t}$, $Q_{2t}$, $Q_{3t}$, $Q_{4t}$, so that

$$Q_{it} = \begin{cases} 1 \text{ in the } i\text{th quarter} \\ 0 \text{ in the other quarters} \end{cases}, i = 1, 2, 3, 4,$$

and rewrite 10.12 as

$$P_t = \alpha_1 Q_{1t} + \alpha_2 Q_{2t} + \alpha_3 Q_{3t} + \alpha_4 Q_{4t} + \beta C_{t-1} + \varepsilon_t. \qquad (10.13)$$

The $Q_i$s, which are usually called 'dummy' variables, have the form

$$Q_1 = (1, 0, 0, 0, 1, 0, 0, 0, 1, \ldots), \qquad Q_2 = (0, 1, 0, 0, 0, 1, 0, 0, 0, 1, \ldots),$$

etc., and they are used, together with $P_t$ and $C_{t-1}$, to estimate the $\alpha$s and $\beta$ and their standard errors in the multiple regression 10.13.

For our example of the production of food, the estimated relationship 10.13 is

$$P_t = -55 \cdot 05 Q_{1t} - 26 \cdot 68 Q_{2t} - 43 \cdot 90 Q_{3t} - 42 \cdot 13 Q_{4t} + 1 \cdot 434 C_{t-1},$$
$$\quad (5 \cdot 49) \qquad (4 \cdot 94) \qquad (5 \cdot 23) \qquad (5 \cdot 42) \qquad (0 \cdot 045)$$
$$R^2 = 0 \cdot 946.$$

In each first quarter $P_t = -55 \cdot 05 + 1 \cdot 434 C_{t-1}$, in each second quarter $P_t = -26 \cdot 68 + 1 \cdot 434 C_{t-1}$, and so on. The lines I–I, II–II, III–III, and IV–IV in Figure 10.7(b) represent these four equations. Comparing this equation with the one reported on p. 212 in which the seasonal variation was ignored, we see that including the seasonal dummy variable has increased the value of $R^2$ from 0·601 to 0·946, that is, we have explained at least a high proportion of the seasonal variation. This is confirmed by the fact that all the estimated coefficients on the dummy variables are statistically significant. The final and most important difference to note is in the coefficient on $C_{t-1}$, which increases from 1·090 to 1·434. This value is almost the same as that obtained by prior seasonal adjustment reported above. However, this is by no means always the case. Prior adjustment by moving averages and seasonal adjustment within a regression model are not exactly equivalent and can lead to different results.

As we are estimating the intercept for each quarter separately, there is no constant term in 10.13.[13] This model can be rewritten to include a constant term if we drop one of the seasonal dummies. For example, if we drop $Q_{4t}$ we may write

$$P_t = \alpha + \beta C_{t-1} + \gamma_1 Q_{1t} + \gamma_2 Q_{2t} + \gamma_3 Q_{3t} + \varepsilon_t. \tag{10.14}$$

Comparing 10.13 and 10.14, as we have only rearranged the constant terms, all measurements in terms of deviations from sample means will be unaffected. Hence $b$ is identical in the two equations, as is its standard error and the value of $R^2$. In the fourth quarter of any year $P_t = \alpha + \beta C_{t-1} + \varepsilon_t$, and comparing this with 10.13 it is clear that $\alpha = \alpha_4$, the intercept for the fourth quarter. In the first quarter $P_t = \alpha + \gamma_1 + \beta C_{t-1} + \varepsilon_t$, so that $\alpha + \gamma_1 = \alpha_1$. That is, $\gamma_1$ measures the shift in the intercept from the fourth quarter level to the first quarter level. The relationship between the two models is shown graphically in Figure 10.8.

For our example on the production of food, 10.14 is estimated as

$$P_t = -42·13 + 1·434C_{t-1} - 12·92Q_{1t} + 15·45Q_{2t} - 1·73Q_{3t}, \qquad R^2 = 0·946.$$
$$\quad\; (5·42) \quad (0·045) \qquad (1·32) \qquad\;\; (1·40) \qquad (1·33)$$

To test the hypotheses $H_0 : \gamma_i = 0, i = 1, 2, 3, H_1 : \gamma_i \neq 0$, we may compute the $t$-ratios $t_1 = -12·92/1·32 = -9·79$, $t_2 = 15·45/1·40 = 11·04$, and $t_3 = -1·73/1·33 = -1·30$. We see that with 68 observations and estimating 5 parameters we have 63 degrees of freedom, so that both $t_1$ and $t_2$ are highly significant while $t_3$ is not significant. These formal tests confirm the visual impression which is obtained from Figure 10.6(b).

Clearly, 10.13 and 10.14 are equivalent, and which form we choose to estimate is a matter of which is more convenient. For example, suppose when we plot the scatter diagram for two variables $Y$ and $X$ (which both contain strong trends) we obtain the result shown in Figure 10.9, the break occurring at the time of a devaluation. From the diagram it would appear that the effect of the devaluation has been to change the intercept, leaving the slope unchanged. A model which incorporates these features is

$$Y_t = \alpha + \beta X_t + \gamma D_t + \varepsilon_y, \tag{10.15}$$

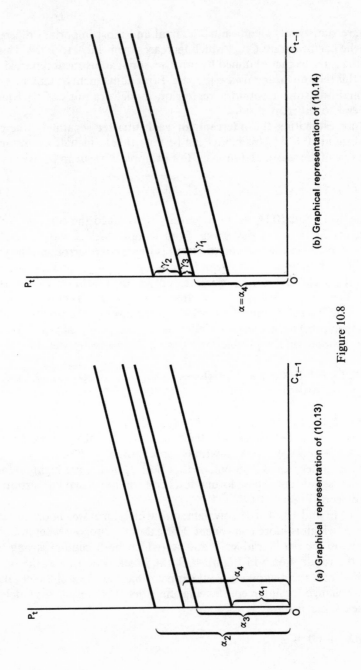

(a) Graphical representation of (10.13)

(b) Graphical representation of (10.14)

Figure 10.8

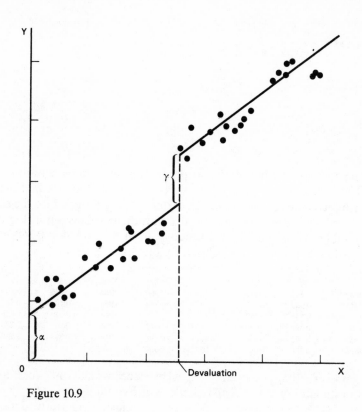

Figure 10.9

where

$$D_t = \begin{cases} 0 \text{ in the period before devaluation} \\ 1 \text{ in the period after devaluation.} \end{cases}$$

By estimating this form of the equation, we may see whether the devaluation produced a shift which was statistically significant or not by testing $H_0 : \gamma = 0$.

# problems

**10.1** Given the data on the industrial production of food which are presented in Table 10.1, calculate the trend in this series by means of the method of moving averages with averages of lengths (a) 2 quarters, (b) 8 quarters. Plot the data and the fitted trends and compare your results with the trend shown in Figure 10.5.

**10.2** The table below contains the residuals from $S_t = 0.554 + 0.368t, (S'_t)$ and from $Y_t = 8.403 + 1.254t, (Y'_t)$ which were plotted in Figure 10.4:

| Year | $S'_t$ | $Y'_t$ | Year | $S'_t$ | $Y'_t$ | Year | $S'_t$ | $Y'_t$ |
|------|--------|--------|------|--------|--------|------|--------|--------|
| 1950 | 0.678 | 1.343 | 1956 | -0.129 | -0.380 | 1962 | -0.136 | -0.604 |
| 1951 | 0.410 | 0.989 | 1957 | -0.397 | -0.734 | 1963 | -0.104 | -0.358 |
| 1952 | 0.343 | 0.535 | 1958 | -0.665 | -1.088 | 1964 | 0.128 | 0.388 |
| 1953 | 0.075 | 0.081 | 1959 | -0.733 | -1.242 | 1965 | 0.560 | 1.234 |
| 1954 | -0.193 | -0.272 | 1960 | -0.400 | -0.996 | 1966 | 0.792 | 1.980 |
| 1955 | -0.261 | -0.326 | 1961 | 0.032 | -0.550 | | | |

Plot $S'_t$ against $Y'_t$. If $S'_t = \gamma + \delta Y'_t + \varepsilon_t$, what values would you expect for $c$ and $d$? Obtain estimates of $\gamma$ and $\delta$ and check your hypothesis.

**10.3** The table below contains seasonally unadjusted quarterly data on the consumption of food:

(a) Plot the data against time and comment on the seasonal variation.

(b) Estimate the trend in the series by means of a moving average of length four quarters. Plot the trend.

(c) Use the deviations from your trend to estimate the seasonal factors and then calculate a seasonally adjusted series for the consumption of food, $C^s_t$.

(d) Plot the seasonally adjusted data on the production of food, $P^s_t$ (given in Table 10.4), against $C^s_{t-1}$.

**10.4** (a) Is a perfect correlation between two explanatory variables in a multiple regression model (i) a necessary or (ii) a sufficient condition for extreme multicollinearity?

(b) Suppose

$$D_{1t} = \begin{cases} 1 & \text{for } t = 1, \ldots, m \\ 0 & \text{for } t = (m + 1), \ldots, T \end{cases}$$

| Year | Quarter | | | |
|------|------|------|------|------|
| | I | II | III | IV |
| 1947 | | | | 106·8 |
| 1948 | 94·0 | 98·4 | 104·4 | 103·2 |
| 1949 | 94·5 | 98·8 | 105·4 | 105·8 |
| 1950 | 97·6 | 100·6 | 107·7 | 109·8 |
| 1951 | 100·9 | 102·1 | 108·4 | 108·8 |
| 1952 | 97·9 | 102·0 | 109·2 | 109·1 |
| 1953 | 99·7 | 106·5 | 111·1 | 113·7 |
| 1954 | 100·5 | 107·3 | 113·7 | 119·0 |
| 1955 | 102·2 | 108·9 | 118·2 | 119·2 |
| 1956 | 106·4 | 110·3 | 116·6 | 123·5 |
| 1957 | 108·4 | 114·3 | 117·5 | 124·5 |
| 1958 | 110·5 | 115·2 | 119·8 | 125·3 |
| 1959 | 110·2 | 122·0 | 123·6 | 127·9 |
| 1960 | 115·8 | 126·0 | 125·0 | 130·2 |
| 1961 | 119·7 | 128·4 | 128·6 | 132·0 |
| 1962 | 119·2 | 128·4 | 127·7 | 135·0 |
| 1963 | 119·2 | 131·0 | 132·2 | 137·0 |
| 1964 | 125·2 | 131·4 | 132·3 | |

and

$$D_{2t} = \begin{cases} 0 & \text{for} \quad t = 1, \ldots, m \\ 1 & \text{for} \quad t = (m+1), \ldots, T. \end{cases}$$

Calculate the correlation between $D_{1t}$ and $D_{2t}$.

(c) If $Y_t = \gamma_1 D_{1t} + \gamma_2 D_{2t} + \varepsilon_t$, $t = 1, \ldots, T$, interpret $\gamma_1$ and $\gamma_2$. How does this model compare with $Y_t = \alpha + \beta D_{2t} + \varepsilon_t$? Estimate $\gamma_1$ and $\gamma_2$. Do the estimates affect your answer to (a) above?

**10.5** An alternative to using dummy variables in our food example would have been to fit a separate regression line to the data for each quarter. Would the results have been the same as those obtained by using dummy variables? If not, which is the more efficient method?

# chapter 11

# the limitations of regression analysis

## 11.1 introduction

In chapter 8 it was shown that, given the assumptions concerning the random errors which were listed in 8.13 to 8.16, the least squares estimates were best, linear and unbiased. Similar results were indicated in chapter 9 for the assumptions listed on p. 180. In this chapter we shall investigate what happens to the properties of the least squares estimators when the various assumptions in the model are relaxed. We shall relax the assumptions one at a time and discuss the effects in the context of the simple linear model as far as possible. Our discussion will consider three things: firstly, the effect of relaxing the assumption; secondly, the possibility of testing for infringement of the assumption; and finally, what can be done if the assumption is untenable.

The first point to note is that the assumptions are not all of equal importance. For example, the sampling theory developed in chapter 8 is based on the assumption that the random errors are drawn from a normal population. However, if the errors are drawn from a non-normal population, provided it is reasonably unskewed, these sampling results hold approximately, the approximation being a good one even for quite small samples.[1] Since this assumption was not necessary to show that the least squares estimators were best, linear and unbiased, these properties are independent of the form of the distribution of the errors.

Secondly, the relaxation of an assumption may have some effect, but it may not be serious from the point of view of economic analysis. For example, suppose we relax the assumption $E(\varepsilon_i) = 0$. As alternatives we might consider either

$$E(\varepsilon_i) = \mu \neq 0 \tag{11.1}$$

or

$$E(\varepsilon_i) = \mu_i \neq 0, \qquad i = 1, \ldots, n, \tag{11.2}$$

the first being that the errors are drawn from a population with a non-zero mean, while the second allows for the possibility that the errors are drawn from different populations with different non-zero means. We may easily show that relaxing the assumption $E(\varepsilon_i) = 0$ can make the least squares estimators biased. To consider

$b$ first, from p. 138 we have

$$b = \beta + \sum k_i \varepsilon_i \tag{8.24}$$

and taking expected values

$$E(b) = \beta + \sum k_i E(\varepsilon_i).$$

Now if 11.1 holds,

$$E(b) = \beta + \mu \sum k_i = \beta,$$

since by 8.21 $\sum k_i = 0$. Thus $b$ remains unbiased. However, if 11.2 holds,

$$E(b) = \beta + \sum k_i \mu_i \neq \beta$$

and $b$ is biased, since there is no reason why $\sum k_i \mu_i$ should be equal to zero.
For $a$, we have (from p. 138)

$$E(a) = \alpha + \frac{1}{n} \sum E(\varepsilon_i) - \overline{X} \sum k_i E(\varepsilon_i)$$

and if 11.1 holds,

$$E(a) = \alpha + \mu \neq \alpha,$$

while if 11.2 holds,

$$E(a) = \alpha + \frac{1}{n} \sum \mu_i - \overline{X} \sum k_i \mu_i \neq \alpha.$$

In either case it is clear that $a$ will be a biased estimator of $\alpha$ while $b$ is an unbiased estimator of $\beta$ if 11.1 holds. However, in economic theory the constant term is usually of no great interest, and hence bias in $a$ is of less concern than bias in $b$. As far as testing the assumption $E(\varepsilon_i) = 0$ goes, there is a major problem, since the $\varepsilon$s are unobservable and it is a property of the method of least squares that the mean of the residuals is always zero.[2] Hence we cannot test this assumption by using the residuals as proxies for the unobservable errors, but must rely on using independent evidence concerning the $\varepsilon$s when it is available.

The relaxation of the remaining assumptions has more serious effects on the properties of the least squares estimators, and these cases will therefore be considered in more detail below. In the next section we shall consider what happens when the variance is not the same for all the $\varepsilon$s, which is known as the problem of *heteroscedasticity*. In section 11.3 we shall discuss the problem that arises when $E(\varepsilon_i \varepsilon_j) \neq 0$, the problem of *autocorrelation*. Finally, in section 11.4 we shall consider what happens if the $X$s are not a set of constants and in particular the implications of the possibility that the $X$s and the $\varepsilon$s may not be independent.

### 11.2 non-constant variances. the problem of heteroscedasticity

If 8.14 holds, the errors are said to be *homoscedastic*. In this section we shall consider what happens when $E(\varepsilon_i^2) \neq \sigma^2$ while assumptions 8.13, 8.15 and 8.16 still hold. To put this problem in an economic context, it is most likely to arise at the micro-level in cross-sectional studies. For example, in a cross-sectional analysis of the consumption function, families with low incomes have less scope to vary their expenditures than families with high incomes. Similarly, in studying a random sample of firms, the range of the random errors in sales may well vary with the size of the firm. Heteroscedasticity can occur in the time series contest if some of the excluded variables change their range of variation as a result of an exogenous disturbance, such as a war, leading to a change in the variance of the random errors in the regression model over time.

In order to investigate the effects of heteroscedasticity on the properties of the least squares estimators, we must replace 8.14 with some alternative assumption. We shall assume extreme heteroscedasticity, that is, we shall consider the case in which the random errors are drawn from distributions which have different variances, so that

$$E(\varepsilon_i^2) = \sigma_i^2, \qquad i = 1, \dots, n. \tag{11.3}$$

Looking back to the derivation of the properties of the least square estimators in section 8.3, it is clear that the new assumption does not affect the linearity and unbiasedness of the estimators. Where it does have an effect is when we consider the variances of the estimators. For example, on p. 139 it was shown that

$$\text{var}(b) = \sum k_i^2 E(\varepsilon_i^2) = \sigma^2 \sum k_i^2,$$

but under the new assumption we now have

$$\text{var}(b) = \sum k_i^2 E(\varepsilon_i^2) = \sum k_i^2 \sigma_i^2 \neq \sigma^2 \sum k_i^2. \tag{11.4}$$

Hence we should not use the usual formula 8.25 to compute the sampling variance of $b$. It could also be shown that the formula for the sampling variance of $a$ which is given in 8.26 is no longer relevant if the errors are heteroscedastic.

If the variances $\sigma_i^2$, $i = 1, \dots, n$, were known, then we could substitute these values into $\sum k_i^2 \sigma_i^2$ and compute the correct value for the sampling variance of $b$.[3] The real problem is that usually the values of the parameters are not known and have to be estimated from the sample data, using the residuals as proxies for the unobservable errors. In this case we have to estimate $n$ variances from $n$ observations, that is, one for each variance, a situation in which estimation is obviously impossible, since we cannot estimate a variance from one observation!

This also raises a problem when we consider how to test whether the errors are homoscedastic or heteroscedastic. We may use the residuals as proxies for the unobservable errors, but with only one observation per distribution we cannot set up any formal test for the difference between variances. At most we might order the observations in some way, for example by the increasing size of the independent variable, and plot them. They may then be examined visually to see whether the range of variation appears to vary over the sample.

In many economic examples our assumption of extreme heteroscedasticity would be unnecessarily general, since the variance may vary in a systematic way across the sample data as a function of the independent variable. An example would arise with a cross-section of firms if the independent variable were the level of sales and the variances of the errors varied with the size of the firm. If this is so, then instead of 11.3 we may propose

$$E(\varepsilon_i^2) = \sigma^2 X_i^\gamma, \qquad i = 1, \ldots, n, \tag{11.5}$$

where $\gamma = 0$ gives errors which are homoscedastic. For simplicity we shall analyse the case in which $\gamma = 2$, so that

$$E(\varepsilon_i^2) = \sigma^2 X_i^2. \tag{11.6}$$

To illustrate this case, consider the following hypothetical data in which $\sigma^2 = 1$, so that $E(\varepsilon_i^2) = \sigma_i^2 = X_i^2$, the $\varepsilon_i$ are $N(0, \sigma_i^2)$, and the $Y$s are generated from $Y_i = 100 + 10X_i + \varepsilon_i$, $i = 1, \ldots, 20$. The data on $Y$ and $X$, ordered by increasing values of $X$, are presented in the first two columns of Table 11.1 and,

Table 11.1. An example of heteroscedasticity

| X | Y | $\hat{Y}$ | e | $\varepsilon$ | X' | Y' |
|---|---|---|---|---|---|---|
| 1·0 | 109·30 | 105·33 | 3·97 | −0·70 | 1·000 | 109·30 |
| 2·0 | 120·00 | 116·10 | 3·90 | −0·00 | 0·500 | 60·00 |
| 3·0 | 130·76 | 126·87 | 3·89 | 0·76 | 0·333 | 43·59 |
| 4·0 | 135·69 | 137·64 | −1·95 | −4·31 | 0·250 | 33·92 |
| 5·0 | 148·01 | 148·41 | −0·40 | −2·00 | 0·200 | 29·60 |
| 6·0 | 168·00 | 159·18 | 8·82 | 8·00 | 0.167 | 28·00 |
| 7·0 | 163·55 | 169·95 | −6·40 | −6·45 | 0·143 | 23·36 |
| 8·0 | 166·91 | 180·72 | −13·81 | −13·09 | 0·125 | 20·86 |
| 9·0 | 179·60 | 191·49 | −11·89 | −10·40 | 0·111 | 19·96 |
| 10·0 | 207·80 | 202·26 | 5·54 | 7·80 | 0·100 | 20·78 |
| 11·0 | 209·18 | 213·03 | −3·85 | −0·82 | 0·091 | 19·02 |
| 12·0 | 245·22 | 223·80 | 21·42 | 25·22 | 0·083 | 20·44 |
| 13·0 | 228·69 | 234·57 | −5·88 | −1·31 | 0·077 | 17·59 |
| 14·0 | 266·26 | 245·34 | 20·92 | 26·26 | 0·071 | 19·02 |
| 15·0 | 246·28 | 256·11 | −9·83 | −3·72 | 0·067 | 16·42 |
| 16·0 | 261·50 | 266·88 | −5·38 | 1·50 | 0·062 | 16·34 |
| 17·0 | 262·57 | 277·65 | −15·08 | −7·43 | 0·059 | 15·44 |
| 18·0 | 305·00 | 288·42 | 16·58 | 25·00 | 0·055 | 16·94 |
| 19·0 | 299·37 | 299·19 | 0·18 | 9·37 | 0·052 | 15·76 |
| 20·0 | 307·00 | 309·96 | −2·96 | 7·00 | 0·050 | 15·35 |

given these data, the least squares estimate of $Y_i = \alpha + \beta X_i + \varepsilon_i$ (if we ignore the heteroscedasticity) is

$$\hat{Y}_i = 94 \cdot 56 + 10 \cdot 77 X_i, \qquad R^2 = 0 \cdot 975.$$
$$\phantom{\hat{Y}_i = 94 \cdot 56 + } (0 \cdot 408)$$

The predicted values of $Y_i$, $\hat{Y}_i$, are presented in the third column of Table 11.1, and by subtracting column 3 from column 2 we obtain the residuals, $e_i = Y_i - \hat{Y}_i$, which are given in column 4. In this case, since the data are hypothetical the $\varepsilon_i$ are known and their values are presented in column 5 for comparison with the es.

The residuals are plotted according to increasing values of $X$ in Figure 11.1(a), and we observe that they do suggest the presence of heteroscedasticity. (The dotted lines show the corresponding values of $X$, that is, they form a range $\pm \sigma$ around the mean of zero.)

A more formal test for the presence of heteroscedasticity may be developed as follows.[4] Suppose we have an even number of observations $n = 2r$; order the residuals by increasing values of the independent variable and divide them into two groups, $e_1, ..., e_r$ and $e_{r+1}, ..., e_n$. Then if 8.14 holds and the errors are homoscedastic, estimates of $\sigma^2$ based on the first $r$ and on the last $r$ values of the residuals should not differ significantly. However, if 11.6 holds we would expect the variance of the last $r$ residuals to be larger than that of the first $r$. However, if the variances are changing slowly, the sharpness of the test may be increased by omitting the central values of the residuals and comparing the variances computed from the first (smallest) residuals and the last (largest) residuals.

The test proposed by Goldfeld and Quandt involves ordering the $n$ observations by the increasing size of $X$ and then omitting the $m$ central values. The relationship $Y_i = \alpha + \beta X_i + \varepsilon_i$ is then estimated separately from the first $(n - m)/2$ observations and the last $(n - m)/2$ observations. If $S_1$ denotes the sum of the squared residuals from the first regression line and $S_2$ the sum of the squared residuals from the second, we form the ratio

$$R = \frac{S_2}{S_1}. \qquad (11.7)$$

Then $R$ follows the $F$–distribution with $[(n - m)/2 - 2, \quad (n - m)2 - 2]$ degrees of freedom.[5]

For our hypothetical data presented in Table 11.1, $n = 20$, and if we omit the six central observations we need to estimate $Y_i = \alpha + \beta X_i + \varepsilon_i$ from the first seven and the last seven observations. We obtain

(first seven observations) $\quad Y_i = 99 \cdot 9 + 9 \cdot 86 X_i; \qquad R^2 = 0 \cdot 956, \qquad S_1 = 124 \cdot 88,$

(last seven observations) $\quad Y_i = 113 \cdot 2 + 9 \cdot 71 X_i; \qquad R^2 = 0 \cdot 718, \qquad S_2 = 1038 \cdot 57,$

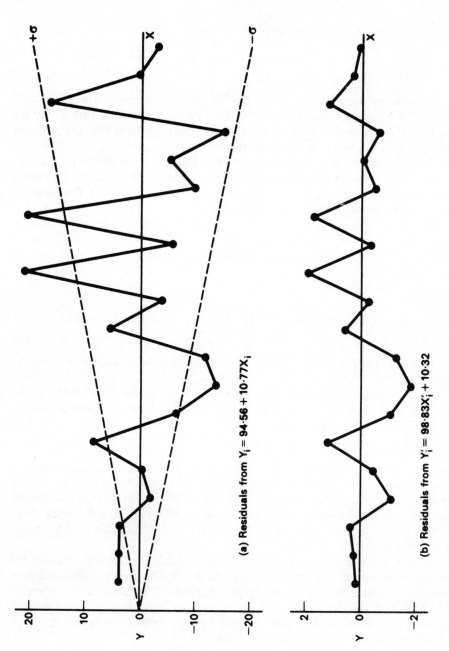

(a) Residuals from $Y_i = 94 \cdot 56 + 10 \cdot 77 X_i$

(b) Residuals from $Y'_i = 98 \cdot 83 X'_i + 10 \cdot 32$

Figure 11.1  An example of heteroscedasticity

and we compute the test statistic

$$R = \frac{1038 \cdot 57}{124 \cdot 88} = 8 \cdot 32.$$

In this example the relevant $F$–distribution has (5,5) degrees of freedom, and choosing the $p = 0.95$ probability level for our test we found from Table A4 that $F = 5.05$. Since $R > 5.05$, we would reject the hypothesis that the errors in our regression model are homoscedastic.

If the test indicates the presence of heteroscedasticity, what can be done about it? If the heteroscedasticity takes the form assumed in 11.6, then it is possible to solve the problem by means of a simple transformation of the data. The equation we are trying to estimate is

$$Y_i = \alpha + \beta X_i + \varepsilon_i,$$

and if we divide both sides of the equation by $X_i$ we obtain

$$\frac{Y_i}{X_i} = \alpha \frac{1}{X_i} + \beta + \frac{\varepsilon_i}{X_i},$$

or defining the new variables $Y_i' = (Y_i/X_i)$, $X_i' = (1/X_i)$ and $\varepsilon_i' = \varepsilon_i/X_i$,

$$Y_i' = \alpha X_i' + \beta + \varepsilon_i' \qquad (11.8)$$

and now $E(\varepsilon_i'^2) = E(\varepsilon_i/X_i)^2 = E(\varepsilon_i^2)/X_i^2 = \sigma^2$, so that the transformed errors are homoscedastic.

If we perform this transformation on the data on $X$ and $Y$ in Table 11.1 we obtain $X'$ and $Y'$, the data presented in the last two columns of Table 11.1. If we use these data to estimate 11.8, we obtain

$$Y_i' = 98 \cdot 83 X_i' + 10 \cdot 32, \qquad R^2 = 0 \cdot 998.$$

The residuals from this regression are plotted in Figure 11.1(b), and while they cannot be compared directly with those plotted in Figure 11.1(a) (because of the transformation of the data), the visual impression they give is that the transformation has largely removed the heteroscedasticity. However, it will be left as an exercise for the reader to conduct a formal test for heteroscedasticity by means of 11.7 (see problem 11.2 below).

We have discussed the problem of heteroscedasticity in some detail in the context of the simple regression model, and the results we have obtained here could be extended to the general regression model without much difficulty.

1. The least squares estimators are linear and unbiased, but the sampling variances are no longer correctly computed by the formulae given in chapter 9 and the minimum variance property is lost.

2. The test for heteroscedasticity is modified in that we now assume

$$E(\varepsilon_i^2) = \sigma^2 X_{ji}^2, \tag{11.9}$$

where $X_{ji}$ is one of the independent variables included in the regression model. Now the observations are ordered by increasing size of $X_{ji}$ and, omitting the central $m$ values, the model

$$Y_i = \beta_1 + \beta_2 X_{2i} + \ldots + \beta_k X_{ki} + \varepsilon_i \tag{11.10}$$

is estimated separately for the first $(n - m)/2$ and last $(n - m)/2$ observations. The ratio of the sums of the squared residuals given in 11.7 is calculated, and $R$ follows the $F$–distribution with $[(n - m - 2k)/2, (n - m - 2k/2]$ degrees of freedom.

3. If the errors are heteroscedastic and the heteroscedasticity takes the form assumed in 11.9, then a transformation which should solve the problem is to divide both sides of 11.10 by $X_{ji}$ to give

$$\frac{Y_i}{X_{ji}} = \beta_1 \frac{1}{X_{ji}} + \beta_2 \frac{X_{2i}}{X_{ji}} + \ldots + \beta_j + \ldots + \beta_k \frac{X_{ki}}{X_{ji}} + \frac{\varepsilon_i}{X_{ji}} \tag{11.11}$$

and estimate the parameters in the transformed equation. We shall not pursue this problem further here, but move on to consider the problem of autocorrelation.[6]

### 11.3 the problem of autocorrelation

In the previous section it was stated that heteroscedasticity most often arises in cross-sectional data owing to the variation in the size of the units being sampled. By contrast, autocorrelation, the problem which arises when assumption 8.15 is relaxed and

$$E(\varepsilon_i \varepsilon_j) \neq 0, \tag{11.12}$$

arises mainly in the context of economic time series when the data are collected in successive periods of time. We shall concentrate on time series at the moment and consider the model $Y_t = \alpha + \beta X_t + \varepsilon_t$, $t = 1, \ldots, T$. In the context of this model we may rewrite 11.12 as

$$E(\varepsilon_t \varepsilon_{t-s}) \neq 0, \qquad s = 1, \ 2, \ldots. \tag{11.13}$$

The absence of autocorrelation requires $E(\varepsilon_t \varepsilon_{t-1}) = 0$, $E(\varepsilon_t \varepsilon_{t-2}) = 0$, etc., that is, the random error in any given period affects the dependent variable in that period only and then dies away completely. However, in many economic situations this is an unlikely assumption. For example, if the dependent variable measures the monthly sales of a firm and the excluded variables produce a positive fluctuation in sales this month, then the firm may still be adjusting to the disturbance next month and possibly in later months as well. In general, the shorter the time period, the more likely this is to be the case.

To see the effects of relaxing assumption 8.15, we may recall that it was not required to establish the linear and unbiased properties of the least squares estimators, so that these properties are not affected. The main effect of auto-correlation is on the sampling variances of the estimators. For $b$ we have (on p. 139)

$$\text{var}(b) = \sum k_t^2 E(\varepsilon_t^2) + \sum k_t k_{t-s} E(\varepsilon_t \varepsilon_{t-s}) \tag{11.14}$$

and under the new assumption

$$\text{var}(b) \neq \sigma^2 \sum k_t^2,$$

since now $\sum k_t k_{t-s} E(\varepsilon_t \varepsilon_{t-s}) \neq 0$. Thus the formula for the variance of $b$ which is given in 8.25 is no longer relevant as it is a biased estimator of var $(b)$, the direction of the bias depending on the sign of the second term in 11.14. This has serious implications for hypothesis testing. In particular, if this term is positive, calculating var $(b)$ by means of 8.14 yields an under-estimate of sampling variance and hence s.e. $(b)$. As a result, when we test the hypothesis that $\beta = 0$ by computing $t = b/\text{s.e.}(b)$, this ratio will be biased *upwards*, that is, towards rejecting the hypothesis.

To fix this idea, let us explore the likely properties of the errors in an economic model when the data are collected over time. We might expect the errors to have some effect for more than one period, but we would also probably expect the effects to die away over time. We might also expect the errors to be in the same direction for a number of consecutive periods, since the excluded variables do not behave in a completely random manner from period to period. For example, if a change in expectations causes a positive fluctuation in sales this month, it may also cause positive fluctuation in future months.

A simple model which has these properties in the *first-order autoregressive process*[7]

$$\varepsilon_t = \rho \varepsilon_{t-1} + \omega_t, \qquad -1 < \rho < 1, \tag{11.15}$$

in which the error is a function of some proportion of the error last period plus the new error this period. If $0 < \rho < 1$ we have positive autocorrelation, negative autocorrelation occurring when $-1 < \rho < 0$. If we lag 11.15 we have $\varepsilon_{t-1} = \rho \varepsilon_{t-2} + \omega_{t-1}$, $\varepsilon_{t-2} = \rho \varepsilon_{t-3} + \omega_{t-2}$, etc., and substituting these expressions into 11.15 gives

$$\varepsilon_t = \rho(\rho \varepsilon_{t-2} + \omega_{t-1}) + \omega_t$$

$$= \omega_t + \rho \omega_{t-1} + \rho^2(\rho \varepsilon_{t-3} + \omega_{t-2})$$

$$= \omega_t + \rho \omega_{t-1} + \rho^2 \omega_{t-2} + \rho^3 \varepsilon_{t-3}$$

$$= \omega_t + \rho \omega_{t-1} + \rho^2 \omega_{t-2} + \rho^3 \omega_{t-3} + \dots, \tag{11.16}$$

that is, $\varepsilon_t$ is composed of the new error in this period, $\omega_t$, plus a weighted sum of past values of $\omega$. The weights are powers of $\rho$, and since $|\rho| < 1$, they decline geometrically, so that the impact of a given $\omega_t$ dies away to zero as we move forward in time.

The statistical properties of the $\varepsilon_t$ depend on those of the $\omega_t$, and we shall explore them under the assumptions that

$$E(\omega_t) = 0 \qquad\qquad\qquad (11.17)$$

$$E(\omega_t^2) = \sigma^2 \qquad\qquad\qquad (11.18)$$

$$E(\omega_t\omega_{t-s}) = 0, \qquad s \neq 0, \qquad\qquad\qquad (11.19)$$

that is, the $\omega_t$ have the properties we assumed for the $\varepsilon_t$ in the non-autocorrelated case. Given these assumptions, it may be shown[8] that

$$E(\varepsilon_t) = 0 \qquad\qquad\qquad (11.20)$$

$$E(\varepsilon_t^2) = \sigma_\varepsilon^2 = \sigma_\omega^2/(1 - \rho^2) \qquad\qquad\qquad (11.21)$$

$$E(\varepsilon_t\varepsilon_{t-s}) = \rho^s\sigma_\varepsilon^2. \qquad\qquad\qquad (11.22)$$

Thus the $\varepsilon$s have a mean of zero and a constant variance, but the covariance between $\varepsilon_t$ and $\varepsilon_{t-s}$ is non-zero, varying as a function of $s$ and tending to zero as $s$ increases.

Figure 11.2 illustrates the appearance of autocorrelated time series. In part (a) we have 30 values of $\omega_t$, chosen at random from the standard normal distribution $N(0, 1)$. These values were used, together with an initial value $\varepsilon_0 = 0$, to generate two autoregressive series. In 11.2(b) we have an example of positive autocorrelation with $\rho = 0.9$, while in 11.2(c) the series exhibits negative autocorrelation, with $\rho = -0.9$. While one cannot infer very much from a single random sample, the diagram does bring out some of the properties of autoregressive series: as compared with the random series in (a), the positively autocorrelated series in (b) does show more pronounced and longer runs of values having the same sign, while the negatively autocorrelated series in (c) appears 'spikier' and has more changes of sign than the random series. This visual impression may be confirmed by counting the number of changes of sign between successive terms which occur for the time series, when we obtain 13 for the random series, 8 for the positively autocorrelated series and 18 for the negatively autocorrelated series.

As we have shown above, the effect of autocorrelation in the errors is to bias the sampling variance of $b$. If the autocorrelation were of the form defined in 11.15 and $\rho$ and $\sigma^2$ were known, then we could substitute for $E(\varepsilon_t\varepsilon_{t-s})$ from 11.22 into the second term in 11.14 and compute the correct sampling variance for $b$. This would be better than ignoring the problem completely, but in the presence of

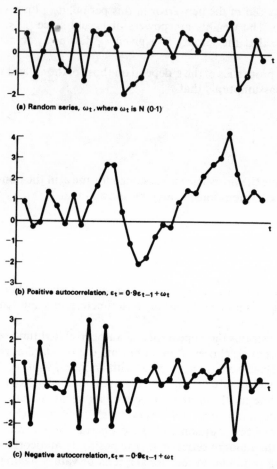

(a) Random series, $\omega_t$, where $\omega_t$ is N (0·1)

(b) Positive autocorrelation, $\varepsilon_t = 0.9\varepsilon_{t-1} + \omega_t$

(c) Negative autocorrelation, $\varepsilon_t = -0.9\varepsilon_{t-1} + \omega_t$

Figure 11.2

autocorrelation it may be shown[9] that $b$ is no longer the minimum variance estimator of $\beta$. We shall discuss possible modifications to the method of estimation below, but before doing so we shall discuss the problem of testing for the presence of autocorrelation.

*The Durbin–Watson test.* The basis for this test lies in the pattern of sign changes in autocorrelated time series which we have already noted. Suppose we have a times series $\varepsilon_1, \ldots, \varepsilon_T$ and calculate

$$\sum_{t=2}^{T} (\varepsilon_t - \varepsilon_{t-1})^2, \qquad\qquad (11.23)$$

how would this sum differ between autocorrelated and random series? Given positive autocorrelation tends to produce longer runs of values having the same sign than we find in a random series, then the sum of squared successive differences will tend to be smaller for the positively autocorrelated series than for the random series. Conversely, successive values in the negatively autocorrelated series tend to change sign more often than is the case in the random series, hence producing larger squared differences.

The formula given in 11.23 is not suitable as a test statistic as it depends on the number of observations and the unit of measurement of the $\varepsilon$s. We may produce a pure number by defining the measure as

$$\frac{\sum_{t=2}^{T} (\varepsilon_t - \varepsilon_{t-1})^2}{\sum_{t=1}^{T} (\varepsilon_t - \bar{\varepsilon})^2}. \qquad\qquad (11.24)$$

To complete the theoretical analysis, we could investigate the distribution of this ratio for series and hence establish significance points for given probability levels such that calculated values of the ratio which were too small (that is, fell in the rejection region in the lower tail) would be taken as evidence of positive auto-correlation. The significance point in the upper tail would provide a test for the presence of negative autocorrelation.

There are two problems in applying the ratio defined in 11.24 to regression analysis. First, the errors are unobservable and we must therefore substitute the residuals as proxies for the errors in 11.24. This yields the test statistic

$$d = \frac{\sum_{t=2}^{T} (e_t - e_{t-1})^2}{\sum_{t=1}^{T} e_t^2}, \qquad\qquad (11.25)$$

usually referred to as the Durbin-Watson statistic. The second problem is that in devising the test we would take account of the available degrees of freedom, a number which varies with the numbers of parameters being estimated.

The rationale behind this test may be seen by expanding the numerator of 11.25, when we obtain

$$d = \frac{\sum e_t^2 + \sum e_{t-1}^2 - 2\sum e_t e_{t-1}}{\sum e_t^2}$$

and, since $\sum e_{t-1}^2$ is approximately equal to $\sum e_t^2$, we may write

$$d = 2 - 2\frac{\sum e_t e_{t-1}}{\sum e_t^2}. \tag{11.26}$$

If the $\varepsilon$s were independent, we would expect $d = 2$, since $\sum e_t e_{t-1}$ would tend to be zero. However, if positive autocorrelation is present, we would expect $\sum e_t e_{t-1}$ to be positive and hence $d$ will be less than 2. (In the case of negative autocorrelation, we would expect $\sum e_t e_{t-1}$ to be negative and hence $d$ to be greater than 2.) Before proceeding, the reader may note that $\sum e_t e_{t-1}/\sum e_t^2$ is approximately equal to $\sum e_t e_{t-1}/\sum e_{t-1}^2$, the least squares estimate of $\rho$ obtained by regressing $e_t$ on $e_{t-1}$. If we denote this estimate by $r$, then we have the approximate result

$$d \approx 2(1 - r), \tag{11.27}$$

a formula which may be convenient in the calculation of $d$.

A rigorous statistical test for autocorrelation can be developed by calculating how large a deviation of $d$ from 2 might be expected on the basis of random sampling fluctuations when the $\varepsilon$s are normally distributed $N(0, \sigma^2)$ and independent. This enables us to tabulate significant values of $d$ for different probability levels, different sample sizes and different numbers of parameters being estimated.

Such values have been tabulated by Durbin and Watson (1950, 1951). The authors suggest that the test may not be appropriate if one of the independent variables consists of lagged values of the dependent variable (for example, if we were estimating a consumption function by regressing $C_t$ on $Y_t$ and $C_{t-1}$),[10] but for the cases to which the test is applicable, 5 per cent and 1 per cent probability levels of $d$ have been tabulated. The test is for first-order autocorrelation, $\varepsilon_t = \rho\varepsilon_{t-1} + \omega_t$, and since economists are mainly interested in positive autocorrelation, $d$ is tabulated to test the hypothesis $H_0: \rho = 0$ against the one-sided alternative $H_1: \rho > 0$ (that is, we have a one-tailed test). In this test the probability of observing given values of $d$ depends on the number of observations in the sample, $n$, and the number of parameters being estimated, $k$.

Unfortunately a complication arises since the residuals are functions of the $X$s and so the distribution of $d$ depends on the values of $X$. As a result the authors were not able to derive exact probability levels for $d$, but for each sample size and number of parameters being estimated they were able to allow for this problem by finding upper and lower bounds within which the value of $d$ must lie. These upper and lower bounds $d_U$ and $d_L$ are tabulated in Table A7 for the 5 per cent and one per cent probability levels. The test procedure is to compute $d$ from the residuals using 11.25 and, having chosen a probability level and given the sample size and $k$, the appropriate values of $d_U$ and $d_L$ may be obtained from Table A7. If the calculated value of $d$ is *less* than $d_L$, the residuals indicate that the null hypothesis that the errors are not autocorrelated is rejected in favour of $H_1:\rho > 0$. If $d$ is *greater* than $d_U$, the null hypothesis is not rejected and we may reasonably accept the standard regression formulae as far as first-order autocorrelation is concerned.

We have concentrated on tests for positive autocorrelation, but modifying the procedure to test for negative autocorrelation is relatively easy since distribution of $d$ is symmetric about 2 with a range from 0 to 4. Hence, given Table A7, we may find the corresponding area in the upper tail be calculating $d'_U = 4 - d_U$ and $d'_L = 4 - d_L$. For example, with $n = 25$, $k = 3$ and $p = 0.95$ we have $d'_U = 4 - 1.55 = 2.45$ and $d'_L = 4 - 1.21 = 2.79$ as the significance points for a test for negative autocorrelation.

A problem arises if the value of $d$ falls between $d_L$ and $d_U$, since in this case the test is indeterminate. Ideally, in this case one would like to draw another sample and repeat the experiment in the hope of resolving the indeterminacy, but this is generally not possible with economic time series data. Given one usually has to make some decision on accepting or rejecting $H_0:\rho = 0$ in this case, one has to weigh up the relative costs of making Type I or Type II errors. We have already shown that ignoring positive autocorrelation (a Type II error) means we are too likely to find significant regression results for a given probability level, while if we treat all values of $d < d_U$ as being a rejection of the null hypothesis, we shall be increasing the size of the Type I error. On balance, since the methodology of testing economic theories is geared towards finding significant results, it could be argued that the Type II error is more important here than the Type I error, so that as a rule of thumb one could resolve the indeterminacy by taking $d_U$ as the relevant point and rejecting $H_0$ if $d < d_U$.

To illustrate the use of the Durbin–Watson statistic, we return to the regressions of aggregate saving and personal disposable income on time, which were reported on p. 199, with the residuals being plotted in Figure 10.4. The reader may recall that a visual inspection of the residuals from the linear trends suggested non-randomness. We may now test the errors for the presence of autocorrelation by computing the Durbin–Watson statistic. For the linear trend equations $n = 17$ and $k = 2$, and choosing the probability level $p = 0.95$ for our test, we find from Table A7 that $d_L = 1.13$ and $d_U = 1.38$. Given the residuals from $S_t = 0.554 + 0.368t$, we compute $d = 0.32$, while for $Y_t = 8.403 + 1.254t$ we calculate $d = 0.22$. In both cases $d < d_L$, indicating the presence of positive autocorrelation and confirming that the linear trend is not a good specification for describing the data.

To see whether the inclusion of the quadratic term in the trend improves the specification, we may calculate the Durbin–Watson statistics. From $S_t = 1.637 + 0.026t + 0.019t^2$ we calculate $d = 1.54$, while from $Y_t = 10.746 + 0.514t + 0.041t^2$ we calculate $d = 0.78$. Now $n = 17$ and $k = 3$, so that from Table A7 we find $d_L = 1.02$ and $d_U = 1.54$. For the saving regression, $d_L < d < d_U$, so that the test becomes indeterminant, although clearly the autocorrelation has been reduced. For the income regression $d < d_L$, so that there is still evidence that the errors are non-random.

For pedagogic purposes it has been convenient to examine the residuals visually in the last chapter and carry out a formal test for autocorrelation here. In practice, most computer regression programmes will compute and print out the residuals and the Durbin–Watson statistic automatically. If the Durbin–Watson statistic is significant, it is well worth examining the residuals carefully to see whether there is any obvious mis-specification of the model of the kind we have noted in this example.[11]

*Removing the effects of autocorrelation.* If we carry out the Durbin–Watson test and detect the presence of autocorrelation, what can be done about this problem? A completely general solution to this problem is beyond the scope of this text as it requires matrix algebra and has therefore been relegated to the appendix to this chapter. Here we shall concentrate on a simple transformation procedure which will remove the effects of first-order autocorrelation as defined in 11.15.

Suppose we wish to estimate

$$Y_t = \alpha + \beta X_t + \varepsilon_t \tag{11.28}$$

where $\varepsilon_t$ is generated by

$$\varepsilon_t = \rho\varepsilon_{t-1} + \omega_t. \tag{11.15}$$

If we knew $\rho$ then lagging 11.28 by one period and multiplying both sides by $\rho$ would give

$$\rho Y_{t-1} = \alpha\rho + \rho\beta X_{t-1} + \rho\varepsilon_{t-1}. \tag{11.29}$$

Subtracting 11.29 from 11.28, we have

$$Y_t - \rho Y_{t-1} = \alpha(1 - \rho) + \beta(X_t - \rho X_{t-1}) + (\varepsilon_t - \rho\varepsilon_{t-1})$$

and denoting $Y_t - Y_{t-1}$ by $Y_t'$ and $X_t - \rho X_{t-1}$ by $X_t'$, we may write the equation equation as

$$Y_t' = \alpha(1 - \rho) + \beta X_t' + \omega_t. \tag{11.30}$$

Since the $\omega$s are not autocorrelated, we may estimate the parameters in 11.30 without having to worry about the problem of autocorrelation.

In general $\rho$ is not known, and so this transformation cannot be performed directly. However, it is possible to estimate $\rho$ from the residuals by regressing $e_t$ on $e_{t-1}$. If $r$ is the resulting least squares estimate of $\rho$, then 11.28 may be approximated by defining $Y'_t = Y_t - rY_{t-1}$, and $X'_t = X_t - rX_{t-1}$, and substituting these variables into 11.30. Whether this transformation has been successful or not may be determined by computing the Durbin–Watson statistic from the new set of residuals and testing $H_0 : \rho = 0$. If the residuals still indicate the presence of autocorrelation, then $\rho$ may be re-estimated from these residuals and we may repeat the exercise of transforming the data and re-estimating 11.30. This process may be repeated until the residuals from 11.30 no longer indicate the presence of autocorrelation.

*An example of the effects of autocorrelation.* In a note on economies of scale, Walters (1963) reports the results of estimating a Cobb–Douglas production function of the form

$$X = AK^{\alpha}L^{\beta}e^{\gamma t}$$

from annual U.S. manufacturing data for the period 1909–49,
where $X$ is gross national product of the private non-farm sector,
$\quad$ $L$ is the number of non-government, non-farm employees,
$\quad$ $K$ in Test (a) is the *net* capital stock,
$\quad$ in Test (b) is the *net* capital stock adjusted for employment, and
$\quad$ in Test (c) is the *gross* capital stock,
and
$\quad$ $e^{\gamma t}$ is a proxy to represent neutral technical progress.

Denoting the estimates of $\alpha$, $\beta$ and $\gamma$ by $a$, $b$ and $c$, the results obtained are presented in the table below, where the figures in parentheses are the standard terms of the estimates.

|          | $a$      | $b$     | $c$      | $(a+b)$  | $t$  | $\bar{R}^2$ | $d$   |
|----------|----------|---------|----------|----------|------|-------------|-------|
| Test (a) | 0·125    | 1·228   | 0·0110   | 1·353    | 2·94 | 0·986       | 1·08  |
|          | (0·077)  | (0·070) | (0·0019) | (0·120)  |      |             |       |
| Test (b) | 0·187    | 1·078   | 0·0123   | 1·265    | 3·73 | 0·987       | 1·13  |
|          | (0·079)  | (0·080) | (0·0012) | (0·071)  |      |             |       |
| Test (c) | 0·151    | 1·224   | 0·0097   | 1·375    | 2·34 | 0·986       | 1·06  |
|          | (0·119)  | (0·072) | (0·0032) | (0·160)  |      |             |       |

Judged by the values of $\bar{R}^2$, all three versions of the production function fit the data well. Here there are $41 - 4 = 37$ degrees of freedom and all the parameter estimates (with the exception of $a$ in Tests (a) and (c)) are significantly different from zero. In our earlier discussion of production functions (pp. 176–7), we developed a test for the constancy of returns to scale by means of 9.51. The $t$-ratio defined by this formula is given in the table and is significant in all three cases. Thus the evidence suggests increasing returns to scale in U.S. production.

However, the final column presents the Durbin–Watson statistic for each equation. Here $n = 41$ and $k = 4$, so that if we choose $p = 0.95$ we find from Table A7 that $d_L = 1.34$ and $d_U = 1.66$. In each case $d < d_L$, indicating the presence of positive autocorrelation. Walters assumed the autocorrelation was generated by the first-order autoregressive process 11.15, and obtained 0.5 as an estimate of $\rho$. He then used the transformation of the data outlined in 11.30 to calculate $\log X_t - 0.5 \log X_{t-1}$,    $\log K_t - 0.5 \log K_{t-1}$, etc., and re-estimated the production function using the transformed data. The results (omitting the estimates of $\gamma$) are presented in the table below.

|          | $a$     | $b$     | $(a + b)$ | $t$  | $\bar{R}^2$ | $d$  |
|----------|---------|---------|-----------|------|-------------|------|
| Test (a) | 0·167   | 1·189   | 1·256     | 1·41 | 0·961       | 2·09 |
|          | (0·130) | (0·095) | (0·181)   |      |             |      |
| Test (b) | 0·227   | 0·993   | 1·220     | 2·32 | 0·963       | 2·17 |
|          | (0·123) | (0·125) | (0·095)   |      |             |      |
| Test (c) | 0·231   | 1·188   | 1·419     | 1·65 | 0·961       | 2·09 |
|          | (0·209) | (0·096) | (0·254)   |      |             |      |

Looking first at the Durbin–Watson statistic, we notice that in all three cases it is very close to 2, the value expected when the errors are random, so that the transformation seems to have solved the problem of autocorrelation. The new parameter estimates differ from those presented in the previous table, but are of the same order of magnitude. The most interesting difference arises when we test the constancy of returns to scale, since now, having corrected for the downward bias in the standard errors caused by ignoring autocorrelation, we would accept $H_0 : \alpha + \beta = 1$ in Tests (a) and (c). The evidence in favour of increasing returns to scale is now considerably reduced.

We have considered this example in some detail since it illustrates neatly the effects of autocorrelation and is a case in which the transformation appears to work very well. Not all autocorrelation is of the kind we have considered here, but a fuller discussion of the problem would take us beyond the level of this text. The reader with a knowledge of matrix algebra will find further discussion of the problem in the appendix to this chapter, but now we shall leave the problem of autocorrelation and move on to consider what happens when we relax the assumption that the $X$s are a set of constants.[12]

### 11.4 the problem of stochastic explanatory variables

In discussing the problem of multicollinearity, we generated the data presented in Table 10.2 (b) from $Y_t = 1 + 2X_{2t} + X_{3t} + \varepsilon_t$, where $X_{2t}$ were chosen from a table of random numbers, $X_{3t} = 1, \ldots, 20$, and $\varepsilon_t$ were chosen at random from $N(0,25)$. Clearly, $X_{3t}$ are a set of constants, and if in repeated samples we now fixed the values of $X_{2t}$ and drew random samples of $\varepsilon_t$, the condition

$$E(X_{2i}\varepsilon_j) = X_{2i}E(\varepsilon_j) = E(X_{3i}\varepsilon_j) = X_{3i}E(\varepsilon_j) = 0, \qquad \text{for all } i,j \qquad (9.14)$$

still holds.

As an alternative we might conduct a sampling experiment in which the $X_{3t}$ are held constant from sample to sample, but now the $X_{2t}$ and $\varepsilon_t$ are drawn at random from their respective populations. Now the $X_{2t}$ can no longer be treated as a set of constants and we must relax assumption 9.14 as far as $X_2$ is concerned. What effect does this have on the properties of the least squares estimators? In this case we are generating both $X_{2t}$ and $\varepsilon_t$ as random samples, and assuming that this randomization ensures that the variables are independent, we may establish that 9.14 holds by using the result obtained on p. 60, that the covariance between two independent variables is zero. *In other words, the properties of the least squares estimators are unaffected, provided the values of the independent variable(s) and the errors are independent of each other.*

However, in much of economic analysis the values of the independent variable(s) are neither a set of constants nor generated by random sampling. This is especially true in the case of time series data, where often the dependent and independent variables are generated by the same stochastic process. For example, even in the simplest Keynesian model both consumption and income are endogenous, a fact whose implications we shall investigate below.

To explore what happens when we relax the assumption that the values of the independent variable are a set of constants, we shall return to the simple model

$$Y_i = \alpha + \beta X_i + \varepsilon_i,$$

assuming that 8.13–8.15 hold, but that now

$$E(X_i\varepsilon_j) \neq 0 \qquad \text{for all } i, j. \qquad (11.31)$$

The reader may recall that for the least squares estimator of $\beta$ (p. 138)

$$b = \beta + \sum k_i\varepsilon_i, \qquad (8.24)$$

where $k_i = x_i / \sum x_i^2$, so that

$$b = \beta + \frac{\sum x_i\varepsilon_i}{\sum x_i^2}. \qquad (11.32)$$

From 11.31 it follows that $E(x_i\varepsilon_i) \neq 0$, but we cannot make use of this result directly to investigate 11.32, since when we take expected values

$$E(b) = \beta + E\left(\frac{\sum x_i \varepsilon_i}{\sum x_i^2}\right) \tag{11.33}$$

and the second term on the right-hand side involves the ratio of two random variables. This causes a problem since, as was demonstrated in problem 3.15, if $X$ and $Y$ are random variables and $Z = X/Y$, then it is possible to find cases in which $E(Z) \neq E(X)/E(Y)$, so that we are not entitled to assume that $E(Z) = E(X)/E(Y)$ in evaluating 11.33.

To proceed, we have to abandon the expected value which holds for all sample sizes, and introduce the large sample concept of a 'limit in probability'. The limit in probability of a random variable is a modification of the mathematical limit obtained by letting the sample size tend to infinity, but assuming that as $n$ tends to infinity the probability that the random variable takes its expected value (if it has one) tends to unity. As an example which is relevant to 11.33, consider $\sum x_i^2$ and assume that the $X$s are drawn from some population with a variance of $\sigma_X^2$. Here we immediately see one restraint that is necessary in defining a limit in probability or, as it is often abbreviated, plim, since

$$\lim_{n \to \infty}\left(\sum_1^n x_i^2\right) = \infty \,,$$

so that if the limit in probability is to be finite, we need to consider

$$\lim_{n \to \infty}\left(\frac{1}{n}\sum_1^n x_i^2\right),$$

that is, the mean square deviation. In this case we would expect the mean square deviation to converge on the population variance, $\sigma_X^2$, as the sample size tends to infinity. That is,

$$\lim_{n \to \infty} p\left(\sigma_X^2 - \delta < \frac{1}{n}\sum x_i^2 < \sigma_X^2 + \delta\right) = 1,$$

where $\delta$ is some arbitrary small positive number.[13] That is,

$$\text{plim}\left(\frac{1}{n}\sum x_i^2\right) = \sigma_X^2. \tag{11.34}$$

In the same way, if we assume that there is a non-zero covariance between $X$ and $\varepsilon$, denoted by $\sigma_{X\varepsilon}$, then

$$\text{plim}\left(\frac{1}{n}\sum x_i \varepsilon_i\right) = \sigma_{X\varepsilon}. \tag{11.35}$$

We may also note that if $\lambda$ is a constant, then $plim(\lambda) = \lambda$.

The relevance of the new concept in dealing with our current problem is that limits in probability have many of the important properties of mathematical limits. In particular, if $X$ and $Y$ are random variables and $Z = X/Y$, then $\text{plim}(Z) = \text{plim}(X)/\text{plim}(Y)$. Thus if we consider the limit in probability of 11.32, we have

$$\text{plim}(b) = \beta + \text{plim}\left(\frac{\frac{1}{n}\sum x_i \varepsilon_i}{\frac{1}{n}\sum x_i^2}\right)$$

$$= \beta + \frac{\text{plim}\left(\frac{1}{n}\sum x_i \varepsilon_i\right)}{\text{plim}\left(\frac{1}{n}\sum x_i^2\right)}$$

$$= \beta + \frac{\sigma_{X\varepsilon}}{\sigma_X^2}, \tag{11.36}$$

using 11.34 and 11.35. Since the second term on the right-hand side is not equal to zero,

$$\text{plim}(b) \neq )\beta,$$

so that $b$ is a biased and inconsistent estimator of $\beta$.

This is a fairly disturbing result, and in addition we may note that, unlike the case of autocorrelation, there is no simple way of devising a statistical test for possibility that the $X$s and $\varepsilon$s are interdependent. This is because one of the algebraic properties of the method of least squares is that $\sum X_i e_i = 0$ (recall problem 8.7 on p. 154), so that we cannot compute the sample covariance between the $X$s and the residuals and use it as a basis for a test. What we must do is to consider the possibility that the $X$s and $\varepsilon$s are interdependent by examining the context in which regression analysis is being applied. We shall explore three potential problem situations here, before considering ways of dealing with the problem.

### 1. *Errors in variables*
At the beginning of chapter 8 it was asserted that for the least squares estimators to be unbiased, it was necessary for the $X$s to be free from errors of measurement (pp. 131–2). To see why this is so, we shall consider the following simple model in which we assume that there are no excluded variables, so that we have an exact relationship

$$Y_i^* = \alpha + \beta X_i^*, \qquad i = 1, \ldots, n, \tag{11.37}$$

between two variables $Y^*$ and $X^*$. However, we assume that when we collect a sample of data, both variables are subject to errors of measurement, so that we observe

$$Y_i = Y_i^* + \omega_{1i}$$

and

$$X_i = X_i^* + \omega_{2i}. \tag{11.38}$$

Substituting for $X_i^*$ and $Y_i^*$ in 11.37, we obtain

$$Y_i - \omega_{1i} = \alpha + \beta (X_i - \omega_{2i}),$$

that is

$$Y_i = \alpha + \beta X_i + \omega_{1i} - \beta \omega_{21},$$

or

$$Y_i = \alpha + \beta X_i + \varepsilon_i, \qquad i = 1, \ldots, n,$$

where $\varepsilon_i = \omega_{1i} - \beta \omega_{2i}$.

We shall also assume that

$$E(\omega_{1i}) = E(\omega_{2i}) = 0, \tag{11.39}$$

$$E(\omega_{1i}^2) = \sigma_1^2,$$

$$E(\omega_{2i}^2) = \sigma_2^2, \tag{11.40}$$

$$E(\omega_{1i}\omega_{2j}) = 0 \qquad \text{for all } i, j \tag{11.41}$$

and that the $X_i^*$ are a set of constants, so that $E(X_i^*\omega_{1j}) = E(X_i^*\omega_{2j}) = 0$, by virtue of 11.39. Now although the $X_i^*$ are constants, it is clear from 11.38 that the $X_i$ are random, so that the second term on the right-hand side of 11.32 involves the ratio of two random variables. Hence to investigate the properties of the least squares estimator we must consider the limit in probability of $b$, in particular plim$(1/n \sum x_i \varepsilon_i)$. From 11.38, $\bar{X} = \bar{X}^* + \bar{\omega}_2$, and denoting deviations from sample means by lower-case letters, so that $x_i = x_i^* + (\omega_{2i} - \bar{\omega}_2)$, we have

$$\text{plim}\left(\frac{1}{n}\sum x_i \varepsilon_i\right) = \text{plim}\left\{\frac{1}{n}\sum [x_i^* + (\omega_{2i} - \bar{\omega}_2)][\omega_{1i} - \beta\omega_{2i}]\right\}$$

$$= -\beta\sigma_2^2,$$

given the assumptions listed in 11.39 to 11.41. So we see that if the $X$s contain errors of measurement, the least squares estimator of $\beta$ is biased and inconsistent.

## 2.  Simultaneous equation bias

It was pointed out above (p. 239) that even in the simplest Keynesian model both consumption, $C_t$, and income, $Y_t$ are endogenous, and we shall now explore the implications of this fact. We shall assume that the consumption function is linear and denote investment, which is assumed to be exogenous, as $I_t$. The simple textbook model may then be written as

$$C_t = \alpha + \beta Y_t + \varepsilon_t, \qquad t = 1, \dots, T, \tag{11.42}$$

$$Y_t = C_t + I_t. \tag{11.43}$$

The implication of the second equation may be seen if we substitute 11.42 in 11.43 to obtain[14]

$$Y_t = \frac{\alpha}{1-\beta} + \frac{1}{1-\beta} I_t + \frac{1}{1-\beta} \varepsilon_t, \tag{11.44}$$

showing that $Y$ is a function of $\varepsilon_t$, so that the independent variable and the error in 11.42 are interdependent. To see how this affects the properties of the least squares estimator of $\beta$ in 11.42, we may write $b$ in terms of deviations from sample means as

$$b = \frac{\sum c_t y_t}{\sum y_t^2}.$$

Substituting for $c_t$ from 11.42 yields

$$b = \beta + \frac{\sum y_t \varepsilon_t}{\sum y_t^2} \tag{11.45}$$

and given 11.44, the second term on the right-hand side of 11.45 involves the ratio of two random variables.

Letting $i_t = I_t - \bar{I}_t$, from 11.44 we have

$$y_t = \frac{1}{1-\beta} i_t + \frac{1}{1-\beta} (\varepsilon_t - \bar{\varepsilon}), \tag{11.46}$$

and substituting this expression for $y_t$ in 11.45 the numerator becomes

$$\frac{1}{1-\beta} \left[ \sum i_t \varepsilon_t + \sum \varepsilon_t (\varepsilon_t - \bar{\varepsilon}) \right].$$

Now if we assume that, being exogenous, $I_t$ is independent of $\varepsilon_t$ and that the $I_t$ come from a population with variance $\sigma_I^2$, while the errors have a mean of zero and a variance $\sigma_\varepsilon^2$, then

$$\text{plim } \frac{1}{1-\beta}\left[\frac{1}{T}\sum i_t\varepsilon_t + \frac{1}{T}\sum \varepsilon_t(\varepsilon_t - \bar{\varepsilon})\right] = \frac{1}{1-\beta}\sigma_\varepsilon^2.$$

Since, from 11.46,

$$\text{plim}\left(\frac{1}{T}\sum y_t^2\right) = \text{plim}\left(\frac{1}{1-\beta}\right)^2\left[\frac{1}{T}\sum i_t^2 + \frac{2}{T}\sum i_t(\varepsilon_t - \varepsilon) + \frac{1}{T}\sum(\varepsilon_t - \varepsilon)^2\right]$$

$$= \left(\frac{1}{1-\beta}\right)^2(\sigma_I^2 + \sigma_\varepsilon^2),$$

we obtain

$$\text{plim }(b) = \beta + \frac{(1-\beta)\sigma_\varepsilon^2}{\sigma_I^2 + \sigma_\varepsilon^2},$$

showing that the least squares estimator is biased and inconsistent.

This is a rather disturbing result and, given most economic models involve the simultaneous determination of a number of endogenous variables, we may expect the interdependence of the explanatory variables and the errors in the equations to arise frequently when regression analysis is applied in economic contexts.

For example, if we change the simple model of 11.42 and 11.43 by making investment endogenous according to the very simple accelerator equation

$$I_t = \gamma + \delta(Y_t - Y_{t-1}) + \omega_t, \tag{11.47}$$

the interdependence of $Y_t$ and $\varepsilon_t$ which we have already discussed would still lead to biased and inconsistent estimates of $\alpha$ and $\beta$ in the consumption function. In addition, substituting 11.42 and 11.47 into 11.43 gives

$$Y_t = \frac{\alpha+\gamma}{1-\beta-\delta} - \frac{\delta}{1-\beta-\delta}Y_{t-1} + \frac{\varepsilon_t + \omega_t}{1-\beta-\delta},$$

and since $Y_t$ is a function of $\omega_t$, the least squares estimators of $\gamma$ and $\delta$ will be biased and inconsistent.

### 3. The problem of the lagged dependent variable
In some economic models the lagged value of the dependent variable appears as an explanatory variable. An example is the 'partial adjustment' model in which

the 'desired' value of the dependent variable is $Y_t^* = \alpha + \beta X_t$, but the actual value of $Y$ only adjusts partially to a discrepancy between the desired and actual values at the beginning of the period, so that

$$Y_t - Y_{t-1} = \gamma(Y_t^* - Y_{t-1}) + \varepsilon_t, 0 < \gamma < 1$$

and substituting for $Y_t^*$ gives

$$Y_t = \alpha\gamma + \beta\gamma X_t + (1 - \gamma) Y_{t-1} + \varepsilon_t. \qquad (11.48)$$

To see how the inclusion of the lagged dependent variable among the explanatory variables may cause problems, we shall analyse the simple model in which

$$Y_t = \alpha + \beta Y_{t-1} + \varepsilon_t. \qquad (11.49)$$

Here the least squares estimator of $\beta$ is

$$b = \frac{\sum y_t y_{t-1}}{\sum y_{t-1}^2}$$

$$= \beta + \frac{\sum y_{t-1}\varepsilon_t}{\sum y_{t-1}^2}, \qquad (11.50)$$

by substituting for $y_t$ from 11.49. Since $Y_{t-1}$ is a random variable (it is a function of $\varepsilon_{t-1}$), the second term on the right-hand side of 11.50 involves the ratio of two random variables, and we must consider limits in probability rather than expectations.

Now provided the $Y$s have a finite variance, $b$ will be a consistent estimator of $\beta$ if

$$\text{plim}\left(\frac{1}{T}\sum y_{t-1}\varepsilon_t\right) = 0. \qquad (11.51)$$

One interesting case in which 11.51 will not hold is when the $\varepsilon$s are autocorrelated, for example if they are generated by 11.15. In that case $\varepsilon_t$ is a function of $\varepsilon_{t-1}$ and, given 11.49, $Y_{t-1}$ is also a function of $\varepsilon_{t-1}$. Hence $Y_{t-1}$ and $\varepsilon_t$ must be related and plim $(1/T\sum y_{t-1}\varepsilon_t) \neq 0$. *In other words, if the lagged dependent variable appears among the explanatory variables and the errors in the model are autocorrelated, the least squares estimators become biased and inconsistent.*

Once we consider stochastic explanatory variables, it is easier to find interesting examples of problems than it is to find solutions to the problems. In some cases no practical solutions have been found, and in general a discussion of the theoretical solutions to the problems would take us well beyond the technical scope of this text. However, to indicate how statisticians have modified the method of least squares to produce consistent parameter estimates in this context, we shall consider the method of *instrumental variables*.

Assume that we are trying to estimate the parameters in the model

$$Y_i = \alpha + \beta X_i + \varepsilon_i, \qquad i = 1, \ldots, n,$$

and $E(X_i \varepsilon_i) \neq 0$ for all $i$. Suppose we can find a new variable, $Z_i$, which has the following properties:

$$\text{plim}\left(\frac{1}{n}\sum z_i \varepsilon_i\right) = 0 \tag{11.52}$$

$$\text{plim}\left(\frac{1}{n}\sum z_i x_i\right) \neq 0, \tag{11.53}$$

where $x_i$ and $z_i$ are deviations from sample means. If we write the model in terms of deviations from sample means as $y_i = \beta x_i + (\varepsilon_i - \bar{\varepsilon})$, multiply through by $z_i$ and sum over the sample observations, we obtain

$$\sum y_i z_i = \beta \sum x_i z_i + \sum z_i (\varepsilon_i - \bar{\varepsilon}).$$

Now consider the limit in probability of this expression,

$$\text{plim}\left(\frac{1}{n}\sum y_i z_i\right) = \beta \,\text{plim}\left(\frac{1}{n}\sum x_i z_i\right) + \text{plim}\left(\frac{1}{n}\sum z_i \left(\varepsilon_i - \bar{\varepsilon}\right)\right)$$

and the plim of the second term on the right-hand side of the equation is zero by virtue of 11.52. Therefore

$$\frac{\text{plim}\left(\dfrac{1}{n}\sum y_i z_i\right)}{\text{plim}\left(\dfrac{1}{n}\sum x_i z_i\right)} = \beta, \tag{11.54}$$

and if we define an estimator of $\beta$ as

$$b^* = \frac{\sum y_i z_i}{\sum x_i z_i}, \tag{11.55}$$

it is clear from 11.54 that $\text{plim}(b^*) = \beta$, so that we have found an estimator of $\beta$ which is consistent. The estimator defined in 11.55 is called the instrumental variable estimator and $Z$ is usually referred to as an instrumental variable, or instrument.

To illustrate the method of instrumental variables with an economic example, consider the simple Keynesian model given in equations 11.42 and 11.43. If we write $i_t = I_t - \bar{I}$, it is clear from 11.43 that plim $(1/T \sum i_t y_t) \neq 0$, and since investment is exogenous, we earlier assumed that plim $(1/T \sum i_t \varepsilon_t) = 0$. Now write the consumption function in deviation form as $c_t = \beta y_t + \varepsilon_t - \bar{\varepsilon}$, multiply through by $i_t$ and sum to obtain

$$\sum c_t i_t = \beta \sum i_t y_t + \sum i_t (\varepsilon_t - \varepsilon)$$

and

$$\frac{\text{plim}\left(\dfrac{1}{T}\sum c_t i_t\right)}{\text{plim}\left(\dfrac{1}{T}\sum i_t y_t\right)} = \beta.$$

Hence, by using $I_t$ as the instrumental variable in this model, we have

$$b^* = \frac{\sum c_t i_t}{\sum i_t y_t}$$

as a consistent estimator of $\beta$.

Unfortunately not all the problems we have raised can be solved as simply as this. The method of instrumental varables is not easy to apply in all situations, as in some cases there may not be any obvious instrumental variables, while in others we may be embarrassed by having an excess of potential instrumental variables and no obvious criteria for making a choice. A full discussion of the problems and the efforts made to solve them is not possible in this text, and the interested reader should consult a text in econometrics at the appropriate level.[15]

## 11.5 conclusions
We conclude with some remarks directed at the reader who, having worked his way through the first ten chapters, has reacted to the problems raised in this chapter by wondering whether his journey was really necessary. The desirable properties of the method of least squares do depend on a number of assumptions that are not always justified in the context of economic analysis. For this reason one should always bear these limitations in mind, test for violations of the assumptions where possible, and introduce appropriate modifications in the method of least squares where necessary. As we have suggested, the problems may not be too difficult if we are dealing with heteroscedasticity or autocorrelation, but may be more severe if the independent variables are stochastic.

However, while we have theoretical solutions to some of the problems that arise if the independent variables are stochastic, the statistical properties of these alternatives to the least squares regression model are dependent on their assumptions. In many cases these assumptions involve a detailed specification of an economic model and, given our current knowledge of the working of an economy, it is very likely that some of these assumptions are going to be violated. This will affect the desirable properties of these alternative methods of estimation, and some of them are less robust to the relaxation of assumptions than is the method of least squares.[16] Given the trade-off in robustness, computational ease and data requirements between the various alternative methods of estimation, at least one major simultaneous equation economic model is currently being estimated by the method of least squares.[17] As our knowledge of the economy and the quality of the data both improve, we shall probably move to more complex methods of estimation as a matter of course, but at present the method of least squares is the estimating technique which still deserves very serious attention.

# appendix to chapter 11

## 11.A1 generalized least squares

The discussion of both heteroscedasticity and autocorrelation may be generalized to the multiple regression model by retaining assumptions 9.A6, 9.A8 and 9.A9 and by changing the assumption made concerning the variance–covariance matrix $E(\varepsilon\varepsilon')$ from

$$E(\varepsilon\varepsilon') = \sigma^2 \mathbf{I} \qquad (9.\text{A}7)$$

to

$$E(\varepsilon\varepsilon') = \sigma^2 \boldsymbol{\Omega}, \qquad (11.\text{A}1)$$

where the elements in $\boldsymbol{\Omega}$ will be discussed below. The new assumption leaves $\mathbf{b}$ as an unbiased estimator of $\boldsymbol{\beta}$, but now when we derive var $(\mathbf{b})$ from 9.A13 we get

$$\begin{aligned}
\text{var}(\mathbf{b}) &= (\mathbf{X}'\mathbf{X})^{-1}\mathbf{X}'E(\varepsilon\varepsilon')\mathbf{X}(\mathbf{X}'\mathbf{X})^{-1} \\
&= \sigma^2 (\mathbf{X}'\mathbf{X})^{-1}\mathbf{X}'\boldsymbol{\Omega}\mathbf{X}(\mathbf{X}'\mathbf{X})^{-1} \qquad (11.\text{A}2) \\
&\neq \sigma^2 (\mathbf{X}'\mathbf{X})^{-1},
\end{aligned}$$

confirming that the effect of either heteroscedasticity or autocorrelation is that the standard formulae for sampling variances are no longer relevant.

If $\boldsymbol{\Omega}$ were known, 11.A2 would enable us to calculate the appropriate sampling variances for $\mathbf{b}$, but the elements of $\mathbf{b}$ are no longer the minimum variance estimators of the $\beta$s. The best estimator of $\boldsymbol{\beta}$ may be developed in this context as follows. Suppose we define a transformation matrix, $\mathbf{T}$, such that

$$\boldsymbol{\Omega} = \mathbf{T}\mathbf{T}'.$$

Then

$$\mathbf{T}^{-1}\boldsymbol{\Omega}\mathbf{T}^{-1'} = \mathbf{T}^{-1}(\mathbf{T}\mathbf{T}')\mathbf{T}^{-1'} = \mathbf{I} \qquad (11.\text{A}3)$$

and

$$\Omega^{-1} = (TT')^{-1} = T^{-1'}T^{-1}. \tag{11.A4}$$

If we premultiply $y = X\beta + \varepsilon$ by $T^{-1}$ we obtain

$$T^{-1}y = T^{-1}X\beta + T^{-1}\varepsilon$$

or

$$y^* = X^*\beta + \varepsilon^*. \tag{11.A5}$$

The least squares estimator of $\beta$ in the transformed model is

$$b^* = (X^{*'}X^*)^{-1}X^{*'}y^* \tag{11.A6}$$

$$= (X'T^{-1'}T^{-1}X)^{-1}X'T^{-1'}T^{-1}y$$

$$= (X'\Omega^{-1}X)^{-1}X'\Omega^{-1}y. \tag{11.A7}$$

Substituting 11.A5 for $y^*$ in 11.A6 gives

$$b^* = (X^{*'}X^*)^{-1}X^{*'}(X^*\beta + \varepsilon^*)$$

$$= \beta + (X^{*'}X^*)^{-1}X^{*'}\varepsilon^*,$$

so that

$$E(b)^* = \beta,$$

given 9.A8 and 9.A6. Thus $b^*$ is an unbiased estimator of $\beta$. In addition,

$$E(\varepsilon^*\varepsilon') = E(T^{-1}\varepsilon\varepsilon'T^{-1'})$$

$$= T^{-1}E(\varepsilon\varepsilon')T^{-1'}$$

$$= T^{-1}\sigma^2\Omega T^{-1'}$$

$$= \sigma^2 T^{-1}\Omega T^{-1'}$$

$$= \sigma^2 I,$$

given 11.A3, so that $b^*$ has all the properties of the least squares estimator $b$ defined in 9.A5. Thus, from 9.A14, we have

$$var(b^*) = \sigma^2 (X^{*'}X^*)^{-1}$$

$$= \sigma^2 (X'T^{-1'}T^{-1}X)^{-1}$$

$$= \sigma^2 (X'\Omega^{-1}X)^{-1}. \tag{11.A8}$$

The estimator defined in 11.A7 is usually called the generalized least squares estimator (GLS), since when $\Omega = \mathbf{I}$, 11.A7 and 11.A8 reduce to 9.A5 and 9.A14 respectively, so that the ordinary least squares (OLS) estimator of $\beta$ is just a special case of generalized least squares model.

## 11.A2 heteroscedasticity
In our earlier discussion of heteroscedasticity we considered the case in which

$$E(\varepsilon_i^2) = \sigma^2 X_i^2. \tag{11.6}$$

If we assume that assumptions 9.A6, 9.A8 and 9.A9 hold and concentrate on the effects of relaxing 9.A7, we may assume that the variances of the errors vary with the values of the $j$th independent variable, so that $E(\varepsilon_i^2) = \sigma^2 X_{ji}^2$. Then for a sample of $n$ observations,

$$E(\varepsilon\varepsilon') = \sigma^2\Omega = \sigma^2 \begin{bmatrix} X_{j1}^2 & 0 & \cdots & 0 \\ 0 & X_{j2}^2 & \cdots & 0 \\ \cdot & \cdot & \cdots & \cdot \\ \cdot & \cdot & \cdots & \cdot \\ \cdot & \cdot & \cdots & \cdot \\ 0 & 0 & \cdots & X_{jn}^2 \end{bmatrix}$$

Here we may define the appropriate transformation matrix as

$$\mathbf{T} = \begin{bmatrix} X_{j1} & 0 & \cdots & 0 \\ 0 & X_{j2} & \cdots & 0 \\ \cdot & \cdot & \cdots & \cdot \\ \cdot & \cdot & \cdots & \cdot \\ \cdot & \cdot & \cdots & \cdot \\ 0 & 0 & \cdots & X_{jn} \end{bmatrix}$$

with

$$\mathbf{T}^{-1} = \begin{bmatrix} 1/X_{j1} & 0 & \cdots & 0 \\ 0 & 1/X_{j2} & \cdots & 0 \\ \cdot & \cdot & \cdots & \cdot \\ \cdot & \cdot & \cdots & \cdot \\ \cdot & \cdot & \cdots & \cdot \\ 0 & 0 & \cdots & 1/X_{jn} \end{bmatrix}.$$

That is, we run the multiple regression after dividing the elements on both sides of the equation $\mathbf{y} = \mathbf{X}\boldsymbol{\beta} + \boldsymbol{\varepsilon}$ by $X_{ji}$.

### 11.A3 autocorrelation

The form of the matrix $\Omega$ depends on the assumptions we make concerning $E\,(\boldsymbol{\varepsilon}\boldsymbol{\varepsilon}')$, and here we shall explore the properties of the simple first-order autoregressive process

$$\varepsilon_t = \rho\varepsilon_{t-1} + \omega_t, \qquad |\rho| < 1, \tag{11.15}$$

where

$$E\,(\omega_t) = 0, \tag{11.17}$$

$$E(\omega_t^2) = \sigma_\omega^2, \tag{11.18}$$

$$E(\omega_t\omega_{t-s}) = 0, \qquad s \neq 0. \tag{11.19}$$

Substituting back successively for $\varepsilon_{t-1}$, $\quad \varepsilon_{t-2}$, etc., in 11.15, we have

$$\varepsilon_t = \omega_t + \rho\omega_{t-1} + \rho^2\omega_{t-2} + \dots$$

$$= \sum_{s=0}^{\infty} \rho^s\omega_{t-s} \tag{11.A9}$$

and taking expected values of both sides of 11.A9,

$$E\,(\varepsilon_t) = E\,\left(\sum \rho^s\omega_{t-s}\right)$$

$$= \sum \rho^s E\,(\omega_{t-s}) = 0 \tag{11.A10}$$

by virtue of 11.17.

Given this result,

$$\text{var}\,(\varepsilon_t) = E\,(\varepsilon_t^2) = E\,\left[\left(\sum \rho^s\omega_{t-s}\right)^2\right]$$

$$= [(\omega_t + \rho\omega_{t-1} + \rho^2\omega_{t-2} + \dots)(\omega_t + \rho\omega_{t-1} + \dots)]$$

$$= E\,[(\omega_t^2 + \rho^2\omega_{t-1}^2 + \rho^4\omega_{t-2}^2 + \dots)],$$

since all the terms involving $\omega_t\omega_{t-s}$ have an expected value of zero by virtue of 11.19, so that

$$\text{var}(\varepsilon_t) = \sigma_\omega^2 (1 + \rho^2 + \rho^4 + \ldots)$$

$$= \frac{\sigma_\omega^2}{1 - \rho^2} = \sigma_\varepsilon^2, \tag{11.A11}$$

since the series of terms inside the parentheses forms a convergent geometric progression.

The covariance between $\varepsilon_t$ and $\varepsilon_{t-1}$ is

$$E(\varepsilon_t \varepsilon_{t-1}) = E[(\omega_t + \rho\omega_{t-1} + \ldots)(\omega_{t-1} + \rho\omega_{t-2} + \ldots)]$$

$$= E[(\rho\omega_{t-1} + \rho^2\omega_{t-2} + \ldots)(\omega_{t-1} + \rho\omega_{t-2} + \ldots)],$$

since $E[\omega_t(\omega_{t-1} + \rho\omega_{t-2} + \ldots)] = 0$, given 11.19. So

$$E(\varepsilon_t \varepsilon_{t-1}) = E[\rho(\omega_{t-1} + \rho\omega_{t-2} + \ldots)(\omega_{t-1} + \rho\omega_{t-2} + \ldots)]$$

$$= \rho E[(\omega_{t-1} + \rho\omega_{t-2} + \ldots)^2]$$

$$= \rho\sigma_\varepsilon^2,$$

given the result obtained in 11.A11. Similarly,

$$E(\varepsilon_t \varepsilon_{t-2}) = E[(\omega_t + \rho\omega_{t-1} + \rho^2\omega_{t-2} + \ldots)(\omega_{t-2} + \rho\omega_{t-3} + \ldots)]$$

$$= \rho^2 E[(\omega_{t-2} + \rho\omega_{t-3} + \ldots)^2] = \rho^2\sigma_\varepsilon^2,$$

and in general

$$E(\varepsilon_t \varepsilon_{t-s}) = \rho^s \sigma_\varepsilon^2. \tag{11.A12}$$

Given the results obtained in 11.A11 and 11.A12,

$$\sigma^2 \Omega = \sigma_\varepsilon^2 \begin{bmatrix} 1 & \rho & \rho^2 & \ldots & \rho^{T-1} \\ \rho & 1 & \rho & \ldots & \rho^{T-2} \\ . & . & . & \ldots & . \\ . & . & . & \ldots & . \\ . & . & . & \ldots & . \\ \rho^{T-1} & \rho^{T-2} & . & \ldots & 1 \end{bmatrix}$$

and it can be shown that

$$\Omega^{-1} = \begin{bmatrix} 1 & -\rho & 0 & \cdots & 0 & 0 & 0 \\ -\rho & 1+\rho^2 & -\rho & \cdots & 0 & 0 & 0 \\ 0 & -\rho & 1+\rho^2 & \cdots & 0 & 0 & 0 \\ \vdots & \vdots & \vdots & \cdots & \vdots & \vdots & \vdots \\ 0 & 0 & 0 & \cdots & -\rho & 1+\rho^2 & -\rho \\ 0 & 0 & 0 & \cdots & 0 & -\rho & 1 \end{bmatrix}$$

So if $\rho$ and $\sigma_\omega^2$ were known, the generalized least squares estimator would be estimated by substituting 11.A13 for $\Omega^{-1}$ in 11.A7.

The appropriate transformation matrix in this case has an inverse

$$\mathbf{T}^{-1} = \begin{bmatrix} \sqrt{1-\rho^2} & 0 & 0 & \cdots & 0 & 0 & 0 \\ -\rho & 1 & 0 & \cdots & 0 & 0 & 0 \\ 0 & -\rho & 1 & \cdots & 0 & 0 & 0 \\ \vdots & \vdots & \vdots & \cdots & \vdots & \vdots & \vdots \\ 0 & 0 & 0 & \cdots & -\rho & 1 & 0 \\ 0 & 0 & 0 & \cdots & 0 & -\rho & 1 \end{bmatrix} \qquad (11.A14)$$

and the reader may verify that

$$E\,(\mathbf{T}^{-1}\boldsymbol{\varepsilon}\boldsymbol{\varepsilon}'\mathbf{T}^{-1\prime}) = \sigma^2\mathbf{I}.$$

The transformation proposed in 11.30 is equivalent to premultiplying $\mathbf{y} = \mathbf{X}\boldsymbol{\beta} + \boldsymbol{\varepsilon}$ by the $(T-1)$ by $T$ matrix

$$\begin{bmatrix} -\rho & 1 & 0 & \cdots & 0 & 0 & 0 \\ 0 & -\rho & 1 & \cdots & 0 & 0 & 0 \\ \vdots & \vdots & \vdots & \cdots & \vdots & \vdots & \vdots \\ 0 & 0 & 0 & \cdots & -\rho & 1 & 0 \\ 0 & 0 & 0 & \cdots & 0 & -\rho & 1 \end{bmatrix}$$

which is equal to the $T$ by $T$ matrix defined in 11.A14 after the first row has been deleted.

## 11.A4 instrumental variables

In terms of matrix algebra, we may consider the limit in probability of 9.A10 and write

$$\text{plim } (\mathbf{b}) = \boldsymbol{\beta} + \text{plim} \left( \frac{1}{n} \mathbf{X'X} \right)^{-1} \text{plim} \left( \frac{1}{n} \mathbf{X'\varepsilon} \right).$$

The problem of finding a consistent estimator of $\boldsymbol{\beta}$ arises when $plim \, (1/n \, \mathbf{X'\varepsilon}) \neq 0$. To apply the method of instrumental variables, we need to find $k$ variables, $Z_{ji}$, $j = 1, ..., k$; $i = 1, ..., n$, or

$$\mathbf{Z} = \begin{bmatrix} Z_{11} \cdots Z_{kl} \\ \cdot \quad \cdots \quad \cdot \\ \cdot \quad \cdots \quad \cdot\cdot \\ \cdot \quad \cdots \quad \cdot \\ Z_{ln} \cdots Z_{kn} \end{bmatrix},$$

such that

$$\text{plim} \left( \frac{1}{n} \mathbf{Z'\varepsilon} \right) = 0$$

$$\text{plim} \left( \frac{1}{n} \mathbf{Z'X} \right) \neq 0$$

$$\text{plim} \left( \frac{1}{n} \mathbf{Z'Z} \right) \neq 0.$$

Then premultiplying $\mathbf{y} = \mathbf{X}\boldsymbol{\beta} + \boldsymbol{\varepsilon}$ by $\mathbf{Z'}$ gives

$$\mathbf{Z'y} = \mathbf{Z'X}\boldsymbol{\beta} + \mathbf{Z'\varepsilon}$$

and the instrumental variable estimator

$$\mathbf{b^*} = (\mathbf{Z'X})^{-1} \mathbf{Z'y} \qquad (11.A15)$$

is a consistent estimator of $\boldsymbol{\beta}$. The sampling variances and covariances for the elements in $\mathbf{b^*}$ may be found from [18]

$$s^2 (\mathbf{Z'X})^{-1} (\mathbf{Z'Z}) (\mathbf{X'Z})^{-1}$$

where

$$s^2 = (\mathbf{y} - \mathbf{Xb^*})' \, (\mathbf{y} - \mathbf{Xb^*}) / (n - k).$$

# problems

**11.1** Consider the model $Y_i = \alpha + \beta X_i + \varepsilon_i$ in which the $X_i$ are free from errors and there are no important excluded variables. $Y$ is income which contains errors of measurement which are (i) caused by people rounding their incomes to the nearest £10, (ii) caused by the fact that people do not know their incomes exactly but always guess within $\pm k$ per cent of the true value. How might these alternative assumptions concerning the random errors affect the properties of the least squares estimators of $\alpha$ and $\beta$?

**11.2** Given the data in Table 11.1 on $Y'$ and $X'$, test for heteroscedasticity by using equation 11.7, letting $m = 0, 2, 4, 6, 8, 10$. Comment on your results.

**11.3*** For the residuals presented in problem 10.2, calculate the Durbin–Watson statistic and comment.

**11.4*** For the data given in Table 10.2 (b), subtract $1 + 2X_{2t} + X_{3t}$ from $Y_t$ to obtain $\varepsilon_t$ and then calculate the correlation between $X_{2t}$ and $\varepsilon_t$.

**11.5** Show that the least squares estimator of $\delta$ in 11.47 is biased and inconsistent.

**11.6** If

$$y_t = \beta y_{t-1} + \varepsilon_t$$

and

$$\varepsilon_t = \rho \varepsilon_{t-2} + \omega_t, \qquad -1 < \rho < 1,$$

where $E(\omega_t) = 0$, $E(\omega_t^2) = \sigma^2$, $E(\omega_t \omega_{t-s}) = 0$, $s \neq 0$, evaluate the covariance between $\varepsilon_t$ and $y_{t-1}$ and comment on the implications for the least squares estimation of $\beta$.

**11.7** For the first-order autoregressive equation

$$\varepsilon_t = \rho \varepsilon_{t-1} + \omega_t, \qquad -1 < \rho < 1,$$

where $E(\omega_t) = 0$, $E(\omega_t^2) = \sigma^2$, $E(\omega_t \omega_{t-s}) = 0$, $s \neq 0$, calculate the expected correlation between $\Delta \varepsilon_t$ and $\Delta \varepsilon_{t-1}$ (where $\Delta \varepsilon_t = \varepsilon_t - \varepsilon_{t-1}$).

**11.8** Given $y_t$ and $\varepsilon_t$ are related by 11.15 and 11.49, show that

$$p\lim\left(\frac{1}{n}\sum y_{t-1}\varepsilon_t\right) \neq 0.$$

What is the variance of $y_t$ in this model?

(N.B. Starred questions involve some computational effort.)

# answers to the odd-numbered numerical problems

## chapter 1

**1.1**  (a) 31;     (b) 20;     (c) $a(X_1 + X_2 + X_3)$;
(d) $4a + b(X_1 + X_2 + X_3 + X_4)$;     (e) $4\sum X_i^2 + 12\sum X_i + 27$;
(f) $4ac + bc\sum X_i + ad\sum Y_i + bd\sum X_i Y_i$

## chapter 2

**2.1**  (a) 2/13;     (b) 1/13;     (c) 10/13;     (d) 7/13.
**2.3**  (a) 1/9;     (b) 1/6;     (c) 2/9;     (d) 5/6.
**2.5**  0·4;     0·6.
**2.7**  (a) 1/120;     (b) 1/5.
**2.9**  2/3.

## chapter 3

**3.1**

| $X$    | 0    | 1    | 2     | 3     | 4    | 5    |
|--------|------|------|-------|-------|------|------|
| $p(X)$ | 1/32 | 5/32 | 10/32 | 10/32 | 5/32 | 1/32 |

**3.3**  Throwing twice, $p$(one six) $= 5/18 = 0\cdot278$. Throwing four times, $p$(two sixes) $= 25/216 = 0\cdot116$

**3.5**

| $X$    | 25   | 35  | 60   | 70  | 85  | 120  |
|--------|------|-----|------|-----|-----|------|
| $p(X)$ | 1/12 | 1/4 | 1/12 | 1/4 | 1/4 | 1/12 |

$E(X) = 64\cdot6$

**3.7**  (a) 0·0668;     (b) 0·1056;     (c) 0·3264;     (d) 0·1628;     (e) 0·0038;
(f) 0·4114.
**3.9**  Lowest $A = 79\cdot2$;     Highest $F = 70\cdot8$.
**3.11**  (a) $p$(19 or more) $= \sum_{x=19}^{25} {}^{25}C_x(\cdot64)^x(\cdot36)^{25-x}$. Using the normal approximation to the binomial distribution, $p$(19 or more) $= 0\cdot1492$
(b) $p(17) = {}^{25}C_{17}(\cdot64)^{17}(\cdot36)^8$. Using normal approximation to the binomial distribution, $p(17) = 0\cdot1515$

**3.13** $p(25 \text{ or more}) = 0{\cdot}1303$. Sample size should be 173 or more.

**3.15** (a) $E(X) = 0$;    $E(Y) = 2$;    $E(X)/E(Y) = 0$;

| Z | − 1 | − 1/3 | 2/9 | 2/3 | 2 |
|------|------|------|------|------|------|
| $p(Z)$ | 1/4 | 1/4 | 3/16 | 1/4 | 1/16 |

$E(Z) = 0$

(b) $E(X) = 1$;    $E(Y) = 0$;    $E(X)/E(Y) = \infty$ ;

| Z | − 3 | 1 | 9 |
|------|------|------|------|
| $p(Z)$ | 1/2 | 1/8 | 3/8 |

$E(Z) = 2$

(c) $E(X) = 0$;    $E(Y) = 0$;    $E(X)/E(Y) = 0/0 = ?$ ;

| Z | − 3 | − 2 | − 2/3 | 1 | 2 | 6 |
|------|------|------|------|------|------|------|
| $p(Z)$ | 3/8 | 1/32 | 3/32 | 1/8 | 9/32 | 3/32 |

$E(Z) = 0$

## chapter 4

**4.3** (a) $p(0{\cdot}10 < \bar{Z} < 0{\cdot}25) = 0{\cdot}0589$;    $p(\bar{Z} < -0{\cdot}20) = 0{\cdot}4207$

(b) $p(0{\cdot}10 < \bar{Z} < 0{\cdot}25) = 0{\cdot}0819$;    $p(\bar{Z} < -0{\cdot}20) = 0{\cdot}3887$

(c) $p(0{\cdot}10 < \bar{Z} < 0{\cdot}25) = 0{\cdot}1122$;    $p(\bar{Z} < -0{\cdot}20) = 0{\cdot}3446$

(d) $p(0{\cdot}10 < \bar{Z} < 0{\cdot}25) = 0{\cdot}1859$;    $p(\bar{Z} < -0{\cdot}20) = 0{\cdot}2119$

(e) $p(0{\cdot}10 < \bar{Z} < 0{\cdot}25) = 0{\cdot}1891$;    $p(\bar{Z} < -0{\cdot}20) = 0{\cdot}0548$

(f) $p(0{\cdot}10 < \bar{Z} < 0{\cdot}25) = 0{\cdot}1525$;    $p(\bar{Z} < -0{\cdot}20) = 0{\cdot}0228$

**4.5** Choose a sample of 385 colour TV sets.

## chapter 5

**5.3** If $\sigma^2$ unknown, use $t$-distribution, $112{\cdot}3 < \mu < 116{\cdot}1$. If $\sigma^2$ known, use standard normal distribution, $112{\cdot}3 < \mu < 116{\cdot}1$.

**5.5** $\bar{X} = 30{\cdot}6, s^2 = 25{\cdot}9$.    $27{\cdot}4 < \mu < 33{\cdot}8$.    $13{\cdot}0 < \sigma^2 < 74{\cdot}6$.

**5.7** $0{\cdot}46 < \pi < 0{\cdot}64$.

## chapter 6

**6.1** Type II error $= 0{\cdot}516$. To obtain Type II error $= 0{\cdot}05$, choose sample of size 420.

**6.3** Ration $\chi^2 = 9(81\cdot3)/125 = 5\cdot85$. Not significant at probability level $p = 0\cdot95$, so accept hypotheses that $\sigma^2 = 125$.

**6.5** Type II error $= 0\cdot55$. Sample size too small for us to discriminate between hypotheses.

**6.7** F-test on variances, $F = 1\cdot25$, which is not significant. Large sample test on the difference of the two population means, $z = 4\cdot37$, which is significant.

**6.9** The difference in proportions is not significant at the probability level $p = 0\cdot95$.

**6.11** F-test, $F = 3\cdot56$, which is significant for $p = 0\cdot95$.

## chapter 7

**7.1** (a) $1\cdot000$;    (b) $0\cdot373$;    (c) $0\cdot903$

**7.3** $Z = 2\cdot03$. Reject hypothesis that $\rho = 0\cdot6$ at probability level $p = 0\cdot95$.

**7.5** (a) $n = 4$;    (b) $n = 5$;    (c) $n = 16$;    (d) $n = 388$;    (e) $n = 38, 420$.

**7.7** Here $s = 1/\sqrt{17} = 0\cdot2425$. $Z = -\cdot205/\cdot2425 = -0\cdot855$. Not significant, so accept hypothesis that $\rho = 0$.

**7.9** Testing difference in proportions passing, $Z = 2\cdot88$. Reject hypothesis that population proportions are equal.

**7.11** $x^2 = 41\cdot83$. Evidence suggests form does affect response.

## chapter 8

**8.1** $Y_i = -4 + 1\cdot33X_i$, $R^2 = 0\cdot53$. To test $\beta = 2$ $t = 2\cdot0$, and with 14 degrees of freedom
             $(0\cdot33)$
and probability level $p = 0\cdot95$ accept the hypothesis that $\beta = 2\cdot0$.

**8.3** $Y_i = -50 + 2\cdot00X_i$,     $R^2 = 0\cdot8$.
           $(0\cdot20)$

To test $H_0 : \beta = 1\cdot5$ against $H_1 : \beta > 1\cdot5$ implies a one-tail test, and $t = 2\cdot5$, so reject in favour of $H_1$.

**8.5** (i) $Q_t = 99\cdot83 - 0\cdot982P_t$,      $R^2 = 0\cdot985$,      $d = 2\cdot29$.
         $(1\cdot45)$   $(0\cdot028)$

     (ii) $Q_t = 3\cdot56 + 0\cdot993P_{t-1}$,      $R^2 = 0\cdot882$,      $d = 1\cdot72$.
         $(4\cdot29)$   $(0\cdot083)$

## chapter 9

**9.1** $Y_i = 30 + 2\cdot0X_{2i} - X_{3i}$, $R^2 = 0\cdot90$. To test $\beta_2 = 1\cdot0$, $t = 7\cdot0$, so reject $H_0 : \beta_2 = 1\cdot0$.

**9.5** $r_{13.2} = -0\cdot261$.

# appendix

# some useful statistical tables

*Table A1.* Areas under the Standard Normal Distribution, $N(0, 1)$ 🔼

| $z$ | 0·00 | 0·01 | 0·02 | 0·03 | 0·04 | 0·05 | 0·06 | 0·07 | 0·08 | 0·09 |
|---|---|---|---|---|---|---|---|---|---|---|
| 0·0 | 0·5000 | 0·5040 | 0·5080 | 0·5120 | 0·5160 | 0·5199 | 0·5239 | 0·5279 | 0·5319 | 0·5359 |
| 0·1 | 0·5398 | 0·5438 | 0·5478 | 0·5517 | 0·5557 | 0·5596 | 0·5636 | 0·5675 | 0·5714 | 0·5753 |
| 0·2 | 0·5793 | 0·5832 | 0·5871 | 0·5910 | 0·5948 | 0·5987 | 0·6026 | 0·6064 | 0·6103 | 0·6141 |
| 0·3 | 0·6179 | 0·6217 | 0·6255 | 0·6293 | 0·6331 | 0·6368 | 0·6406 | 0·6443 | 0·6480 | 0·6517 |
| 0·4 | 0·6554 | 0·6591 | 0·6628 | 0·6664 | 0·6700 | 0·6736 | 0·6772 | 0·6808 | 0·6844 | 0·6879 |
| 0·5 | 0·6915 | 0·6950 | 0·6985 | 0·7019 | 0·7054 | 0·7088 | 0·7123 | 0·7157 | 0·7190 | 0·7224 |
| 0·6 | 0·7257 | 0·7291 | 0·7324 | 0·7357 | 0·7389 | 0·7422 | 0·7454 | 0·7486 | 0·0717 | 0·7549 |
| 0·7 | 0·7580 | 0·7611 | 0·7642 | 0·7673 | 0·7703 | 0·7734 | 0·7764 | 0·7794 | 0·7823 | 0·7852 |
| 0·8 | 0·7881 | 0·7910 | 0·7939 | 0·7967 | 0·7995 | 0·8023 | 0·8051 | 0·8078 | 0·8106 | 0·8133 |
| 0·9 | 0·8159 | 0·8186 | 0·8212 | 0·8238 | 0·8264 | 0·8289 | 0·8315 | 0·8340 | 0·8365 | 0·8389 |
| 1·0 | 0·8413 | 0·8438 | 0·8461 | 0·8485 | 0·8508 | 0·8531 | 0·8554 | 0·8577 | 0·8599 | 0·8621 |
| 1·1 | 0·8643 | 0·8665 | 0·8686 | 0·8708 | 0·8729 | 0·8749 | 0·8770 | 0·8790 | 0·8810 | 0·8830 |
| 1·2 | 0·8849 | 0·8869 | 0·8888 | 0·8907 | 0·8925 | 0·8944 | 0·8962 | 0·8980 | 0·8997 | 0·90147 |
| 1·3 | 0·90320 | 0·90490 | 0·90658 | 0·90824 | 0·90988 | 0·91149 | 0·91309 | 0·91466 | 0·91621 | 0·91774 |
| 1·4 | 0·91924 | 0·92073 | 0·92220 | 0·92364 | 0·92507 | 0·92647 | 0·92785 | 0·92922 | 0·93056 | 0·93189 |
| 1·5 | 0·93319 | 0·93448 | 0·93574 | 0·93699 | 0·93822 | 0·93943 | 0·94062 | 0·94179 | 0·94295 | 0·94408 |
| 1·6 | 0·94520 | 0·94630 | 0·94738 | 0·94845 | 0·94950 | 0·95053 | 0·95154 | 0·95254 | 0·95352 | 0·95449 |
| 1·7 | 0·95543 | 0·95637 | 0·95728 | 0·95818 | 0·95907 | 0·95994 | 0·96080 | 0·96164 | 0·96246 | 0·96327 |
| 1·8 | 0·96407 | 0·96485 | 0·96562 | 0·96638 | 0·96712 | 0·96784 | 0·96856 | 0·96926 | 0·96995 | 0·97062 |
| 1·9 | 0·97128 | 0·97193 | 0·97257 | 0·97320 | 0·97381 | 0·97441 | 0·97500 | 0·97558 | 0·97615 | 0·97670 |
| 2·0 | 0·97725 | 0·97778 | 0·97831 | 0·97882 | 0·97932 | 0·97982 | 0·98030 | 0·98077 | 0·98124 | 0·98169 |
| 2·1 | 0·98214 | 0·98257 | 0·98300 | 0·98341 | 0·98382 | 0·98422 | 0·98461 | 0·98500 | 0·98537 | 0·98574 |
| 2·2 | 0·98610 | 0·98645 | 0·98679 | 0·98713 | 0·98745 | 0·98778 | 0·98809 | 0·98840 | 0·98870 | 0·98899 |
| 2·3 | 0·98928 | 0·98956 | 0·98983 | $0·9^2 0097$ | $0·9^2 0358$ | $0·9^2 0613$ | $0·9^2 0863$ | $0·9^2 1106$ | $0·9^2 1344$ | $0·9^2 1576$ |
| 2·4 | $0·9^2 1802$ | $0·9^2 2024$ | $0·9^2 2240$ | $0·9^2 2451$ | $0·9^2 2656$ | $0·9^2 2857$ | $0·9^2 3053$ | $0·9^2 3244$ | $0·9^2 3431$ | $0·9^2 3613$ |
| 2·5 | $0·9^2 3790$ | $0·9^2 3963$ | $0·9^2 4132$ | $0·9^2 4297$ | $0·9^2 4457$ | $0·9^2 4614$ | $0·9^2 4766$ | $0·9^2 4915$ | $0·9^2 5060$ | $0·8^2 5201$ |
| 2·6 | $0·9^2 5339$ | $0·9^2 5473$ | $0·9^2 5604$ | $0·9^2 5731$ | $0·9^2 5855$ | $0·9^2 5975$ | $0·9^2 6093$ | $0·9^2 6207$ | $0·9^2 6319$ | $0·9^2 6427$ |
| 2·7 | $0·9^2 6533$ | $0·9^2 6636$ | $0·9^2 6736$ | $0·9^2 6833$ | $0·9^2 6928$ | $0·9^2 7020$ | $0·9^2 7110$ | $0·9^2 7197$ | $0·9^2 7282$ | $0·9^2 7365$ |
| 2·8 | $0·9^2 7445$ | $0·9^2 7523$ | $0·9^2 7599$ | $0·9^2 7673$ | $0·9^2 7744$ | $0·9^2 7814$ | $0·9^2 7882$ | $0·9^2 7948$ | $0·9^2 8012$ | $0·9^2 8074$ |
| 2·9 | $0·9^2 8134$ | $0·9^2 8193$ | $0·9^2 8250$ | $0·9^2 8305$ | $0·9^2 8359$ | $0·9^2 8411$ | $0·9^2 8462$ | $0·9^2 8511$ | $0·9^2 8559$ | $0·9^2 8605$ |
| 3·0 | $0·9^2 8650$ | $0·9^2 8694$ | $0·9^2 8736$ | $0·9^2 8777$ | $0·9^2 8817$ | $0·9^2 8856$ | $0·9^2 8893$ | $0·9^2 8930$ | $0·9^2 8965$ | $0·9^2 8999$ |
| 3·1 | $0·9^3 0324$ | $0·9^3 0646$ | $0·9^3 0957$ | $0·9^3 1260$ | $0·9^3 1553$ | $0·9^3 1836$ | $0·9^3 2112$ | $0·9^3 2378$ | $0·9^3 2636$ | $0·9^3 2886$ |
| 3·2 | $0·9^3 3129$ | $0·9^3 3363$ | $0·9^3 3590$ | $0·9^3 3810$ | $0·9^3 4024$ | $0·9^3 4230$ | $0·9^3 4429$ | $0·9^3 4623$ | $0·9^3 4810$ | $0·9^3 4991$ |
| 3·3 | $0·9^3 5166$ | $0·9^3 5335$ | $0·9^3 5499$ | $0·9^3 5658$ | $0·9^3 5811$ | $0·9^3 5959$ | $0·9^3 6103$ | $0·9^3 6242$ | $0·9^3 6376$ | $0·9^3 6505$ |
| 3·4 | $0·9^3 6631$ | $0·9^3 6752$ | $0·9^3 6869$ | $0·9^3 6982$ | $0·9^3 7091$ | $0·9^3 7197$ | $0·9^3 7299$ | $0·9^3 7398$ | $0·9^3 7493$ | $0·9^3 7585$ |
| 3·5 | $0·9^3 7674$ | $0·9^3 7759$ | $0·9^3 7842$ | $0·9^3 7922$ | $0·9^3 7999$ | $0·9^3 8074$ | $0·9^3 8146$ | $0·9^3 8215$ | $0·9^3 8282$ | $0·9^3 8347$ |
| 3·6 | $0·9^3 8409$ | $0·9^3 8469$ | $0·9^3 8527$ | $0·9^3 8583$ | $0·9^3 8637$ | $0·9^3 8689$ | $0·9^3 8739$ | $0·9^3 8787$ | $0·9^3 8834$ | $0·9^3 8879$ |
| 3·7 | $0·9^3 8922$ | $0·9^3 8964$ | $0·9^4 0039$ | $0·9^4 0426$ | $0·9^4 0799$ | $0·9^4 1158$ | $0·9^4 1504$ | $0·9^4 1838$ | $0·9^4 2159$ | $0·9^4 2468$ |
| 3·8 | $0·9^4 2765$ | $0·9^4 3052$ | $0·9^4 3327$ | $0·9^4 3593$ | $0·9^4 3848$ | $0·9^4 4094$ | $0·9^4 4331$ | $0·9^4 4558$ | $0·9^4 4777$ | $0·9^4 4988$ |
| 3·9 | $0·9^4 5190$ | $0·9^4 5385$ | $0·9^4 5573$ | $0·9^4 5753$ | $0·9^4 5926$ | $0·9^4 6092$ | $0·9^4 6253$ | $0·9^4 6406$ | $0·9^4 6554$ | $0·9^4 6696$ |
| 4·0 | $0·9^4 6833$ | $0·9^4 6964$ | $0·9^4 7090$ | $0·9^4 7211$ | $0·9^4 7327$ | $0·9^4 7439$ | $0·9^4 7546$ | $0·9^4 7649$ | $0·9^4 7748$ | $0·9^4 7843$ |

For example: $F(3.82) = 0·9^4 3327 = 0·99993327 = 1 - 0·00006673 = 1 - F(-3.82)$.

From *Statistical Tables and Formulas*, by A. Hald, John Wiley and Sons, New York, 1952; reproduced by permission of Professor A. Hald and the publishers.

*Table A2.* Critical values for the *t*-distribution

| Degrees of Freedom | Probability of a Larger Value, Sign Ignored | | | | | | | | |
|---|---|---|---|---|---|---|---|---|---|
| | 0·500 | 0·400 | 0·200 | 0·100 | 0·050 | 0·025 | 0·010 | 0·005 | 0·001 |
| 1 | 1·000 | 1·376 | 3·078 | 6·314 | 12·706 | 25·452 | 63·657 | | |
| 2 | 0·816 | 1·061 | 1·886 | 2·920 | 4·303 | 6·205 | 9·925 | 14·089 | 31·598 |
| 3 | ·765 | 0·978 | 1·638 | 2·353 | 3·182 | 4·176 | 5·841 | 7·453 | 12.941 |
| 4 | ·741 | ·941 | 1·533 | 2·132 | 2·776 | 3·495 | 4·604 | 5·598 | 8·610 |
| 5 | ·727 | ·920 | 1·476 | 2·015 | 2·571 | 3·163 | 4·032 | 4·773 | 6·859 |
| 6 | ·718 | ·906 | 1·440 | 1·943 | 2·447 | 2·969 | 3·707 | 4·317 | 5·959 |
| 7 | ·711 | ·896 | 1·415 | 1·895 | 2·365 | 2·841 | 3·499 | 4·029 | 5·405 |
| 8 | ·706 | ·889 | 1·397 | 1·860 | 2·306 | 2·752 | 3·355 | 3·832 | 5·041 |
| 9 | ·703 | ·883 | 1·383 | 1·833 | 2·262 | 2·685 | 3·250 | 3·690 | 4·781 |
| 10 | ·700 | ·879 | 1·372 | 1·812 | 2·228 | 2·634 | 3·169 | 3·581 | 4·587 |
| 11 | ·697 | ·876 | 1·363 | 1·796 | 2·201 | 2·593 | 3·106 | 3·497 | 4·437 |
| 12 | ·695 | ·873 | 1·356 | 1·782 | 2·179 | 2·560 | 3·055 | 3·428 | 4·318 |
| 13 | ·694 | ·870 | 1·350 | 1·771 | 2·160 | 2·533 | 3·012 | 3·372 | 4·221 |
| 14 | ·692 | ·868 | 1·345 | 1·761 | 2·145 | 2·510 | 2·077 | 3·326 | 4·140 |
| 15 | ·691 | ·866 | 1·341 | 1·753 | 2·131 | 2·490 | 2·947 | 3·286 | 4·073 |
| 16 | ·690 | ·865 | 1·337 | 1·746 | 2·120 | 2·473 | 2·921 | 3·252 | 4·015 |
| 17 | ·689 | ·863 | 1·333 | 1·740 | 2·110 | 2·458 | 2·898 | 3·222 | 3·965 |
| 18 | ·688 | ·862 | 1·330 | 1·734 | 2·101 | 2·445 | 2·878 | 3·197 | 3·922 |
| 19 | ·688 | ·861 | 1·328 | 1·729 | 2·093 | 2·433 | 2·861 | 3·174 | 3·883 |
| 20 | ·687 | ·860 | 1·325 | 1·725 | 2·086 | 2·423 | 2·845 | 3·153 | 3·850 |
| 21 | ·686 | ·859 | 1·323 | 1·721 | 2·080 | 2·414 | 2·831 | 3·135 | 3·819 |
| 22 | ·686 | ·858 | 1·321 | 1·717 | 2·074 | 2·406 | 2·819 | 3·119 | 3·792 |
| 23 | ·685 | ·858 | 1·319 | 1·714 | 2·069 | 2·398 | 2·807 | 3·104 | 3·767 |
| 24 | ·685 | ·857 | 1·318 | 1·711 | 2·064 | 2·391 | 2·797 | 3·090 | 3·745 |
| 25 | ·684 | ·856 | 1·316 | 1·708 | 2·060 | 2·385 | 2·787 | 3·078 | 3·725 |
| 26 | ·684 | ·856 | 1·315 | 1·706 | 2·056 | 2·379 | 2·779 | 3·067 | 3·707 |
| 27 | ·684 | ·855 | 1·314 | 1·703 | 2·052 | 2·373 | 2·771 | 3·056 | 3·690 |
| 28 | ·683 | ·855 | 1·313 | 1·701 | 2·048 | 2·368 | 2·763 | 3·047 | 3·674 |
| 29 | ·683 | ·854 | 1·311 | 1·699 | 2·045 | 2·364 | 2·756 | 3·038 | 3·659 |
| 30 | ·683 | ·854 | 1·310 | 1·697 | 2·042 | 2·360 | 2·750 | 3·030 | 3·646 |
| 35 | ·682 | ·852 | 1·306 | 1·690 | 2·030 | 2·342 | 2·724 | 2·996 | 3·591 |
| 40 | ·681 | ·851 | 1·303 | 1·684 | 2·021 | 2·329 | 2·704 | 2·971 | 3·551 |
| 45 | ·680 | ·850 | 1·301 | 1·680 | 2·014 | 2·319 | 2·690 | 2·952 | 3·520 |
| 50 | ·680 | ·849 | 1·299 | 1·676 | 2·008 · | 2·310 | 2·678 | 2·937 | 3·496 |
| 55 | ·679 | ·849 | 1·297 | 1·673 | 2·004 | 2·304 | 2·669 | 2·925 | 3·476 |
| 60 | ·679 | ·848 | 1·296 | 1·671 | 2·000 | 2·299 | 2·660 | 2·915 | 3·460 |
| 70 | ·678 | ·847 | 1·294 | 1·667 | 1·994 | 2·290 | 2·648 | 2·899 | 3·435 |
| 80 | ·678 | ·847 | 1·293 | 1·665 | 1·989 | 2·284 | 2·638 | 2·887 | 3·416 |
| 90 | ·678 | ·846 | 1·291 | 1·662 | 1·986 | 2·279 | 2·631 | 2·878 | 3·402 |
| 100 | ·677 | ·846 | 1·290 | 1·661 | 1·982 | 2·276 | 2·625 | 2·871 | 3·390 |
| 120 | ·677 | ·845 | 1·289 | 1·658 | 1·980 | 2·270 | 2·617 | 2·860 | 3·373 |
| ∞ | ·6745 | ·8416 | 1·2816 | 1·6448 | 1·9600 | 2·2414 | 2·5758 | 2·8070 | 3·2905 |

Parts of this table are reprinted by permission from R. A. Fisher's *Statistical Methods for Research Workers*, published by Oliver and Boyd, Edinburgh (1925–50); from Maxine Merrington's 'Table of Percentage Points of the *t*-Distribution', *Biometrika*, **32** (1942) 300; and from Bernard Ostle's *Statistics in Research*, Iowa State University Press (1954).

*Table A3.* Critical values for the chi-square distribution

Probability of a greater value

| Degrees of freedom | 0·995 | 0·990 | 0·975 | 0·950 | 0·900 | 0·750 | 0·500 | 0·250 | 0·100 | 0·050 | 0·025 | 0·010 | 0·005 |
|---|---|---|---|---|---|---|---|---|---|---|---|---|---|
| 1 | — | — | — | — | 0·02 | 0·10 | 0·45 | 1·32 | 2·71 | 3·84 | 5·02 | 6·63 | 7·88 |
| 2 | 0·01 | 0·02 | 0·05 | 0·10 | 0·21 | 0·58 | 1·39 | 2·77 | 4·61 | 5·99 | 7·38 | 9·21 | 10·60 |
| 3 | 0·07 | 0·11 | 0·22 | 0·35 | 0·58 | 1·21 | 2·37 | 4·11 | 6·25 | 7·81 | 9·35 | 11·34 | 12·84 |
| 4 | 0·21 | 0·30 | 0·48 | 0·71 | 1·06 | 1·92 | 3·36 | 5·39 | 7·78 | 9·49 | 11·14 | 13·28 | 14·86 |
| 5 | 0·41 | 0·55 | 0·83 | 1·15 | 1·61 | 2·67 | 4·35 | 6·63 | 9·24 | 11·07 | 12·83 | 15·09 | 16·75 |
| 6 | 0·68 | 0·87 | 1·24 | 1·64 | 2·20 | 3·45 | 5·35 | 7·84 | 10·64 | 12·59 | 14·45 | 16·81 | 18·55 |
| 7 | 0·99 | 1·24 | 1·69 | 2·17 | 2·83 | 4·25 | 6·35 | 9·04 | 12·02 | 14·07 | 16·01 | 18·48 | 20·28 |
| 8 | 1·34 | 1·65 | 2·18 | 2·73 | 3·49 | 5·07 | 7·34 | 10·22 | 13·36 | 15·51 | 17·53 | 20·09 | 21·96 |
| 9 | 1·73 | 2·09 | 2·70 | 3·33 | 4·17 | 5·90 | 8·34 | 11·39 | 14·68 | 16·92 | 19·02 | 21·67 | 23·59 |
| 10 | 2·16 | 2·56 | 3·25 | 3·94 | 4·87 | 6·74 | 9·34 | 12·55 | 15·99 | 18·31 | 20·48 | 23·21 | 25·19 |
| 11 | 2·60 | 3·05 | 3·82 | 4·57 | 5·58 | 7·58 | 10·34 | 13·70 | 17·28 | 19·68 | 21·92 | 24·72 | 26·76 |
| 12 | 3·07 | 3·57 | 4·40 | 5·23 | 6·30 | 8·44 | 11·34 | 14·85 | 18·55 | 21·03 | 23·34 | 26·22 | 28·30 |
| 13 | 3·57 | 4·11 | 5·01 | 5·89 | 7·04 | 9·30 | 12·34 | 15·98 | 19·81 | 22·36 | 24·74 | 27·69 | 29·82 |
| 14 | 4·07 | 4·66 | 5·63 | 6·57 | 7·70 | 10·17 | 13·34 | 17·12 | 21·06 | 23·68 | 26·12 | 29·14 | 31·32 |
| 15 | 4·60 | 5·23 | 6·27 | 7·26 | 8·55 | 11·04 | 14·34 | 18·25 | 22·31 | 25·00 | 27·49 | 30·58 | 32·80 |
| 16 | 5·14 | 5·81 | 6·91 | 7·96 | 9·31 | 11·91 | 15·34 | 19·37 | 23·54 | 26·30 | 28·85 | 32·00 | 34·27 |
| 17 | 5·70 | 6·41 | 7·56 | 8·67 | 10·09 | 12·79 | 16·34 | 20·49 | 24·77 | 27·59 | 30·19 | 33·41 | 35·72 |
| 18 | 6·26 | 7·01 | 8·23 | 9·39 | 10·86 | 13·68 | 17·34 | 21·60 | 25·99 | 28·87 | 31·53 | 34·81 | 37·16 |
| 19 | 6·84 | 7·63 | 8·91 | 10·12 | 11·65 | 14·56 | 18·34 | 22·72 | 27·20 | 30·14 | 32·85 | 36·19 | 38·58 |
| 20 | 7·43 | 8·26 | 9·59 | 10·85 | 12·44 | 15·45 | 19·34 | 23·83 | 28·41 | 31·41 | 34·17 | 37·57 | 40·00 |

Table A3. Critical values for the chi-distribution (continued)

| Degrees of freedom | Probability of a greater value | | | | | | | | | | | | |
|---|---|---|---|---|---|---|---|---|---|---|---|---|---|
| | 0·995 | 0·990 | 0·975 | 0·950 | 0·900 | 0·750 | 0·500 | 0·250 | 0·100 | 0·050 | 0·025 | 0·010 | 0·005 |
| 21 | 8·03 | 8·90 | 10·28 | 11·59 | 13·24 | 16·34 | 20·34 | 24·93 | 29·62 | 32·67 | 35·48 | 38·93 | 41·40 |
| 22 | 8·64 | 9·54 | 10·98 | 12·34 | 14·04 | 17·24 | 21·34 | 26·04 | 30·81 | 33·92 | 36·78 | 40·29 | 42·80 |
| 23 | 9·26 | 10·20 | 11·69 | 13·09 | 14·85 | 18·14 | 22·34 | 27·14 | 32·01 | 35·17 | 38·08 | 41·64 | 44·18 |
| 24 | 9·89 | 10·86 | 12·40 | 13·85 | 15·66 | 19·04 | 23·34 | 28·24 | 33·20 | 36·42 | 39·36 | 42·98 | 45·56 |
| 25 | 10·52 | 11·52 | 13·12 | 14·61 | 16·47 | 19·94 | 24·34 | 29·34 | 34·38 | 37·65 | 40·65 | 44·31 | 46·93 |
| 26 | 11·16 | 12·20 | 13·84 | 15·38 | 17·29 | 20·84 | 25·34 | 30·43 | 35·56 | 38·89 | 41·92 | 45·64 | 48·29 |
| 27 | 11·81 | 12·88 | 14·57 | 16·15 | 18·11 | 21·75 | 26·34 | 31·53 | 36·74 | 40·11 | 43·19 | 46·96 | 49·64 |
| 28 | 12·46 | 13·56 | 15·31 | 16·93 | 18·94 | 22·66 | 27·34 | 32·62 | 37·92 | 41·34 | 44·46 | 48·28 | 50·99 |
| 29 | 13·12 | 14·26 | 16·05 | 17·71 | 19·77 | 23·57 | 28·34 | 33·71 | 39·09 | 42·56 | 45·72 | 49·59 | 52·34 |
| 30 | 13·79 | 14·95 | 16·79 | 18·49 | 20·60 | 24·48 | 29·34 | 34·80 | 40·26 | 43·77 | 46·98 | 50·89 | 53·67 |
| 40 | 20·71 | 22·16 | 24·43 | 26·51 | 29·05 | 33·66 | 39·34 | 45·62 | 51·80 | 55·76 | 59·34 | 63·69 | 66·77 |
| 50 | 27·99 | 29·71 | 32·36 | 34·76 | 37·69 | 42·94 | 49·33 | 56·33 | 63·17 | 67·50 | 71·42 | 76·15 | 79·49 |
| 60 | 35·53 | 37·48 | 40·48 | 43·19 | 46·46 | 52·29 | 59·33 | 66·98 | 74·40 | 79·08 | 83·30 | 88·38 | 91·95 |
| 70 | 43·28 | 45·44 | 48·76 | 51·74 | 55·33 | 61·70 | 69·33 | 77·58 | 85·53 | 90·53 | 95·02 | 100·42 | 104·22 |
| 80 | 51·17 | 53·54 | 57·15 | 60·39 | 64·28 | 71·14 | 79·33 | 88·13 | 96·58 | 101·88 | 106·63 | 112·33 | 116·32 |
| 90 | 59·20 | 61·75 | 65·65 | 69·13 | 73·29 | 80·62 | 89·33 | 98·64 | 107·56 | 113·14 | 118·14 | 124·12 | 128·30 |
| 100 | 67·33 | 70·06 | 74·22 | 77·93 | 82·36 | 90·13 | 99·33 | 109·14 | 118·50 | 124·34 | 129·56 | 135·81 | 140·17 |

Condensed from table 6 with significant figures by Catherine M. Thompson, by permission of the Editor of *Biometrika*.

## Table A4. Critical values for the F-Distribution (continued)

5% (Roman type) and 1% (bold face type) points for the distribution of F

$f_1$ degrees of freedom (for greater mean square)

| $f_2$ | 1 | 2 | 3 | 4 | 5 | 6 | 7 | 8 | 9 | 10 | 11 | 12 | 14 | 16 | 20 | 24 | 30 | 40 | 50 | 75 | 100 | 200 | 500 | ∞ | $f_2$ |
|---|---|---|---|---|---|---|---|---|---|---|---|---|---|---|---|---|---|---|---|---|---|---|---|---|---|
| 1 | 161 **4,052** | 200 **4,999** | 216 **5,403** | 225 **5,625** | 230 **5,764** | 234 **5,859** | 237 **5,928** | 239 **5,981** | 241 **6,022** | 242 **6,056** | 243 **6,082** | 244 **6,106** | 245 **6,142** | 246 **6,169** | 248 **6,208** | 249 **6,234** | 250 **6,261** | 251 **6,286** | 252 **6,302** | 253 **6,323** | 253 **6,334** | 254 **6,352** | 254 **6,361** | 254 **6,366** | 1 |
| 2 | 18.51 **98.49** | 19.00 **99.00** | 19.16 **99.17** | 19.25 **99.25** | 19.30 **99.30** | 19.33 **99.33** | 19.36 **99.36** | 19.37 **99.37** | 19.38 **99.39** | 19.39 **99.40** | 19.40 **99.41** | 19.41 **99.42** | 19.42 **99.43** | 19.43 **99.44** | 19.44 **99.45** | 19.45 **99.46** | 19.46 **99.47** | 19.47 **99.48** | 19.47 **99.48** | 19.48 **99.49** | 19.49 **99.49** | 19.49 **99.49** | 19.50 **99.50** | 19.50 **99.50** | 2 |
| 3 | 10.13 **34.12** | 9.55 **30.82** | 9.28 **29.46** | 9.12 **28.71** | 9.01 **28.24** | 8.94 **27.91** | 8.88 **27.67** | 8.84 **27.49** | 8.81 **27.34** | 8.78 **27.23** | 8.76 **27.13** | 8.74 **27.05** | 8.71 **26.92** | 8.69 **26.83** | 8.66 **26.69** | 8.64 **26.60** | 8.62 **26.50** | 8.60 **26.41** | 8.58 **26.35** | 8.57 **26.27** | 8.56 **26.23** | 8.54 **26.18** | 8.54 **26.14** | 8.53 **26.12** | 3 |
| 4 | 7.71 **21.20** | 6.94 **18.00** | 6.59 **16.69** | 6.39 **15.98** | 6.26 **15.52** | 6.16 **15.21** | 6.09 **14.98** | 6.04 **14.80** | 6.00 **14.66** | 5.96 **14.54** | 5.93 **14.45** | 5.91 **14.37** | 5.87 **14.24** | 5.84 **14.15** | 5.80 **14.02** | 5.77 **13.93** | 5.74 **13.83** | 5.71 **13.74** | 5.70 **13.69** | 5.68 **13.61** | 5.66 **13.57** | 5.65 **13.52** | 5.64 **13.48** | 5.63 **13.46** | 4 |
| 5 | 6.61 **16.26** | 5.79 **13.27** | 5.41 **12.06** | 5.19 **11.39** | 5.05 **10.97** | 4.95 **10.67** | 4.88 **10.45** | 4.82 **10.29** | 4.78 **10.15** | 4.74 **10.05** | 4.70 **9.96** | 4.68 **9.89** | 4.64 **9.77** | 4.60 **9.68** | 4.56 **9.55** | 4.53 **9.47** | 4.50 **9.38** | 4.46 **9.29** | 4.44 **9.24** | 4.42 **9.17** | 4.40 **9.13** | 4.38 **9.07** | 4.37 **9.04** | 4.36 **9.02** | 5 |
| 6 | 5.99 **13.74** | 5.14 **10.92** | 4.76 **9.78** | 4.53 **9.15** | 4.39 **8.75** | 4.28 **8.47** | 4.21 **8.26** | 4.15 **8.10** | 4.10 **7.98** | 4.06 **7.87** | 4.03 **7.79** | 4.00 **7.72** | 3.96 **7.60** | 3.92 **7.52** | 3.87 **7.39** | 3.84 **7.31** | 3.81 **7.23** | 3.77 **7.14** | 3.75 **7.09** | 3.72 **7.02** | 3.71 **6.99** | 3.69 **6.94** | 3.68 **6.90** | 3.67 **6.88** | 6 |
| 7 | 5.59 **12.25** | 4.74 **9.55** | 4.35 **8.45** | 4.12 **7.85** | 3.97 **7.46** | 3.87 **7.19** | 3.79 **7.00** | 3.73 **6.84** | 3.68 **6.71** | 3.63 **6.62** | 3.60 **6.54** | 3.57 **6.47** | 3.52 **6.35** | 3.49 **6.27** | 3.44 **6.15** | 3.41 **6.07** | 3.38 **5.98** | 3.34 **5.90** | 3.32 **5.85** | 3.29 **5.78** | 3.28 **5.75** | 3.25 **5.70** | 3.24 **5.67** | 3.23 **5.65** | 7 |
| 8 | 5.32 **11.26** | 4.46 **8.65** | 4.07 **7.59** | 3.84 **7.01** | 3.69 **6.63** | 3.58 **6.37** | 3.50 **6.19** | 3.44 **6.03** | 3.39 **5.91** | 3.34 **5.82** | 3.31 **5.74** | 3.28 **5.67** | 3.23 **5.56** | 3.20 **5.48** | 3.15 **5.36** | 3.12 **5.28** | 3.08 **5.20** | 3.05 **5.11** | 3.03 **5.06** | 3.00 **5.00** | 2.98 **4.96** | 2.96 **4.91** | 2.94 **4.88** | 2.93 **4.86** | 8 |
| 9 | 5.12 **10.56** | 4.26 **8.02** | 3.86 **6.99** | 3.63 **6.42** | 3.48 **6.06** | 3.37 **5.80** | 3.29 **5.62** | 3.23 **5.47** | 3.18 **5.35** | 3.13 **5.26** | 3.10 **5.18** | 3.07 **5.11** | 3.02 **5.00** | 2.98 **4.92** | 2.93 **4.80** | 2.90 **4.73** | 2.86 **4.64** | 2.82 **4.56** | 2.80 **4.51** | 2.77 **4.45** | 2.76 **4.41** | 2.73 **4.36** | 2.72 **4.33** | 2.71 **4.31** | 9 |
| 10 | 4.96 **10.04** | 4.10 **7.56** | 3.71 **6.55** | 3.48 **5.99** | 3.33 **5.64** | 3.22 **5.39** | 3.14 **5.21** | 3.07 **5.06** | 3.02 **4.95** | 2.97 **4.85** | 2.94 **4.78** | 2.91 **4.71** | 2.86 **4.60** | 2.82 **4.52** | 2.77 **4.41** | 2.74 **4.33** | 2.70 **4.25** | 2.67 **4.17** | 2.64 **4.12** | 2.61 **4.05** | 2.59 **4.01** | 2.56 **3.96** | 2.55 **3.93** | 2.54 **3.91** | 10 |
| 11 | 4.84 **9.65** | 3.98 **7.20** | 3.59 **6.22** | 3.36 **5.67** | 3.20 **5.32** | 3.09 **5.07** | 3.01 **4.88** | 2.95 **4.74** | 2.90 **4.63** | 2.86 **4.54** | 2.82 **4.46** | 2.79 **4.40** | 2.74 **4.29** | 2.70 **4.21** | 2.65 **4.10** | 2.61 **4.02** | 2.57 **3.94** | 2.53 **3.86** | 2.50 **3.80** | 2.47 **3.74** | 2.45 **3.70** | 2.42 **3.66** | 2.41 **3.62** | 2.40 **3.60** | 11 |
| 12 | 4.75 **9.33** | 3.88 **6.93** | 3.49 **5.95** | 3.26 **5.41** | 3.11 **5.06** | 3.00 **4.82** | 2.92 **4.65** | 2.85 **4.50** | 2.80 **4.39** | 2.76 **4.30** | 2.72 **4.22** | 2.69 **4.16** | 2.64 **4.05** | 2.60 **3.98** | 2.54 **3.86** | 2.50 **3.78** | 2.46 **3.70** | 2.42 **3.61** | 2.40 **3.56** | 2.36 **3.49** | 2.35 **3.46** | 2.32 **3.41** | 2.31 **3.38** | 2.30 **3.36** | 12 |
| 13 | 4.67 **9.07** | 3.80 **6.70** | 3.41 **5.74** | 3.18 **5.20** | 3.02 **4.86** | 2.92 **4.62** | 2.84 **4.44** | 2.77 **4.30** | 2.72 **4.19** | 2.67 **4.10** | 2.63 **4.02** | 2.60 **3.96** | 2.55 **3.85** | 2.51 **3.78** | 2.46 **3.67** | 2.42 **3.59** | 2.38 **3.51** | 2.34 **3.42** | 2.32 **3.37** | 2.28 **3.30** | 2.26 **3.27** | 2.24 **3.21** | 2.22 **3.18** | 2.21 **3.16** | 13 |

*Table A4.* Critical values for the *F*-Distribution (*continued*)

$f_1$ degrees of freedom (for greater mean square)

| $f_2$ | 1 | 2 | 3 | 4 | 5 | 6 | 7 | 8 | 9 | 10 | 11 | 12 | 14 | 16 | 20 | 24 | 30 | 40 | 50 | 75 | 100 | 200 | 500 | ∞ | $f_2$ |
|---|---|---|---|---|---|---|---|---|---|---|---|---|---|---|---|---|---|---|---|---|---|---|---|---|---|
| 14 | 4·60 / 8·86 | 3·74 / 6·51 | 3·34 / 5·56 | 3·11 / 5·03 | 2·96 / 4·69 | 2·85 / 4·46 | 2·77 / 4·28 | 2·70 / 4·14 | 2·65 / 4·03 | 2·60 / 3·94 | 2·56 / 3·86 | 2·53 / 3·80 | 2·48 / 3·70 | 2·44 / 3·62 | 2·39 / 3·51 | 2·35 / 3·43 | 2·31 / 3·34 | 2·27 / 3·26 | 2·24 / 3·21 | 2·21 / 3·14 | 2·19 / 3·11 | 2·16 / 3·06 | 2·14 / 3·02 | 2·13 / 3·00 | 14 |
| 15 | 4·54 / 8·68 | 3·68 / 6·36 | 3·29 / 5·42 | 3·06 / 4·89 | 2·90 / 4·56 | 2·79 / 4·32 | 2·70 / 4·14 | 2·64 / 4·00 | 2·59 / 3·89 | 2·55 / 3·80 | 2·51 / 3·73 | 2·48 / 3·67 | 2·43 / 3·56 | 2·39 / 3·48 | 2·33 / 3·36 | 2·29 / 3·29 | 2·25 / 3·20 | 2·21 / 3·12 | 2·18 / 3·07 | 2·15 / 3·00 | 2·12 / 2·97 | 2·10 / 2·92 | 2·08 / 2·89 | 2·07 / 2·87 | 15 |
| 16 | 4·49 / 8·53 | 3·63 / 6·23 | 3·24 / 5·29 | 3·01 / 4·77 | 2·85 / 4·44 | 2·74 / 4·20 | 2·66 / 4·03 | 2·59 / 3·89 | 2·54 / 3·78 | 2·49 / 3·69 | 2·45 / 3·61 | 2·42 / 3·55 | 2·37 / 3·45 | 2·33 / 3·37 | 2·28 / 3·25 | 2·24 / 3·18 | 2·20 / 3·10 | 2·16 / 3·01 | 2·13 / 2·96 | 2·09 / 2·89 | 2·07 / 2·86 | 2·04 / 2·80 | 2·02 / 2·77 | 2·01 / 2·75 | 16 |
| 17 | 4·45 / 8·40 | 3·59 / 6·11 | 3·20 / 5·18 | 2·96 / 4·67 | 2·81 / 4·34 | 2·70 / 4·10 | 2·62 / 3·93 | 2·55 / 3·79 | 2·50 / 3·68 | 2·45 / 3·59 | 2·41 / 3·52 | 2·38 / 3·45 | 2·33 / 3·35 | 2·29 / 3·27 | 2·23 / 3·16 | 2·19 / 3·08 | 2·15 / 3·00 | 2·11 / 2·92 | 2·08 / 2·86 | 2·04 / 2·79 | 2·02 / 2·76 | 1·99 / 2·70 | 1·97 / 2·67 | 1·96 / 2·65 | 17 |
| 18 | 4·41 / 8·28 | 3·55 / 6·01 | 3·16 / 5·09 | 2·93 / 4·58 | 2·77 / 4·25 | 2·66 / 4·01 | 2·58 / 3·85 | 2·51 / 3·71 | 2·46 / 3·60 | 2·41 / 3·51 | 2·37 / 3·44 | 2·34 / 3·37 | 2·29 / 3·27 | 2·25 / 3·19 | 2·19 / 3·07 | 2·15 / 3·00 | 2·11 / 2·91 | 2·07 / 2·83 | 2·04 / 2·78 | 2·00 / 2·71 | 1·98 / 2·68 | 1·95 / 2·62 | 1·93 / 2·59 | 1·92 / 2·57 | 18 |
| 19 | 4·38 / 8·18 | 3·52 / 5·93 | 3·13 / 5·01 | 2·90 / 4·50 | 2·74 / 4·17 | 2·63 / 3·94 | 2·55 / 3·77 | 2·48 / 3·63 | 2·43 / 3·52 | 2·38 / 3·43 | 2·34 / 3·36 | 2·31 / 3·30 | 2·26 / 3·19 | 2·21 / 3·12 | 2·15 / 3·00 | 2·11 / 2·92 | 2·07 / 2·84 | 2·02 / 2·76 | 2·00 / 2·70 | 1·96 / 2·63 | 1·94 / 2·60 | 1·91 / 2·54 | 1·90 / 2·51 | 1·88 / 2·49 | 19 |
| 20 | 4·35 / 8·10 | 3·49 / 5·85 | 3·10 / 4·94 | 2·87 / 4·43 | 2·71 / 4·10 | 2·60 / 3·87 | 2·52 / 3·71 | 2·45 / 3·56 | 2·40 / 3·45 | 2·35 / 3·37 | 2·31 / 3·30 | 2·28 / 3·23 | 2·23 / 3·13 | 2·18 / 3·05 | 2·12 / 2·94 | 2·08 / 2·86 | 2·04 / 2·77 | 1·99 / 2·69 | 1·96 / 2·63 | 1·92 / 2·56 | 1·90 / 2·53 | 1·87 / 2·47 | 1·85 / 2·44 | 1·84 / 2·42 | 20 |
| 21 | 4·32 / 8·02 | 3·47 / 5·78 | 3·07 / 4·87 | 2·84 / 4·37 | 2·68 / 4·04 | 2·57 / 3·81 | 2·49 / 3·65 | 2·42 / 3·51 | 2·37 / 3·40 | 2·32 / 3·31 | 2·28 / 3·24 | 2·25 / 3·17 | 2·20 / 3·07 | 2·15 / 2·99 | 2·09 / 2·88 | 2·05 / 2·80 | 2·00 / 2·72 | 1·96 / 2·63 | 1·93 / 2·58 | 1·89 / 2·51 | 1·87 / 2·47 | 1·84 / 2·42 | 1·82 / 2·38 | 1·81 / 2·36 | 21 |
| 22 | 4·30 / 7·94 | 3·44 / 5·72 | 3·05 / 4·82 | 2·82 / 4·31 | 2·66 / 3·99 | 2·55 / 3·76 | 2·47 / 3·59 | 2·40 / 3·45 | 2·35 / 3·35 | 2·30 / 3·26 | 2·26 / 3·18 | 2·23 / 3·12 | 2·18 / 3·02 | 2·13 / 2·94 | 2·07 / 2·83 | 2·03 / 2·75 | 1·98 / 2·67 | 1·93 / 2·58 | 1·91 / 2·53 | 1·87 / 2·46 | 1·84 / 2·42 | 1·81 / 2·37 | 1·80 / 2·33 | 1·78 / 2·31 | 22 |
| 23 | 4·28 / 7·88 | 3·42 / 5·66 | 3·03 / 4·76 | 2·80 / 4·26 | 2·64 / 3·94 | 2·53 / 3·71 | 2·45 / 3·54 | 2·38 / 3·41 | 2·32 / 3·30 | 2·28 / 3·21 | 2·24 / 3·14 | 2·20 / 3·07 | 2·14 / 2·97 | 2·10 / 2·89 | 2·04 / 2·78 | 2·00 / 2·70 | 1·96 / 2·62 | 1·91 / 2·53 | 1·88 / 2·48 | 1·84 / 2·41 | 1·82 / 2·37 | 1·79 / 2·32 | 1·77 / 2·28 | 1·76 / 2·26 | 23 |
| 24 | 4·26 / 7·82 | 3·40 / 5·61 | 3·01 / 4·72 | 2·78 / 4·22 | 2·62 / 3·90 | 2·51 / 3·67 | 2·43 / 3·50 | 2·36 / 3·36 | 2·30 / 3·25 | 2·26 / 3·17 | 2·22 / 3·09 | 2·18 / 3·03 | 2·13 / 2·93 | 2·09 / 2·85 | 2·02 / 2·74 | 1·98 / 2·66 | 1·94 / 2·58 | 1·89 / 2·49 | 1·86 / 2·44 | 1·82 / 2·36 | 1·80 / 2·33 | 1·76 / 2·27 | 1·74 / 2·23 | 1·73 / 2·21 | 24 |
| 25 | 4·24 / 7·77 | 3·38 / 5·57 | 2·99 / 4·68 | 2·76 / 4·18 | 2·60 / 3·86 | 2·49 / 3·63 | 2·41 / 3·46 | 2·34 / 3·32 | 2·28 / 3·21 | 2·24 / 3·13 | 2·20 / 3·05 | 2·16 / 2·99 | 2·11 / 2·89 | 2·06 / 2·81 | 2·00 / 2·70 | 1·96 / 2·62 | 1·92 / 2·54 | 1·87 / 2·45 | 1·84 / 2·40 | 1·80 / 2·32 | 1·77 / 2·29 | 1·74 / 2·23 | 1·72 / 2·19 | 1·71 / 2·17 | 25 |
| 26 | 4·22 / 7·72 | 3·37 / 5·53 | 2·98 / 4·64 | 2·74 / 4·14 | 2·59 / 3·82 | 2·47 / 3·59 | 2·39 / 3·42 | 2·32 / 3·29 | 2·27 / 3·17 | 2·22 / 3·09 | 2·18 / 3·02 | 2·15 / 2·96 | 2·10 / 2·86 | 2·05 / 2·77 | 1·99 / 2·66 | 1·95 / 2·58 | 1·90 / 2·50 | 1·85 / 2·41 | 1·82 / 2·36 | 1·78 / 2·28 | 1·76 / 2·25 | 1·72 / 2·19 | 1·70 / 2·15 | 1·69 / 2·13 | 26 |

Table A4. Critical values for the F-Distribution (continued)

$f_1$ degrees of freedom (for greater mean square)

Each cell shows two critical values (upper row over lower row).

| $f_2$ | 1 | 2 | 3 | 4 | 5 | 6 | 7 | 8 | 9 | 10 | 11 | 12 | 14 | 16 | 20 | 24 | 30 | 40 | 50 | 75 | 100 | 200 | 500 | ∞ | $f_2$ |
|---|---|---|---|---|---|---|---|---|---|---|---|---|---|---|---|---|---|---|---|---|---|---|---|---|---|
| 27 | 4·21 / 7·68 | 3·35 / 5·49 | 2·96 / 4·60 | 2·73 / 4·11 | 2·57 / 3·79 | 2·46 / 3·56 | 2·37 / 3·39 | 2·30 / 3·26 | 2·25 / 3·14 | 2·20 / 3·06 | 2·16 / 2·98 | 2·13 / 2·93 | 2·08 / 2·83 | 2·03 / 2·74 | 1·97 / 2·63 | 1·93 / 2·55 | 1·88 / 2·47 | 1·84 / 2·38 | 1·80 / 2·33 | 1·76 / 2·25 | 1·74 / 2·21 | 1·71 / 2·16 | 1·68 / 2·12 | 1·67 / 2·10 | 27 |
| 28 | 4·20 / 7·64 | 3·34 / 5·45 | 2·95 / 4·57 | 2·71 / 4·07 | 2·56 / 3·76 | 2·44 / 3·53 | 2·36 / 3·36 | 2·29 / 3·23 | 2·24 / 3·11 | 2·19 / 3·03 | 2·15 / 2·95 | 2·12 / 2·90 | 2·06 / 2·80 | 2·02 / 2·71 | 1·96 / 2·60 | 1·91 / 2·52 | 1·87 / 2·44 | 1·81 / 2·35 | 1·78 / 2·30 | 1·75 / 2·22 | 1·72 / 2·18 | 1·69 / 2·13 | 1·67 / 2·09 | 1·65 / 2·06 | 28 |
| 29 | 4·18 / 7·60 | 3·33 / 5·42 | 2·93 / 4·54 | 2·70 / 4·04 | 2·54 / 3·73 | 2·43 / 3·50 | 2·35 / 3·33 | 2·28 / 3·20 | 2·22 / 3·08 | 2·18 / 3·00 | 2·14 / 2·92 | 2·10 / 2·87 | 2·05 / 2·77 | 2·00 / 2·68 | 1·94 / 2·57 | 1·90 / 2·49 | 1·85 / 2·41 | 1·80 / 2·32 | 1·77 / 2·27 | 1·73 / 2·19 | 1·71 / 2·15 | 1·68 / 2·10 | 1·65 / 2·06 | 1·64 / 2·03 | 29 |
| 30 | 4·17 / 7·56 | 3·32 / 5·39 | 2·92 / 4·51 | 2·69 / 4·02 | 2·53 / 3·70 | 2·42 / 3·47 | 2·34 / 3·30 | 2·27 / 3·17 | 2·21 / 3·06 | 2·16 / 2·98 | 2·12 / 2·90 | 2·09 / 2·84 | 2·04 / 2·74 | 1·99 / 2·66 | 1·93 / 2·55 | 1·89 / 2·47 | 1·84 / 2·38 | 1·79 / 2·29 | 1·76 / 2·24 | 1·72 / 2·16 | 1·69 / 2·13 | 1·66 / 2·07 | 1·64 / 2·03 | 1·62 / 2·01 | 30 |
| 32 | 4·15 / 7·50 | 3·30 / 5·34 | 2·90 / 4·46 | 2·67 / 3·97 | 2·51 / 3·66 | 2·40 / 3·42 | 2·32 / 3·25 | 2·25 / 3·12 | 2·19 / 3·01 | 2·14 / 2·94 | 2·10 / 2·86 | 2·07 / 2·80 | 2·02 / 2·70 | 1·97 / 2·62 | 1·91 / 2·51 | 1·86 / 2·42 | 1·82 / 2·34 | 1·76 / 2·25 | 1·74 / 2·20 | 1·69 / 2·12 | 1·67 / 2·08 | 1·64 / 2·02 | 1·61 / 1·98 | 1·59 / 1·96 | 32 |
| 34 | 4·13 / 7·44 | 3·28 / 5·29 | 2·88 / 4·42 | 2·65 / 3·93 | 2·49 / 3·61 | 2·38 / 3·38 | 2·30 / 3·21 | 2·23 / 3·08 | 2·17 / 2·97 | 2·12 / 2·89 | 2·08 / 2·82 | 2·05 / 2·76 | 2·00 / 2·66 | 1·95 / 2·58 | 1·89 / 2·47 | 1·84 / 2·38 | 1·80 / 2·30 | 1·74 / 2·21 | 1·71 / 2·15 | 1·67 / 2·08 | 1·64 / 2·04 | 1·61 / 1·98 | 1·59 / 1·94 | 1·57 / 1·91 | 34 |
| 36 | 4·11 / 7·39 | 3·26 / 5·25 | 2·86 / 4·38 | 2·63 / 3·89 | 2·48 / 3·58 | 2·36 / 3·35 | 2·28 / 3·18 | 2·21 / 3·04 | 2·15 / 2·94 | 2·10 / 2·86 | 2·06 / 2·78 | 2·03 / 2·72 | 1·98 / 2·62 | 1·93 / 2·54 | 1·87 / 2·43 | 1·82 / 2·35 | 1·78 / 2·26 | 1·72 / 2·17 | 1·69 / 2·12 | 1·65 / 2·04 | 1·62 / 2·00 | 1·59 / 1·94 | 1·56 / 1·90 | 1·55 / 1·87 | 36 |
| 38 | 4·10 / 7·35 | 3·25 / 5·21 | 2·85 / 4·34 | 2·62 / 3·86 | 2·46 / 3·54 | 2·35 / 3·32 | 2·26 / 3·15 | 2·19 / 3·02 | 2·14 / 2·91 | 2·09 / 2·82 | 2·05 / 2·75 | 2·02 / 2·69 | 1·96 / 2·59 | 1·92 / 2·51 | 1·85 / 2·40 | 1·80 / 2·32 | 1·76 / 2·22 | 1·71 / 2·14 | 1·67 / 2·08 | 1·63 / 2·00 | 1·60 / 1·97 | 1·57 / 1·90 | 1·54 / 1·86 | 1·53 / 1·84 | 38 |
| 40 | 4·08 / 7·31 | 3·23 / 5·18 | 2·84 / 4·31 | 2·61 / 3·83 | 2·45 / 3·51 | 2·34 / 3·29 | 2·25 / 3·12 | 2·18 / 2·99 | 2·12 / 2·88 | 2·07 / 2·80 | 2·04 / 2·73 | 2·00 / 2·66 | 1·95 / 2·56 | 1·90 / 2·49 | 1·84 / 2·37 | 1·79 / 2·29 | 1·74 / 2·20 | 1·69 / 2·11 | 1·66 / 2·05 | 1·61 / 1·97 | 1·59 / 1·94 | 1·55 / 1·88 | 1·53 / 1·84 | 1·51 / 1·81 | 40 |
| 42 | 4·07 / 7·27 | 3·22 / 5·15 | 2·83 / 4·29 | 2·59 / 3·80 | 2·44 / 3·49 | 2·32 / 3·26 | 2·24 / 3·10 | 2·17 / 2·96 | 2·11 / 2·86 | 2·06 / 2·77 | 2·02 / 2·70 | 1·99 / 2·64 | 1·94 / 2·54 | 1·89 / 2·46 | 1·82 / 2·35 | 1·78 / 2·26 | 1·73 / 2·17 | 1·68 / 2·08 | 1·64 / 2·02 | 1·60 / 1·94 | 1·57 / 1·91 | 1·54 / 1·85 | 1·51 / 1·80 | 1·49 / 1·78 | 42 |
| 44 | 4·06 / 7·24 | 3·21 / 5·12 | 2·82 / 4·26 | 2·58 / 3·78 | 2·43 / 3·46 | 2·31 / 3·24 | 2·23 / 3·07 | 2·16 / 2·94 | 2·10 / 2·84 | 2·05 / 2·75 | 2·01 / 2·68 | 1·98 / 2·62 | 1·92 / 2·52 | 1·88 / 2·44 | 1·81 / 2·32 | 1·76 / 2·24 | 1·72 / 2·15 | 1·66 / 2·06 | 1·63 / 2·00 | 1·58 / 1·92 | 1·56 / 1·88 | 1·52 / 1·82 | 1·50 / 1·78 | 1·48 / 1·75 | 44 |
| 46 | 4·05 / 7·21 | 3·20 / 5·10 | 2·81 / 4·24 | 2·57 / 3·76 | 2·42 / 3·44 | 2·30 / 3·22 | 2·22 / 3·05 | 2·14 / 2·92 | 2·09 / 2·82 | 2·04 / 2·73 | 2·00 / 2·66 | 1·97 / 2·60 | 1·91 / 2·50 | 1·87 / 2·42 | 1·80 / 2·30 | 1·75 / 2·22 | 1·71 / 2·13 | 1·65 / 2·04 | 1·62 / 1·98 | 1·57 / 1·90 | 1·54 / 1·86 | 1·51 / 1·80 | 1·48 / 1·76 | 1·46 / 1·72 | 46 |
| 48 | 4·04 / 7·19 | 3·19 / 5·08 | 2·80 / 4·22 | 2·56 / 3·74 | 2·41 / 3·42 | 2·30 / 3·20 | 2·21 / 3·04 | 2·14 / 2·90 | 2·08 / 2·80 | 2·03 / 2·71 | 1·99 / 2·64 | 1·96 / 2·58 | 1·90 / 2·48 | 1·86 / 2·40 | 1·79 / 2·28 | 1·74 / 2·20 | 1·70 / 2·11 | 1·64 / 2·02 | 1·61 / 1·96 | 1·56 / 1·88 | 1·53 / 1·84 | 1·50 / 1·78 | 1·47 / 1·73 | 1·45 / 1·70 | 48 |

269

Table A4. Critical values for the F-Distribution. *(continued)*

$f_1$ degrees of freedom (for greater mean square)

| $f_2$ | 1 | 2 | 3 | 4 | 5 | 6 | 7 | 8 | 9 | 10 | 11 | 12 | 14 | 16 | 20 | 24 | 30 | 40 | 50 | 75 | 100 | 200 | 500 | ∞ |
|---|---|---|---|---|---|---|---|---|---|---|---|---|---|---|---|---|---|---|---|---|---|---|---|---|
| 50 | 4·03 / 7·17 | 3·18 / 5·06 | 2·79 / 4·20 | 2·56 / 3·72 | 2·40 / 3·41 | 2·29 / 3·18 | 2·20 / 3·02 | 2·13 / 2·88 | 2·07 / 2·78 | 2·02 / 2·70 | 1·98 / 2·62 | 1·95 / 2·56 | 1·90 / 2·46 | 1·85 / 2·39 | 1·78 / 2·26 | 1·74 / 2·18 | 1·69 / 2·10 | 1·63 / 2·00 | 1·60 / 1·94 | 1·55 / 1·86 | 1·52 / 1·82 | 1·48 / 1·76 | 1·46 / 1·71 | 1·44 / 1·68 |
| 55 | 4·02 / 7·12 | 3·17 / 5·01 | 2·78 / 4·16 | 2·54 / 3·68 | 2·38 / 3·37 | 2·27 / 3·15 | 2·18 / 2·98 | 2·11 / 2·85 | 2·05 / 2·75 | 2·00 / 2·66 | 1·97 / 2·59 | 1·93 / 2·53 | 1·88 / 2·43 | 1·83 / 2·35 | 1·76 / 2·23 | 1·72 / 2·15 | 1·67 / 2·06 | 1·61 / 1·96 | 1·58 / 1·90 | 1·52 / 1·82 | 1·50 / 1·78 | 1·46 / 1·71 | 1·43 / 1·66 | 1·41 / 1·64 |
| 60 | 4·00 / 7·08 | 3·15 / 4·98 | 2·76 / 4·13 | 2·52 / 3·65 | 2·37 / 3·34 | 2·25 / 3·12 | 2·17 / 2·95 | 2·10 / 2·82 | 2·04 / 2·72 | 1·99 / 2·63 | 1·95 / 2·56 | 1·92 / 2·50 | 1·86 / 2·40 | 1·81 / 2·32 | 1·75 / 2·20 | 1·70 / 2·12 | 1·65 / 2·03 | 1·59 / 1·93 | 1·56 / 1·87 | 1·50 / 1·79 | 1·48 / 1·74 | 1·44 / 1·68 | 1·41 / 1·68 | 1·39 / 1·60 |
| 65 | 3·99 / 7·04 | 3·14 / 4·95 | 2·75 / 4·10 | 2·51 / 3·62 | 2·36 / 3·31 | 2·24 / 3·09 | 2·15 / 2·93 | 2·08 / 2·79 | 2·02 / 2·70 | 1·98 / 2·61 | 1·94 / 2·54 | 1·90 / 2·47 | 1·85 / 2·37 | 1·80 / 2·30 | 1·73 / 2·18 | 1·68 / 2·09 | 1·63 / 2·00 | 1·57 / 1·90 | 1·54 / 1·84 | 1·49 / 1·76 | 1·46 / 1·71 | 1·42 / 1·64 | 1·39 / 1·60 | 1·37 / 1·56 |
| 70 | 3·98 / 7·01 | 3·13 / 4·92 | 2·74 / 4·08 | 2·50 / 3·60 | 2·35 / 3·29 | 2·23 / 3·07 | 2·14 / 2·91 | 2·07 / 2·77 | 2·01 / 2·67 | 1·97 / 2·59 | 1·93 / 2·51 | 1·89 / 2·45 | 1·84 / 2·35 | 1·79 / 2·28 | 1·72 / 2·15 | 1·67 / 2·07 | 1·62 / 1·98 | 1·56 / 1·88 | 1·53 / 1·82 | 1·47 / 1·74 | 1·45 / 1·69 | 1·40 / 1·62 | 1·37 / 1·56 | 1·35 / 1·53 |
| 80 | 3·96 / 6·96 | 3·11 / 4·88 | 2·72 / 4·04 | 2·48 / 3·56 | 2·33 / 3·25 | 2·21 / 3·04 | 2·12 / 2·87 | 2·05 / 2·74 | 1·99 / 2·64 | 1·95 / 2·55 | 1·91 / 2·48 | 1·88 / 2·41 | 1·82 / 2·32 | 1·77 / 2·24 | 1·70 / 2·11 | 1·65 / 2·03 | 1·60 / 1·94 | 1·54 / 1·84 | 1·51 / 1·78 | 1·45 / 1·70 | 1·42 / 1·65 | 1·38 / 1·57 | 1·35 / 1·52 | 1·32 / 1·49 |
| 100 | 3·94 / 6·90 | 3·09 / 4·82 | 2·70 / 3·98 | 2·46 / 3·51 | 2·30 / 3·20 | 2·19 / 2·99 | 2·10 / 2·82 | 2·03 / 2·69 | 1·97 / 2·59 | 1·92 / 2·51 | 1·88 / 2·43 | 1·85 / 2·36 | 1·79 / 2·26 | 1·75 / 2·19 | 1·68 / 2·06 | 1·63 / 1·98 | 1·57 / 1·89 | 1·51 / 1·79 | 1·48 / 1·73 | 1·42 / 1·64 | 1·39 / 1·59 | 1·34 / 1·51 | 1·30 / 1·46 | 1·28 / 1·43 |
| 125 | 3·92 / 6·84 | 3·07 / 4·78 | 2·68 / 3·94 | 2·44 / 3·47 | 2·29 / 3·17 | 2·17 / 2·95 | 2·08 / 2·79 | 2·01 / 2·65 | 1·95 / 2·56 | 1·90 / 2·47 | 1·86 / 2·40 | 1·83 / 2·33 | 1·77 / 2·23 | 1·72 / 2·15 | 1·65 / 2·03 | 1·60 / 1·94 | 1·55 / 1·85 | 1·49 / 1·75 | 1·45 / 1·68 | 1·39 / 1·59 | 1·36 / 1·54 | 1·31 / 1·46 | 1·27 / 1·40 | 1·25 / 1·37 |
| 150 | 3·91 / 6·81 | 3·06 / 4·75 | 2·67 / 3·91 | 2·43 / 3·44 | 2·27 / 3·14 | 2·16 / 2·92 | 2·07 / 2·76 | 2·00 / 2·62 | 1·94 / 2·53 | 1·89 / 2·44 | 1·85 / 2·37 | 1·82 / 2·30 | 1·76 / 2·20 | 1·71 / 2·12 | 1·64 / 2·00 | 1·59 / 1·91 | 1·54 / 1·83 | 1·47 / 1·72 | 1·44 / 1·66 | 1·37 / 1·56 | 1·34 / 1·51 | 1·29 / 1·43 | 1·25 / 1·37 | 1·22 / 1·33 |
| 200 | 3·89 / 6·76 | 3·04 / 4·71 | 2·65 / 3·88 | 2·41 / 3·41 | 2·26 / 3·11 | 2·14 / 2·90 | 2·05 / 2·73 | 1·98 / 2·60 | 1·92 / 2·50 | 1·87 / 2·41 | 1·83 / 2·34 | 1·80 / 2·28 | 1·74 / 2·17 | 1·69 / 2·09 | 1·62 / 1·97 | 1·57 / 1·88 | 1·52 / 1·79 | 1·45 / 1·69 | 1·42 / 1·62 | 1·35 / 1·53 | 1·32 / 1·48 | 1·26 / 1·39 | 1·22 / 1·33 | 1·19 / 1·28 |
| 400 | 3·86 / 6·70 | 3·02 / 4·66 | 2·62 / 3·83 | 2·39 / 3·36 | 2·23 / 3·06 | 2·12 / 2·85 | 2·03 / 2·69 | 1·96 / 2·55 | 1·90 / 2·46 | 1·85 / 2·37 | 1·81 / 2·29 | 1·78 / 2·23 | 1·72 / 2·12 | 1·67 / 2·04 | 1·60 / 1·92 | 1·54 / 1·84 | 1·49 / 1·74 | 1·42 / 1·64 | 1·38 / 1·57 | 1·32 / 1·47 | 1·28 / 1·42 | 1·22 / 1·32 | 1·16 / 1·24 | 1·13 / 1·19 |
| 1000 | 3·85 / 6·66 | 3·00 / 4·62 | 2·61 / 3·80 | 2·38 / 3·34 | 2·22 / 3·04 | 2·10 / 2·82 | 2·02 / 2·66 | 1·95 / 2·53 | 1·89 / 2·43 | 1·84 / 2·34 | 1·80 / 2·26 | 1·76 / 2·20 | 1·70 / 2·09 | 1·65 / 2·01 | 1·58 / 1·89 | 1·53 / 1·81 | 1·47 / 1·71 | 1·41 / 1·61 | 1·36 / 1·54 | 1·30 / 1·44 | 1·26 / 1·38 | 1·19 / 1·28 | 1·13 / 1·19 | 1·08 / 1·11 |
| ∞ | 3·84 / 6·64 | 2·99 / 4·60 | 2·60 / 3·78 | 2·37 / 3·32 | 2·21 / 3·02 | 2·09 / 2·80 | 2·01 / 2·64 | 1·94 / 2·51 | 1·88 / 2·41 | 1·83 / 2·32 | 1·79 / 2·24 | 1·75 / 2·18 | 1·69 / 2·07 | 1·64 / 1·99 | 1·57 / 1·87 | 1·52 / 1·79 | 1·46 / 1·69 | 1·40 / 1·59 | 1·35 / 1·52 | 1·28 / 1·41 | 1·24 / 1·36 | 1·17 / 1·25 | 1·11 / 1·15 | 1·00 / 1·00 |

The function, $F = e$ with exponent $2z$, is computed in part from Fisher's Table VI (7). Additional entries are by interpolation, mostly graphical. From *Statistical Methods* by G. W. Snedecor and W. G. Cochran, Iowa State University Press, USA, 6th edition, pp. 560–3.

*Table A5.* Transformation of the correlation coefficient. $z = \frac{1}{2}\log_e\left[\dfrac{1+r}{1-r}\right]$

| r | z | r | z | r | z | r | z | r | z | r | z |
|---|---|---|---|---|---|---|---|---|---|---|---|
| 0·00 | $0{\cdot}000_{20}$ | 0·40 | $0{\cdot}424_{24}$ | 0·80 | $1{\cdot}099_{28}$ | 0·940 | $1{\cdot}738_{9}$ | 0·960 | $1{\cdot}946_{13}$ | 0·980 | $2{\cdot}298_{25}$ |
| ·02 | $\cdot020_{20}$ | ·42 | $\cdot448_{24}$ | ·81 | $\cdot127_{30}$ | ·941 | $\cdot747_{9}$ | ·961 | $\cdot959_{13}$ | ·981 | $\cdot323_{28}$ |
| ·04 | $\cdot040_{20}$ | ·44 | $\cdot472_{25}$ | ·82 | $\cdot157_{31}$ | ·942 | $\cdot756_{8}$ | ·962 | $\cdot972_{14}$ | ·982 | $\cdot351_{29}$ |
| ·06 | $\cdot060_{20}$ | ·46 | $\cdot497_{26}$ | ·83 | $\cdot188_{33}$ | ·943 | $\cdot764_{10}$ | ·963 | $1{\cdot}986_{14}$ | ·983 | $\cdot380_{30}$ |
| ·08 | $\cdot080_{20}$ | ·48 | $\cdot523_{26}$ | ·84 | $\cdot221_{35}$ | ·944 | $\cdot774_{9}$ | ·964 | $2{\cdot}000_{14}$ | ·984 | $\cdot410_{33}$ |
| 0·10 | $0{\cdot}100_{21}$ | 0·50 | $0{\cdot}549_{27}$ | 0·85 | $1{\cdot}256_{37}$ | 0·945 | $1{\cdot}783_{9}$ | 0·965 | $2{\cdot}014_{15}$ | 0·985 | $2{\cdot}443_{34}$ |
| ·12 | $\cdot121_{20}$ | ·52 | $\cdot576_{28}$ | ·86 | $\cdot293_{40}$ | ·946 | $\cdot792_{10}$ | ·966 | $\cdot029_{15}$ | ·986 | $\cdot477_{38}$ |
| ·14 | $\cdot141_{20}$ | ·54 | $\cdot604_{29}$ | ·87 | $\cdot333_{43}$ | ·947 | $\cdot802_{10}$ | ·967 | $\cdot044_{16}$ | ·987 | $\cdot515_{40}$ |
| ·16 | $\cdot161_{21}$ | ·56 | $\cdot633_{29}$ | ·88 | $\cdot376_{46}$ | ·948 | $\cdot812_{10}$ | ·968 | $\cdot060_{16}$ | ·988 | $\cdot555_{44}$ |
| ·18 | $\cdot182_{21}$ | ·58 | $\cdot662_{31}$ | ·89 | $\cdot422_{50}$ | ·949 | $\cdot822_{10}$ | ·969 | $\cdot076_{16}$ | ·989 | $\cdot599_{48}$ |
| 0·20 | $0{\cdot}203_{21}$ | 0·60 | $0{\cdot}693_{32}$ | 0·90 | $1{\cdot}472_{56}$ | 0·950 | $1{\cdot}832_{10}$ | 0·970 | $2{\cdot}092_{18}$ | 0·990 | $2{\cdot}647_{53}$ |
| ·22 | $\cdot224_{21}$ | ·62 | $\cdot725_{33}$ | ·91 | $\cdot528_{61}$ | ·951 | $\cdot842_{11}$ | ·971 | $\cdot110_{17}$ | ·991 | $\cdot700_{59}$ |
| ·24 | $\cdot245_{21}$ | ·64 | $\cdot758_{35}$ | ·92 | $\cdot589_{69}$ | ·952 | $\cdot853_{10}$ | ·972 | $\cdot127_{19}$ | ·992 | $\cdot759_{67}$ |
| ·26 | $\cdot266_{22}$ | ·66 | $\cdot793_{36}$ | ·93 | $\cdot658_{80}$ | ·953 | $\cdot863_{11}$ | ·973 | $\cdot146_{19}$ | ·993 | $\cdot826_{77}$ |
| ·28 | $\cdot288_{22}$ | ·68 | $\cdot829_{38}$ | ·94 | $\cdot738$ | ·954 | $\cdot874_{12}$ | ·974 | $\cdot165_{20}$ | ·994 | $\cdot903$ |
| 0·30 | $0{\cdot}310_{22}$ | 0·70 | $0{\cdot}867_{41}$ | 0·95 | $1{\cdot}832$ | 0·955 | $1{\cdot}886_{11}$ | 0·975 | $2{\cdot}185_{20}$ | 0·995 | $2{\cdot}994$ |
| ·32 | $\cdot332_{22}$ | ·72 | $\cdot908_{42}$ | ·96 | $1{\cdot}946$ | ·956 | $\cdot897_{12}$ | ·976 | $\cdot205_{22}$ | ·996 | $3{\cdot}106$ |
| ·34 | $\cdot354_{23}$ | ·74 | $\cdot950_{46}$ | ·97 | $2{\cdot}092$ | ·957 | $\cdot909_{12}$ | ·977 | $\cdot227_{22}$ | ·997 | $\cdot250$ |
| ·36 | $\cdot377_{23}$ | ·76 | $0{\cdot}996_{49}$ | ·98 | $\cdot298$ | ·958 | $\cdot921_{12}$ | ·978 | $\cdot249_{24}$ | ·998 | $\cdot453$ |
| ·38 | $\cdot400_{24}$ | ·78 | $1{\cdot}045_{54}$ | ·99 | $2{\cdot}647$ | ·959 | $\cdot933_{13}$ | ·979 | $\cdot273_{25}$ | ·999 | $3{\cdot}800$ |
| 0·40 | $0{\cdot}424$ | 0·80 | $1{\cdot}099$ | 1·00 | $\infty$ | 0·960 | $1{\cdot}946$ | 0·980 | $2{\cdot}298$ | 1·000 | $\infty$ |

(see next columns)

The function tabulated is $z = \tanh^{-1}r = \frac{1}{2}\ln\dfrac{1+r}{1-r} = 1.1513\log\dfrac{1+r}{1-r}$.

If $r$ is a partial correlation coefficient, after $s$ variables have been eliminated, in a sample of size $n$ from a multivariate normal population with the corresponding partial correlation coefficient $\rho$, then $z$ is approximately normally distributed with mean $\tanh^{-1}\rho + \rho/2(n - s - 1)$ and variance $1/(n - s - 3)$. For $s = 0$ we have the ordinary correlation coefficient.

From *Cambridge Elementary Statistical Tables* by D. V. Lindley and J. C. P. Miller, Cambridge University Press, 1953, p. 6.

*Table A6.* 500 Random Numbers

| | | | | | | | | | |
|---|---|---|---|---|---|---|---|---|---|
| 12159 | 66144 | 05091 | 13446 | 45653 | 13684 | 66024 | 91410 | 51351 | 22772 |
| 30156 | 90519 | 95785 | 47544 | 66735 | 35754 | 11088 | 67310 | 19720 | 08379 |
| 59069 | 01722 | 53338 | 41942 | 65118 | 71236 | 01932 | 70343 | 25812 | 62275 |
| 54107 | 58081 | 82470 | 59407 | 13475 | 95872 | 16268 | 78436 | 39251 | 64247 |
| 99681 | 81295 | 06315 | 28212 | 45029 | 57701 | 96327 | 85436 | 33614 | 29070 |
| | | | | | | | | | |
| 27252 | 37875 | 53679 | 01889 | 35714 | 63534 | 63791 | 76342 | 47717 | 73684 |
| 93259 | 74585 | 11863 | 78985 | 03881 | 46567 | 93696 | 93521 | 54970 | 37607 |
| 84068 | 43759 | 75814 | 32261 | 12728 | 09636 | 22336 | 75629 | 01017 | 45503 |
| 68582 | 97054 | 28251 | 63787 | 57285 | 18854 | 35006 | 16343 | 51867 | 67979 |
| 60646 | 11298 | 19680 | 10087 | 66391 | 70853 | 24423 | 73007 | 74958 | 29020 |
| | | | | | | | | | |
| 97437 | 52922 | 80739 | 59178 | 50628 | 61017 | 51652 | 40915 | 94696 | 67843 |
| 58009 | 20681 | 98823 | 50979 | 01237 | 70152 | 13711 | 73916 | 87902 | 84759 |
| 77211 | 70110 | 93803 | 60135 | 22881 | 13423 | 30999 | 07104 | 27400 | 25414 |
| 54256 | 84591 | 65302 | 99257 | 92970 | 28924 | 36632 | 54044 | 91798 | 78018 |
| 37493 | 69330 | 94069 | 39544 | 14050 | 03476 | 25804 | 49350 | 92525 | 87941 |
| | | | | | | | | | |
| 87569 | 22661 | 55970 | 52623 | 35419 | 76660 | 42394 | 63210 | 62626 | 00581 |
| 22896 | 62237 | 39635 | 63725 | 10463 | 87944 | 92075 | 90914 | 30599 | 35671 |
| 02697 | 33230 | 64527 | 97210 | 41359 | 79399 | 13941 | 88378 | 68503 | 33609 |
| 20080 | 15652 | 37216 | 00679 | 02088 | 34138 | 13953 | 68939 | 05630 | 27653 |
| 20550 | 95151 | 60557 | 57449 | 77115 | 87372 | 02574 | 07851 | 22428 | 39189 |
| | | | | | | | | | |
| 72771 | 11672 | 67492 | 42904 | 64647 | 94354 | 45994 | 42538 | 54885 | 15983 |
| 38472 | 43379 | 76295 | 69406 | 96510 | 16529 | 83500 | 28590 | 49787 | 29822 |
| 24511 | 56510 | 72654 | 13277 | 45031 | 42235 | 96502 | 25567 | 23653 | 36707 |
| 01054 | 06674 | 58283 | 82831 | 97048 | 42983 | 06471 | 12350 | 49990 | 04809 |
| 94437 | 94907 | 95274 | 26487 | 60496 | 78222 | 43032 | 04276 | 70800 | 17378 |
| | | | | | | | | | |
| 97842 | 69095 | 25982 | 03484 | 25173 | 05982 | 14624 | 31653 | 17170 | 92785 |
| 53047 | 13486 | 69712 | 33567 | 82313 | 87631 | 03197 | 02438 | 12374 | 40329 |
| 40770 | 47013 | 63306 | 48154 | 80970 | 87976 | 04939 | 21233 | 20572 | 31013 |
| 52733 | 66251 | 69661 | 58387 | 72096 | 21355 | 51659 | 19003 | 75556 | 33095 |
| 41749 | 46502 | 18378 | 83141 | 63920 | 85516 | 75743 | 66317 | 45428 | 45940 |
| | | | | | | | | | |
| 10271 | 85184 | 46468 | 38860 | 24039 | 80949 | 51211 | 35411 | 40470 | 16070 |
| 98791 | 48848 | 68129 | 51024 | 53044 | 55039 | 71290 | 26484 | 70682 | 56255 |
| 30196 | 09295 | 47685 | 56768 | 29285 | 06272 | 98789 | 47188 | 35063 | 24158 |
| 93373 | 64343 | 92433 | 06388 | 65713 | 35386 | 43370 | 19254 | 55014 | 98621 |
| 27782 | 27552 | 42156 | 23239 | 46823 | 91077 | 06306 | 17756 | 84459 | 92513 |
| | | | | | | | | | |
| 67791 | 35910 | 56921 | 51976 | 78475 | 15336 | 92544 | 82601 | 17996 | 72268 |
| 64018 | 44004 | 08136 | 56129 | 77024 | 82650 | 18163 | 29158 | 33935 | 94262 |
| 79715 | 33859 | 10835 | 94936 | 02857 | 87486 | 70613 | 41909 | 80667 | 52176 |
| 20190 | 40737 | 82688 | 07099 | 65255 | 52767 | 65930 | 45861 | 32575 | 93731 |
| 82421 | 01208 | 49762 | 66360 | 00231 | 87540 | 88302 | 62686 | 38456 | 25872 |
| | | | | | | | | | |
| 00083 | 81269 | 35320 | 72064 | 10472 | 92080 | 80447 | 15259 | 62654 | 70882 |
| 56558 | 09762 | 20813 | 48719 | 35530 | 96437 | 96343 | 21212 | 32567 | 34305 |
| 41183 | 20460 | 08608 | 75283 | 43401 | 25888 | 73405 | 35639 | 92114 | 48006 |
| 39977 | 10603 | 35052 | 53751 | 64219 | 36235 | 84687 | 42091 | 42587 | 16996 |
| 29310 | 84031 | 03052 | 51356 | 44747 | 19678 | 14619 | 03600 | 08066 | 93899 |
| | | | | | | | | | |
| 47360 | 03571 | 95657 | 85065 | 80919 | 14890 | 97623 | 57375 | 77855 | 15735 |
| 48481 | 98262 | 50414 | 41929 | 05977 | 78903 | 47602 | 52154 | 47901 | 84523 |
| 48097 | 56362 | 16342 | 75261 | 27751 | 28715 | 21871 | 37943 | 17850 | 90999 |
| 20648 | 30751 | 96515 | 51581 | 43877 | 94494 | 80164 | 02115 | 09738 | 51938 |
| 60704 | 10107 | 59220 | 64220 | 23944 | 34684 | 83696 | 82344 | 19020 | 84834 |

From *A Million Random Digits*, The Rand Corporation, Glencoe, Illinois, 1955, p. 150.

*Table A7.* Critical values for the Durbin–Watson test statistic.

Lower and upper bounds of the 5% points of the Durbin–Watson Test Statistic[a]

| n | K = 2 | | K = 3 | | K = 4 | | K = 5 | | K = 6 | |
|---|---|---|---|---|---|---|---|---|---|---|
| | $d_L$ | $d_U$ | $d_L$ | $d_U$ | $d_L$ | $d_U$ | $d_L$ | $d_U$ | $d_L$ | $d_U$ |
| 15 | 1·08 | 1·36 | ·95 | 1·54 | ·82 | 1·75 | ·69 | 1·97 | ·56 | 2·21 |
| 16 | 1·10 | 1·37 | ·98 | 1·54 | ·86 | 1·73 | ·74 | 1·93 | ·62 | 2·15 |
| 17 | 1·13 | 1·38 | 1·02 | 1·54 | ·90 | 1·71 | ·78 | 1·90 | ·67 | 2·10 |
| 18 | 1·16 | 1·39 | 1·05 | 1·53 | ·93 | 1·69 | ·82 | 1·87 | ·71 | 2·06 |
| 19 | 1·18 | 1·40 | 1·08 | 1·53 | ·97 | 1·68 | ·86 | 1·85 | ·75 | 2·02 |
| 20 | 1·20 | 1·41 | 1·10 | 1·54 | 1·00 | 1·68 | ·90 | 1·83 | ·79 | 1·99 |
| 21 | 1·22 | 1·42 | 1·13 | 1·54 | 1·03 | 1·67 | ·93 | 1·81 | ·83 | 1·96 |
| 22 | 1·24 | 1·43 | 1·15 | 1·54 | 1·05 | 1·66 | ·96 | 1·80 | ·86 | 1·94 |
| 23 | 1·26 | 1·44 | 1·17 | 1·54 | 1·08 | 1·66 | ·99 | 1·79 | ·90 | 1·92 |
| 24 | 1·27 | 1·45 | 1·19 | 1·55 | 1·10 | 1·66 | 1·01 | 1·78 | ·93 | 1·90 |
| 25 | 1·29 | 1·45 | 1·21 | 1·55 | 1·12 | 1·66 | 1·04 | 1·77 | ·95 | 1·89 |
| 26 | 1·30 | 1·46 | 1·22 | 1·55 | 1·14 | 1·65 | 1·06 | 1·76 | ·98 | 1·88 |
| 27 | 1·32 | 1·47 | 1·24 | 1·56 | 1·16 | 1·65 | 1·08 | 1·76 | 1·01 | 1·86 |
| 28 | 1·33 | 1·48 | 1·26 | 1·56 | 1·18 | 1·65 | 1·10 | 1·75 | 1·03 | 1·85 |
| 29 | 1·34 | 1·48 | 1·27 | 1·56 | 1·20 | 1·65 | 1·12 | 1·74 | 1·05 | 1·84 |
| 30 | 1·35 | 1·49 | 1·28 | 1·57 | 1·21 | 1·65 | 1·14 | 1·74 | 1·07 | 1·83 |
| 31 | 1·36 | 1·50 | 1·30 | 1·57 | 1·23 | 1·65 | 1·16 | 1·74 | 1·09 | 1·83 |
| 32 | 1·37 | 1·50 | 1·31 | 1·57 | 1·24 | 1·65 | 1·18 | 1·73 | 1·11 | 1·82 |
| 33 | 1·38 | 1·51 | 1·32 | 1·58 | 1·26 | 1·65 | 1·19 | 1·73 | 1·13 | 1·81 |
| 34 | 1·39 | 1·51 | 1·33 | 1·58 | 1·27 | 1·65 | 1·21 | 1·73 | 1·15 | 1·81 |
| 35 | 1·40 | 1·52 | 1·34 | 1·58 | 1·28 | 1·65 | 1·22 | 1·73 | 1·16 | 1·80 |
| 36 | 1·41 | 1·52 | 1·35 | 1·59 | 1·29 | 1·65 | 1·24 | 1·73 | 1·18 | 1·80 |
| 37 | 1·42 | 1·53 | 1·36 | 1·59 | 1·31 | 1·66 | 1·25 | 1·72 | 1·19 | 1·80 |
| 38 | 1·43 | 1·54 | 1·37 | 1·59 | 1·32 | 1·66 | 1·26 | 1·72 | 1·21 | 1·79 |
| 39 | 1·43 | 1·54 | 1·38 | 1·60 | 1·33 | 1·66 | 1·27 | 1·72 | 1·22 | 1·79 |
| 40 | 1·44 | 1·54 | 1·39 | 1·60 | 1·34 | 1·66 | 1·29 | 1·72 | 1·23 | 1·79 |
| 45 | 1·48 | 1·57 | 1·43 | 1·62 | 1·38 | 1·67 | 1·34 | 1·72 | 1·29 | 1·78 |
| 50 | 1·50 | 1·59 | 1·46 | 1·63 | 1·42 | 1·67 | 1·38 | 1·72 | 1·34 | 1·77 |
| 55 | 1·53 | 1·60 | 1·49 | 1·64 | 1·45 | 1·68 | 1·41 | 1·72 | 1·38 | 1·77 |
| 60 | 1·55 | 1·62 | 1·51 | 1·65 | 1·48 | 1·69 | 1·44 | 1·73 | 1·41 | 1·77 |
| 65 | 1·57 | 1·63 | 1·54 | 1·66 | 1·50 | 1·70 | 1·47 | 1·73 | 1·44 | 1·77 |
| 70 | 1·58 | 1·64 | 1·55 | 1·67 | 1·52 | 1·70 | 1·49 | 1·74 | 1·46 | 1·77 |
| 75 | 1·60 | 1·65 | 1·57 | 1·68 | 1·54 | 1·71 | 1·51 | 1·74 | 1·49 | 1·77 |
| 80 | 1·61 | 1·66 | 1·59 | 1·69 | 1·56 | 1·72 | 1·53 | 1·74 | 1·51 | 1·77 |
| 85 | 1·62 | 1·67 | 1·60 | 1·70 | 1·57 | 1·72 | 1·55 | 1·75 | 1·52 | 1·77 |
| 90 | 1·63 | 1·68 | 1·61 | 1·70 | 1·59 | 1·73 | 1·57 | 1·75 | 1·54 | 1·78 |
| 95 | 1·64 | 1·69 | 1·62 | 1·71 | 1·60 | 1·73 | 1·58 | 1·75 | 1·56 | 1·78 |
| 100 | 1·65 | 1·69 | 1·63 | 1·72 | 1·61 | 1·74 | 1·59 | 1·76 | 1·57 | 1·78 |

This table is taken from Durbin and Watson (1951) with the kind permission of *Biometrika*, the publisher, and the authors.

[a] The value of K at the head of each pair of columns is the number of elements of the parameter vector, and it is assumed that one of these elements is a constant term.

*Table A*7. Critical values for the Durbin–Watson test statistic (*continued*)

Lower and upper bounds of the 1% points of the Durbin–Watson Test Statistic.

| n | K = 2 $d_L$ | K = 2 $d_L$ | K = 3 $d_L$ | K = 3 $d_U$ | K = 4 $d_L$ | K = 4 $d_U$ | K = 5 $d_L$ | K = 5 $d_U$ | K = 6 $d_L$ | K = 6 $d_U$ |
|---|---|---|---|---|---|---|---|---|---|---|
| 15 | ·81 | 1·07 | ·70 | 1·25 | ·59 | 1·46 | ·49 | 1·70 | ·39 | 1·96 |
| 16 | ·84 | 1·09 | ·74 | 1·25 | ·63 | 1·44 | ·53 | 1·66 | ·44 | 1·90 |
| 17 | ·87 | 1·10 | ·77 | 1·25 | ·67 | 1·43 | ·57 | 1·63 | ·48 | 1·85 |
| 18 | ·90 | 1·12 | ·80 | 1·26 | ·71 | 1·42 | ·61 | 1·60 | ·52 | 1·80 |
| 19 | ·93 | 1·13 | ·83 | 1·26 | ·74 | 1·41 | ·65 | 1·58 | ·56 | 1·77 |
| 20 | ·95 | 1·15 | ·86 | 1·27 | ·77 | 1·41 | ·68 | 1·57 | ·60 | 1·74 |
| 21 | ·97 | 1·16 | ·89 | 1·27 | ·80 | 1·41 | ·72 | 1·55 | ·63 | 1·71 |
| 22 | 1·00 | 1·17 | ·91 | 1·28 | ·83 | 1·40 | ·75 | 1·54 | ·66 | 1·69 |
| 23 | 1·02 | 1·19 | ·94 | 1·29 | ·86 | 1·40 | ·77 | 1·53 | ·70 | 1·67 |
| 24 | 1·04 | 1·20 | ·96 | 1·30 | ·88 | 1·41 | ·80 | 1·53 | ·72 | 1·66 |
| 25 | 1·05 | 1·21 | ·98 | 1·30 | ·90 | 1·41 | ·83 | 1·52 | ·75 | 1·65 |
| 26 | 1·07 | 1·22 | 1·00 | 1·31 | ·93 | 1·41 | ·85 | 1·52 | ·78 | 1·64 |
| 27 | 1·09 | 1·23 | 1·02 | 1·32 | ·95 | 1·41 | ·88 | 1·51 | ·81 | 1·63 |
| 28 | 1·10 | 1·24 | 1·04 | 1·32 | ·97 | 1·41 | ·90 | 1·51 | ·83 | 1·62 |
| 29 | 1·12 | 1·25 | 1·05 | 1·33 | ·99 | 1·42 | ·92 | 1·51 | ·85 | 1·61 |
| 30 | 1·13 | 1·26 | 1·07 | 1·34 | 1·01 | 1·42 | ·94 | 1·51 | ·88 | 1·61 |
| 31 | 1·15 | 1·27 | 1·08 | 1·34 | 1·02 | 1·42 | ·96 | 1·51 | ·90 | 1·60 |
| 32 | 1·16 | 1·28 | 1·10 | 1·35 | 1·04 | 1·43 | ·98 | 1·51 | ·92 | 1·60 |
| 33 | 1·17 | 1·29 | 1·11 | 1·36 | 1·05 | 1·43 | 1·00 | 1·51 | ·94 | 1·59 |
| 34 | 1·18 | 1·30 | 1·13 | 1·36 | 1·07 | 1·43 | 1·01 | 1·51 | ·95 | 1·59 |
| 35 | 1·19 | 1·31 | 1·14 | 1·37 | 1·08 | 1·44 | 1·03 | 1·51 | ·97 | 1·59 |
| 36 | 1·21 | 1·32 | 1·15 | 1·38 | 1·10 | 1·44 | 1·04 | 1·51 | ·99 | 1·59 |
| 37 | 1·22 | 1·32 | 1·16 | 1·38 | 1·11 | 1·45 | 1·06 | 1·51 | 1·00 | 1·59 |
| 38 | 1·23 | 1·33 | 1·18 | 1·39 | 1·12 | 1·45 | 1·07 | 1·52 | 1·02 | 1·58 |
| 39 | 1·24 | 1·34 | 1·19 | 1·39 | 1·14 | 1·45 | 1·09 | 1·52 | 1·03 | 1·58 |
| 40 | 1·25 | 1·34 | 1·20 | 1·40 | 1·15 | 1·46 | 1·10 | 1·52 | 1·05 | 1·58 |
| 45 | 1·29 | 1·38 | 1·24 | 1·42 | 1·20 | 1·48 | 1·16 | 1·53 | 1·11 | 1·58 |
| 50 | 1·32 | 1·40 | 1·28 | 1·45 | 1·24 | 1·49 | 1·20 | 1·54 | 1·16 | 1·59 |
| 55 | 1·36 | 1·43 | 1·32 | 1·47 | 1·28 | 1·51 | 1·25 | 1·55 | 1·21 | 1·59 |
| 60 | 1·38 | 1·45 | 1·35 | 1·48 | 1·32 | 1·52 | 1·28 | 1·56 | 1·25 | 1·60 |
| 65 | 1·41 | 1·47 | 1·38 | 1·50 | 1·35 | 1·53 | 1·31 | 1·57 | 1·28 | 1·61 |
| 70 | 1·43 | 1·49 | 1·40 | 1·52 | 1·37 | 1·55 | 1·34 | 1·58 | 1·31 | 1·61 |
| 75 | 1·45 | 1·50 | 1·42 | 1·53 | 1·39 | 1·56 | 1·37 | 1·59 | 1·34 | 1·62 |
| 80 | 1·47 | 1·52 | 1·44 | 1·54 | 1·42 | 1·57 | 1·39 | 1·60 | 1·36 | 1·62 |
| 85 | 1·48 | 1·53 | 1·46 | 1·55 | 1·43 | 1·58 | 1·41 | 1·60 | 1·39 | 1·63 |
| 90 | 1·50 | 1·54 | 1·47 | 1·56 | 1·45 | 1·59 | 1·43 | 1·61 | 1·41 | 1·64 |
| 95 | 1·51 | 1·55 | 1·49 | 1·57 | 1·47 | 1·60 | 1·45 | 1·62 | 1·42 | 1·64 |
| 100 | 1·52 | 1·56 | 1·50 | 1·58 | 1·48 | 1·60 | 1·46 | 1·63 | 1·44 | 1·65 |

This table is taken from Durbin and Watson (1951) with the kind permission of *Biometrika*, the publisher, and the authors.

# notes

## chapter 1

1 Reference to works cited will be by author's name and the date of publication. Full details may be obtained from the bibliography.

2 The reader who consults other statistical texts may be puzzled by the fact that many authors refer to the expression for $s'$ as the standard deviation and its square as the variance. If he reads on he will find that some of them change their definition of the standard deviation half-way through the text. Rather than change definitions in mid-stream we have chosen to use the 'old fashioned' terms 'root mean square deviation' and 'mean square deviation' at this stage, and will introduce the sample standard deviation, defined as

$$s = \sqrt{\frac{1}{n-1}\sum(X_i - \overline{X})^2},$$

and the sample variance ( $= s^2$) in chapter 5 in the context of estimation.

3 In cases where the range of summation is clear we shall write $\sum$ rather than $\sum_{i=1}^{n}$.

## chapter 2

1 The reader should be clear that mutually exclusive events are different from independent events. It may help if he recalls that we met mutually exclusive events in connection with the addition rule and independent events in the context of the multiplication rule.

2 A proof of the Binomial Theorem may be found in a number of algebra texts. See, for example, O'Brien and Garcia (1971) pp. 102–6.

## chapter 3

1 A more general definition of the rectangular distribution would be for a discrete random variable $X$ whose range is the set of $k$ consecutive, equally spaced values $m, (m + 1), \ldots, (n - 1), n$, with $p(X) = 1/k$.

2 So called because the probabilities of $X$ successes, $X = 0, \ldots, n$, are given by the terms obtained when $[(1 - p) + p]^n$ is expanded using the Binomial Theorem.

3 The median and the mode were discussed as representative values in chapter 1, but the corresponding concepts for a theoretical distribution will not be developed here. They could be defined, but the population mean has considerable theoretical advantages which will be developed in the next chapter.

4 Equation 3.5 defines the expected value of a *discrete* random variable. The definition will be modified for *continuous* random variables below.

5   The quartile deviation and the range will not be developed, since the standard deviation has considerable theoretical advantages which will be discussed in the next chapter.
6   For a proof of this result, see Hoel (1971) pp 60–1.
7   We shall not prove this result here. See, for example, Suits (1966) p. 50, n. 11.

**chapter 4**
1   For example, ERNIE, the Electronic Random Number Indicator Equipment, is a computer designed to generate the random numbers which are used to select Premium Bond prize winners.
2   Sampling theory is a specialized branch of statistical theory which will not be explored here as we shall be dealing mainly with simple random samples. For a fuller discussion, see Kish (1965).
3   This is a difficult result to prove formally, but see Hoel (1971) pp. 121–3.

**chapter 5**
1   For a more extended discussion of this topic see, for example, Kmenta (1971) pp. 154–71.
2   This discussion is based on Kendall and Stuart, vol. I (1958) p. 327.
3   See, for example, ibid., pp. 255–6.
4   The numerator of 5.10, being a standard normal variable, is also the square root of a chi-square variable, in this case with one degree of freedom. Hence $t^2$ is a function of the ratio of two chi-square variables, a result we shall refer to in chapter 6.
5   In some cases if variables do not come from a normal population it is possible to apply a reasonably simple transformation which makes them normal. We shall meet such a transformation below in chapter 7, section 7.3.
6   We shall use $\hat{\pi}$ to denote the sample proportion rather than $p$ to avoid confusion between proportions and probabilities in what follows.

**chapter 6**
1   There are two degrees of freedom lost here, since if we explore the distribution of $t$ defined in 6.15 holding $\bar{X}_1$ and $\bar{X}_2$ constant, we have $(n_1 - 1)$ degrees of freedom for the first sample and $(n_2 - 1)$ degrees of freedom for the second.
2   This distribution was named the $F$-distribution in honour of R. A. Fisher, the statistician who first explored its properties.
3   The results we obtain do not depend on the samples being of equal size, but this assumption simplifies the algebra. It will be relaxed subsequently.

**chapter 7**
1   The observant reader may be surprised by the use of the biased estimator of the standard deviation in the definition of $r$. The answer is that $r$ was developed during the late nineteenth century when statisticians were not interested either in bias or small samples. Obviously, if one works with large samples the problem of bias does not arise.
2   See below, p. 148.

**chapter 8**
1   Usually the true relationship is unknown, but in order to provide an illustrative example

the author constructed the data by using the relationship $Y_i = 100 + 5X_i + \varepsilon_i$. The $\varepsilon_i$ were obtained by drawing twenty values from a table of random standard normal deviates and multiplying each value by ten, so that the distribution of $\varepsilon_i$ is $N(0, 100)$. We may assume that the true relationship is not known to the sales manager.

2  The reader should be clear on the distinction between the random errors, $\varepsilon_i$, and the residuals, $e_i$. The former measure deviations from the *true* line, are fixed once our sample of $Y$s has been drawn, and are unobservable. The residuals measure the deviations from the *fitted* line, vary with the position of the fitted line, and may be evaluated from 8.3. We may think of the $e_i$ as being estimates of the $\varepsilon_i$.

3  Strictly speaking, $Y_i = a + bX_i + e_i$ and so we should not write $Y_i = a + bX_i$, but introduce a new symbol for the value of $Y$ explained by the regression line, say $\hat{Y}_i = a + bX_i$. However, in reporting their empirical results most economists follow the line of least notation and write $Y_i = a + bX_i$. We shall follow their example.

4  Estimators which have the minimum variance property are often called 'best' estimators. Since the least squares estimators are best, linear and unbiased estimators, they are referred to as BLUE in some texts.

5  In order to keep down the amount of algebraic detail in the main text, a proof of this result, together with the derivation of a simple method of calculating $\sum e_i^2$, is given in an appendix to this chapter (see p. 150).

### chapter 9

1  A simpler notation with mnemonic appeal would denote the quantity demanded as $Q$, the price as $P$ and income as $Y$, writing the demand equation as $Q_i = \alpha + \beta P_i + \gamma Y_i + \varepsilon_i$. However, there are only a limited number of Greek letters and the notation introduced in 9.1 extends very easily to the general regression model to be discussed below.

2  The subscript $i$ will be dropped in the summations that follow where its omission does not lead to confusion.

3  The interpretation for this assumption is the same as that presented on p. 136. In our economic example we could imagine repeating the set of prices of the commodity, $X_{2i}$, and incomes, $X_{3i}$, from sample to sample, observing different values for the quantity demanded, $Y_i$, depending on the behaviour of the excluded variables.

4  The extension from the result obtained in chapter 8 is clear. We lose three degrees of freedom here since three parameters have been estimated.

5  See the discussion of the problem of multicollinearity in section 10.3 below.

6  The data are taken from Salter (1966) p. 167, Table 14.

7  There will be no constant term in this equation since $\bar{e}_{(1.3)} = \bar{e}_{(2.3)} = 0$.

8  The sampling theory of the partial correlation coefficient is too difficult to be developed in this text. If $n$ is large, an approximate test of the hypothesis that $\rho_{ij.k} = 0$ may be obtained by treating $r_{ij.k}$ as though it were a simple correlation coefficient and using the $z$-transformation described in section 7.3. See Kendall and Stuart, vol. II (1961) pp. 329–33.

9  We shall not prove this result or the one quoted in the next paragraph. The reader is recommended to consult Johnston (1972) chap. 3.

10  See Allen (1938) pp. 342–3. The formulae for the partial derivatives are given in exercise 9.10 at the end of this chapter.

11   See Goldberger (1964) pp. 118–22.

12   We have deliberately chosen a hypothetical example here in order to illustrate calculations and tests without getting involved in unnecessary arithmetic. We shall give examples of estimates of production functions based on actual economic data below.

13   If we wish to estimate a linear combination of parameters, say $\sum_{i=1}^{k} \lambda_k \beta_k$, then it may be proved that $\sum_{i=1}^{k} \lambda_k b_k$ is the best, linear, unbiased estimator. See Goldberger (1964) pp. 167–8.

14   But it is beyond the scope of this text. See Goldberger (1964) p. 178.

15   In practice it is very unlikely that the reader will be required to make calculations involving two or more independent variables, so that he does not need to know the algebraic details. He is more likely to be faced with the numerical values of parameter estimates and their standard errors as produced by a computer, and his problem will be one of interpretation. We shall concentrate on interpretation in the rest of this section.

16   If $n = 1$ and $k = 2$, we would be attempting to estimate the slope and intercept of a straight line from one observation, an impossible task. If $n = 2$ and $k = 2$, we would have two points and would be forced to draw our estimated line through the two points.

17   However, an increase in $\bar{R}^2$ does not necessarily imply that the latest variable is statistically significant, since $\bar{R}^2$ will increase provided the $t$-ratio for the additional variable is greater than one. This means that one should still perform a $t$-test on the parameter estimates as well as looking at $\bar{R}^2$.

18   We shall discuss time trends and their interpretation in the next chapter. At this point the reader need only note that $t$ is a set of consecutive numbers with $1952 = 1$.

19   We shall suggest an explanation for this result in terms of high correlations among the independent variables in the next chapter – see section 10.3 below.

20   See Evans (1969) pp. 31–2.

21   Matrices and vectors will be printed in bold type. The usual mathematical convention concerning matrices is that the first subscript refers to the row and the second to the column. However, the more natural choice for statisticians is to use the first subscript for the variable and the second subscript for the observation, thus reversing the mathematical convention. We shall follow the statisticians' convention here.

22   We shall not prove that $s^2$ is an unbiased estimator of $\sigma^2$ nor that $\mathbf{b}$ is the best estimator of $\beta$. For proofs of these results, see Johnston (1972) pp. 126–9.

### chapter 10

1   Marx is alleged to have said that history repeats itself, once as a tragedy and once as a farce, but even if this were true, it hardly constitutes repeated sampling.

2   This argument allows us to suggest a way in which saving could vary in our hypothetical samples, but unless these variations are exactly offset by variations in some other component(s) in disposable income, this variable must change from sample to sample and infringe the assumption that the independent variable is a set of constants. Relaxing this assumption can cause serious problems, but we shall defer their discussion until the next chapter (see section 11.4 below).

3   For a discussion of the 'sunspot' theory of cycles, see the paper by W. S. Jevons, reprinted

in Hansen and Clemence (1953) pp. 83–95. For a revival of the theory that there are regular cycles everywhere, see Dewey and Dakin (1947).

4 Some economists would argue that the sophisticated analysis of business cycles associated with the National Bureau of Economic Research constituted an important exception to this generalization. See Moore (1961) for an account of N.B.E.R. methods, while Koopmans (1967) suggests some criticisms of the approach.

5 As an example of such an examination see Evans (1969) chap. 15 and 21.

6 We shall develop a formal test for randomness in section 11.3 below. For the monent we shall rely on visual impressions gained from plotting residuals against time.

7 See Kendall and Stuart, vol. III (1966) pp. 366–9.

8 More formally, the effects of excluding variables from the equation to be estimated may be to produce estimates which are biased and inconsistent. We shall explore this result in section 11.4 below.

9 This is a simplified version of the method of seasonal adjustment actually used by many government agencies. See, for example, O.E.C.D. (1961).

10 The adjustment in this case is very small, but can be important in some cases and is therefore worth checking in every case.

11 At this point the reader may find it worth while to break off and complete problem 10.3 at the end of this chapter, since the data given in this problem are used immediately below in the text.

12 For a discussion of this relationship, see Klein *et al.* (1961) pp. 52–3. A simple explanation would be that food supply responds to demand, but only after a lapse of time.

13 The reader who has followed the matrix presentation of the multiple regression model in the appendix to Chapter 9 may note that, if we had put a constant term in 10.13, the $X'X$ matrix would become singular.

**chapter 11**

1 When the properties of an estimator are not much affected by relaxing the assumptions on which they are based, the estimator is said to be robust. The least squares estimators are robust with respect to the normality assumption in the same way that the sample mean is (see section 4.4 above).

2 Remember problem 8.7 above.

3 This would give us the correct formula for calculating the sampling variance of $b$, but unfortunately, when the errors are heteroscedastic, $b$ is no longer the minimum variance estimator of $\beta$. See Johnston (1972) chap. 7.3.

4 This test was suggested by Goldfeld and Quandt (1965).

5 Each regression is based on $(n - m)/2$ observations and 2 parameters are being estimated.

6 For a more extensive discussion of heteroscedasticity the reader should consult Kmenta (1971) chap. 8.1 or, if he has a knowledge of matrix algebra, Johnston (1972) chap. 7.3.

7 The process is said to be autoregressive since it relates $\varepsilon$ to itself. It is first-order since only $\varepsilon_{t-1}$ appears. A second-order process would contain $\varepsilon_{t-2}$. Such a model is often referred to as a *stochastic* (that is, non-random) process.

8 These results are derived in the appendix at the end of this chapter (see pp. 252–3 below).

9    See Johnston (1972) chap. 8.2.

10    We shall return to this point again in the next section (see p. 244 below).

11    For a further example of this use of the Durbin–Watson statistic, see Nerlove (1965) chap. 6.

12    For a further discussion of the problem of autocorrelation, see Kmenta (1971) chap. 8.2 or Johnston (1972) chap. 8.

13    The reader may find it helpful here to review some basic results concerning mathematical limits; see, for example, O'Brien and Garcia (1971) pp. 350–1.

14    It is usual to refer to the equations that comprise an economic model as *structural* equations. The equations that express each of the endogenous variables in terms of exogenous variables only are called *reduced-form* equations.

15    An excellent textbook in econometrics which does not really require a knowledge of matrix algebra is Kmenta (1971). For the reader with some knowledge of matrix algebra, Wallis (1972) is short, lucid and provides a good introduction to Christ (1966). For the reader whose knowledge of matrix algebra is good, Johnston (1972) is still probably the best text in econometrics.

16    It is difficult to make theoretical comparisons between alternative methods of estimation for small samples of observations. For this reason, most of our knowledge of the relative properties and sensitivity of different estimating techniques is based on simulation experiments. A good summary of some of these studies may be found in Johnston (1972) chap. 13.8.

17    This is the London Business School Quarterly econometric model of the U.K economy. The equations in the model presented in Surrey (1971) are also estimated by least squares.

18    See Johnston (1972) chap. 9.2.

# bibliography

R G D ALLEN   *Mathematical Analysis for Economists* London: Macmillan 1938

C BLYTH   *The Use of Economic Statistics* London: Allen and Unwin 1960

C F CHRIST   *Econometric Models and Methods* New York: Wiley 1966

E R DEWEY and E F DAKIN *Cycles: The Science of Prediction* New York: Holt 1947

J DURBIN and G S WATSON 'Testing for Serial Correlation in Least Squares Regression', *Biometrika* **37** (Dec. 1950) 409–28; **38** (June 1951) 159–78.

M C EVANS   *An Econometric Model of the French Economy* Paris: O E C D, Mar. 1969

M C EVANS   *Macroeconomic Activity* New York: Harper and Row 1969

A S GOLDBERGER   *Economic Theory* New York: Wiley 1964

S M GOLDFELD and R E QUANDT 'Some Tests for Homoscedasticity', *Journal of the American Statistical Association* **60** (1965) 539–48.

A H HANSEN and R V CLEMENCE   *Readings in Business Cycles and National Income* London: Allen and Unwin 1953

P G HOEL   *Introduction to Mathematical Statistics* Third Edition New York: Wiley 1971

J JOHNSTON   *Econometric Methods* Second Edition New York: McGraw-Hill 1972

M G KENDALL and A STUART   *The Advanced Theory of Statistics* vol. I London: Griffin 1958

M G KENDALL and A STUART   *The Advanced Theory of Statistics* vol. II London: Griffin 1961

M G KENDALL and A STUART   *The Advanced Theory of Statistics* vol. III London: Griffin 1966

L KISH   *Survey Sampling* New York: Wiley 1965

L R KLEIN, R J BALL, A HAZELWOOD and P VANDOME   *An Econometric Model of the United Kingdom* Oxford: Oxford University Press 1961

J KMENTA   *Elements of Econometrics* New York: Macmillan 1971

T C KOOPMANS   'Measurement without Theory' *Review of Economics and Statistics* **29** (1947) 161–72.

G H MOORE (ed.)   *Business Cycle Indicators* vol. I Princeton: Princeton University Press 1961

O MORGENSTERN   *On the Accuracy of Economic Observations* Princeton: Princeton University Press 1950

M NERLOVE   *Estimation and Identification of Cobb–Douglas Production Functions* Amsterdam: North-Holland Publishing Company 1965

R J NICHOLSON    *Economic Statistics and Economic Problems* London: McGraw-Hill 1969

R J O'BRIEN and G G GARCIA    *Mathematics for Economists and Social Scientists* London: Macmillan 1971

O E C D    *Seasonal Adjustments on Electronic Computers* Paris: O E C D 1961

W E G SALTER    *Productivity and Technical Change* Cambridge: Cambridge University Press 1966

R M SOLOW 'Technical Change and the Aggregate Production Function', *Review of Economics and Statistics* **39** (1957) 312–20.

D B SUITS    *Statistics: An Introduction to Quantitative Economic Research* London: John Murray 1966

M J C SURREY    *The Analysis and Forecasting of the British Economy* Cambridge: Cambridge University Press 1971

K F WALLIS    *Introductory Econometrics* London: Gray-Mills 1972

A A WALTERS    'A Note on Economies of Scale', *Review of Economics and Statistics* **45** (1963) 425–7.

# index